Mastering Muscles & Movement

A Brain-Friendly System for Learning Musculoskeletal Anatomy and Basic Kinesiology

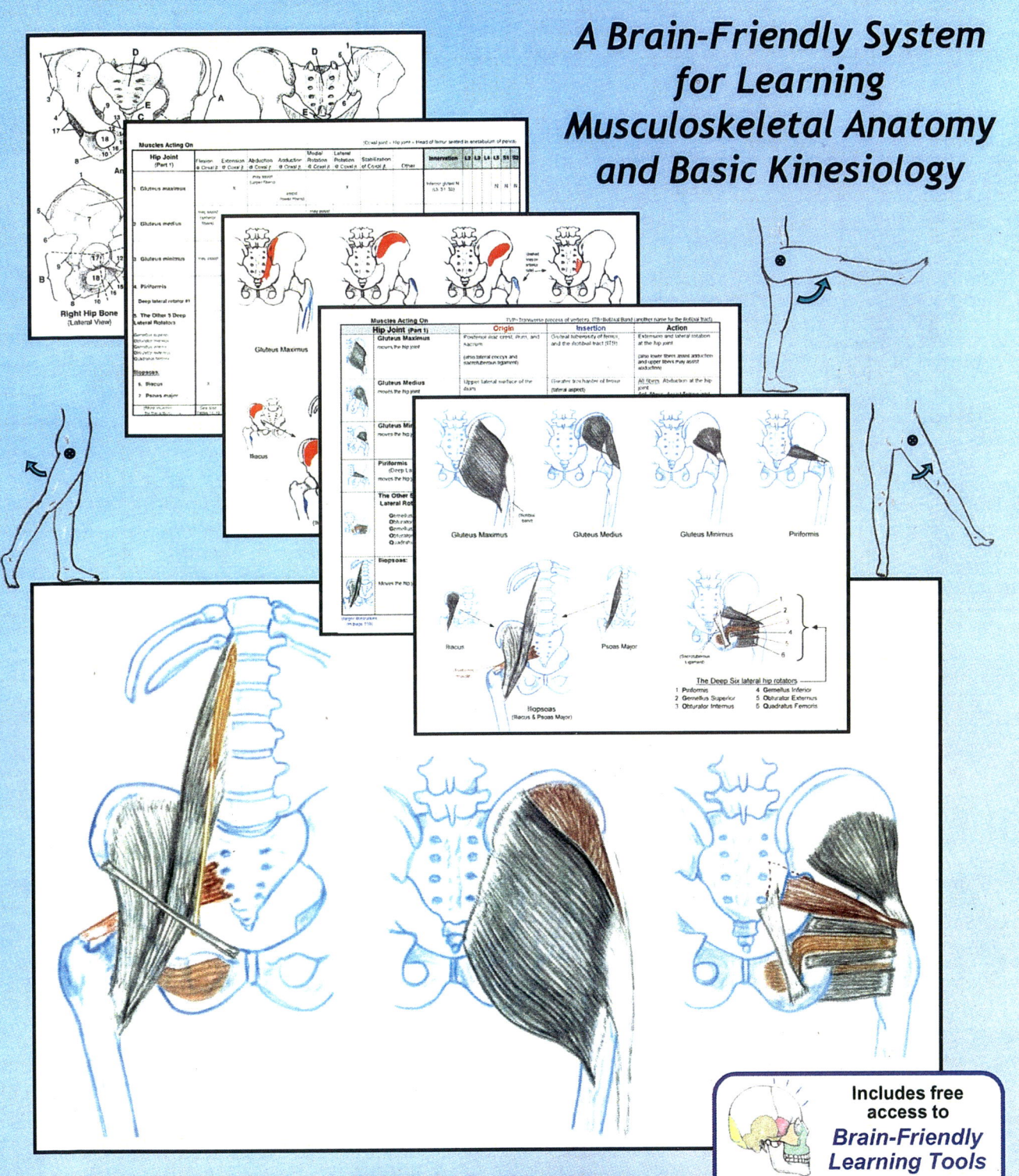

David M. Campbell

Includes free access to *Brain-Friendly Learning Tools*
www.StudyMuscles.com

test #1:

Terminology:

- **anatomical position** — feet, face, palms forward, erect posture, arms @ side, fingers open

  - **Direction & Position:** superior ↑ / inferior ↓, Anterior / Posterior
    (Front)   (back)
    medial / lateral, proximal / distal, supine, prone
    (mid) (sides)   (closer) (farther)   (face up) (face down)

Planes: (AXIS)

- **Sagittal** — a horizontal, front to back line. ANTERIOR to POSTERIOR AXIS!
  - front 2 back, flexion/extension
- **Coronal (frontal)** — Adduction, Abduction, Lateral flexion
  "side 2 side"
- **Transverse (Longitudinal)** — lying @ right angles of transverse
  - Rotations   — a vertical top to bottom line

Action Pairs:

Sagittal → *1st pair* — Flexion & Extension :
   front 2 back movement

- Flexion
- 2 opposite movements in a specific plane
- extension
- moving around an axis
  ↳ @ a specific joint

Coronal (frontal) → *2nd pair* — side to side

Transverse → *3rd pair* : external ← lateral
                          internal ← medial   Rotations

Mastering Muscles & Movement

*A Brain-Friendly System
for Learning
Musculoskeletal Anatomy
and Basic Kinesiology*

#1 Types of joint: 3 major categories by **structure** & **function**

1.) function = Synarthrotic (immovable) ex: teeth, eye socket, cranial

2.) function = AMPHIARTHROTIC (slightly movable) sternum, pubic bone, intervertebral disc

3.) function = DIARTHROTIC (fully moveable) → ligaments, bursae, cartilage pads

Part #2 of joints: Synovial joints (diarthrotic joints)
L>GO OVER! all have same components
 - bones
 - cartilage
 - joint capsule
 - synovial membrane
 - joint cavity filled w/ synovial fluid

6 types:
1. Ball & Socket
2. Pivot
3. Hinge
4. Ellipsoid / Condyloid
5. Gliding
6. Saddle

Mastering Muscles & Movement

A Brain-Friendly System for Learning Musculoskeletal Anatomy and Basic Kinesiology

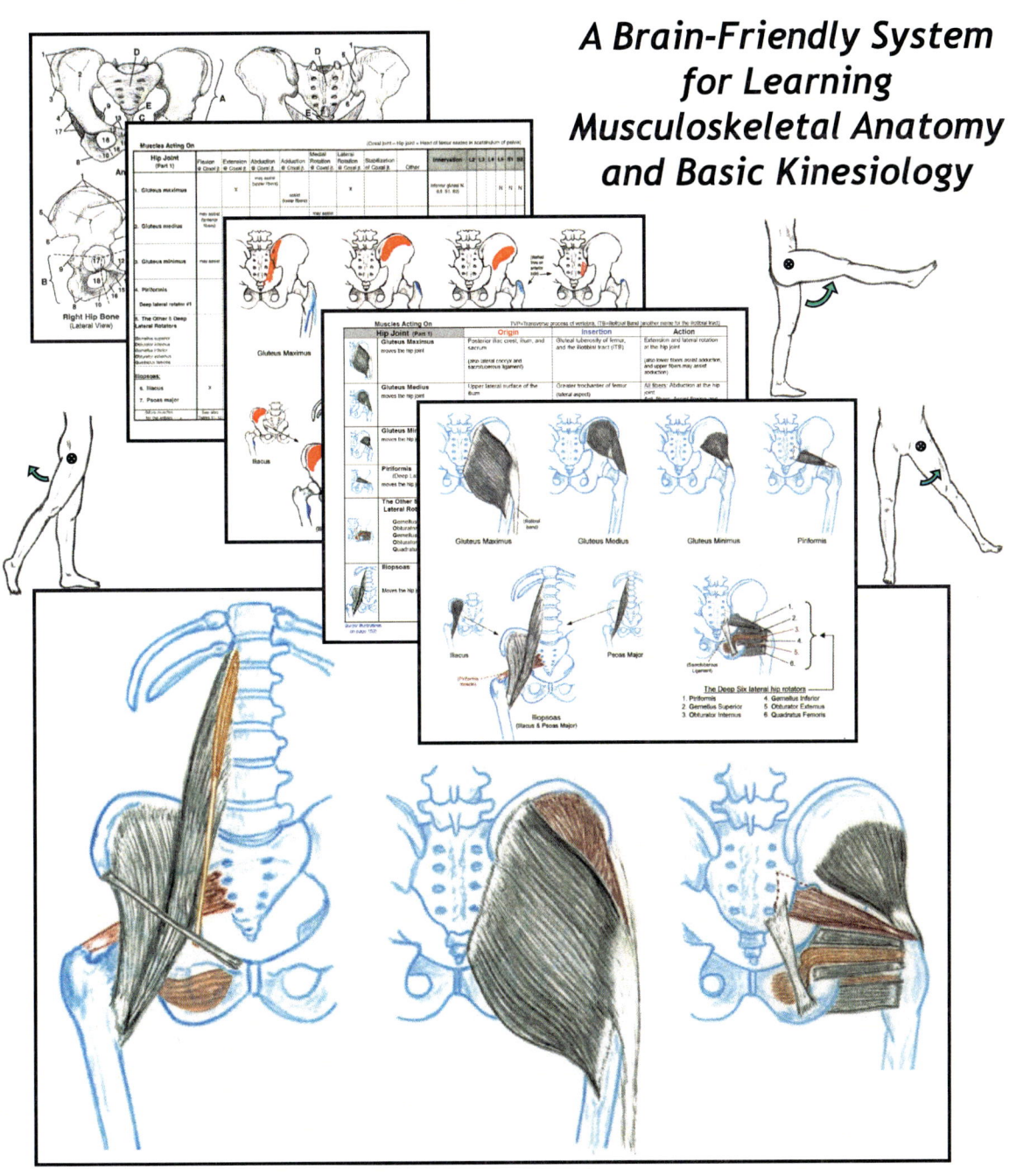

Written and illustrated by

David M. Campbell

Copyright © 2007, text, illustrations by David M. Campbell. All rights reserved. No part of this book, except for brief review, may be reproduced, stored in a retrieval system, or transmitted in any form or by any means, electronic, mechanical, photocopying, recording, or otherwise, without prior permission in writing from the publisher.

Published by Bodylight Books, Bellingham, WA

Library of Congress Cataloging-in-Publication Data

Campbell, David M.
    Mastering Muscles & Movement : a brain-friendly system for learning musculoskeletal anatomy and basic kinesiology / David M. Campbell.
       p. cm.
    Includes bibliographical references and index.
    ISBN-10: 0-9788664-0-1
    ISBN-13: 978-0-9788664-0-2
    1. Human anatomy 2. Kinesiology 3. Human locomotion 4. Musculoskeletal system – anatomy

Contact Information

Dave Campbell, LMP, CTP
The Natural Health Clinic
1707 F Street
Bellingham, WA 98225
(360) 734-1560

www.bodylightbooks.com

Preface

Mastering Muscles & Movement – A Brain-Friendly System for Learning Musculoskeletal Anatomy and Basic Kinesiology, provides a unique and well-organized approach for learning the muscles, bones, joints, and movements of the human body. As the subtitle implies, research in brain-based learning has been richly applied in the design of this book to facilitate understanding, memorization, and mastery of this body of knowledge.

Approach

This book provides a basic set of knowledge for the study of musculoskeletal anatomy and kinesiology. While there are many books available that cover the subject, **Mastering Muscles & Movement** presents a fresh *approach* that is designed to leverage the natural ways the brain observes, learns, and recalls this type of information. Rather than employing the usual one-muscle-per-page format, this book treats *groups* of muscles as "movement families" and presents them in a way that provides a rich visual, verbal, and relational learning environment.

The result is a truly **brain-friendly** experience for the student. The myriad details and interrelationships are easily recognized in simple and natural ways by the innovative arrangement of the muscle information on each page and from page-to-page. The reader comfortably stays aware of the bigger picture while studying any one item, easily compares and contrasts related features and facts, and is enabled to structure study time to play to strengths or to eliminate weaknesses. Please see Chapter 3 – How To Use the Muscle Chapters for a full description of this approach.

Some benefits of this approach are:

- Isolates and supports learning and repetition from many directions: visual, verbal, relational.
- Supports the brain in doing what it does best: Consistently encourages the reader, simply by the way the material is laid out and sized on the pages, to compare and contrast, see patterns, perceive interrelationships, an "come at" the information from different directions.
- Muscle and bone information are arranged to allow easy and repetitive self-testing while studying.
- Anchors the information in the brain with multiple "hooks", providing rich cross-neuronal connections that are important for easy recall of information.
- Clearly organized and has visual cues that always keep the reader aware of where they are within the greater body of knowledge contained in the book.
- Precise and uncluttered presentation clarifies common misunderstandings, and illuminates facts and relationships that are often overlooked.

Audience

This book will serve as a course <u>textbook</u> in some educational programs, and will be valuable as <u>supplemental material</u> in others (depending on the level of specialization required for the course). In addition, this book serves as an easily accessible <u>reference</u> on the shelf of practicing professionals. Finally, because of its clarity of organization and simplicity of approach, it is an excellent <u>quick-review</u> book for students who are preparing for examination, and for practicing professionals who want to refresh their knowledge before attending continuing education classes.

Additional Materials Available

Several supplemental materials based on the text and drawings in *Mastering Muscles & Movement* are available. Please see Chapter 8 – Study Tools for more information about these materials.

- Flash cards – can be used to practice and test recall from *verbal* and *visual* directions
    - *MusclePlus⁺* Flashcards
    - Muscle Tickets
    - Bony Landmark Flashcards

- Laminated skeleton drawings, to be used with dry erase markers to wipe off and practice repeatedly

- Card games to understand muscle families and study synergist and antagonist relationships

- Bony landmark practice sheets and Synergist/Antagonist practice sheets

---

"Mastering Muscles and Movement presents an innovative and practical learning tool for students of Anatomy and Kinesiology. It is the first book that I have seen that actually demonstrates the steps to successful memorization and information retention. It gives you the sense that you are looking through the notes of the best student in class and learning their secret code. The simple, quick access to detailed content and the excellent selection of study tools will make this a book that students reach for first."

Ellen K. Geary MS, LAc

"Mastering Muscles & Movement makes human anatomy and kinesiology highly accessible to the reader. The information is presented in a format that accelerates the learning process and creates a long-term functional memory for the student. Accompanied with the study cards, Mastering Muscles & Movement makes learning fun and simple. I highly recommend this book for undergraduate to post-graduate students as well as practitioners in the field of healing arts and musculoskeletal medicine."

Avilio Halme, MPT

"It motivates you to pay attention, clearly modeling not only how to study this subject, but why it is important and how it is interconnected. The details of the pictures and organized details of the charts enhance the learning process completely and will allow for easier referencing upon completion of training."

Diane Brampton, LMP, OTA

Mastering Muscles & Movement

Contents

Preface i

Muscles – List by Group v

Muscles – Alphabetic List vi

About the Information in This Book viii

Chapter 1 – Basic Information 1
    Introduction 2
    Anatomical Terms 2
    Skeletal System – The Bones 11
    Articular System – The Joints 14
    Muscular System – The Muscles 19
    Nervous System – The Nerves 22
    Kinesiology Concepts 26

Chapter 2 – Bones, Joints and Bony Landmarks 31
    Introduction 32
    Bones 33
    Joints 36
    Bony Landmarks 40

Chapter 3 – How to Use the Muscle Chapters 51
    Introduction 53
    How To Use the General Information Pages 54
    How To Use the Tables and Figures 56
    About Mastering the Muscles 62
    Summary and Generalizations 65

Chapter 4 – Muscles That Move the Upper Extremity 67
    Introduction 68
    Movement of the Scapula/Clavicle .... (Muscle Group 1) 69
    Movement of the Shoulder Joint .... (Muscle Group 2) 77
    Movement of the Elbow and Forearm .... (Muscle Group 3) 85
    Movement of the Wrist, Hand, and Fingers .... (Muscle Group 4) 93
    Movement of the Thumb .... (Muscle Group 5) 101
    Intrinsic Muscles of the Hand 108

Chapter 5 – Muscles That Move the Axial Skeleton 109
    Introduction 110
    Movement of the Face and Jaw .... (Muscle Group 6) 113
    Movement of the Neck and Head .... (Muscle Group 7) 121
    Movement of the Spine .... (Muscle Group 8) 129
    Movement of the Thorax, Abdomen, Breathing .... (Muscle Group 9) 137

Chapter 6 – Muscles That Move the Lower Extremity 151
Introduction 152
Movement of the Hip Joint (Part 1) .... (Muscle Group 10) 153
Movement of the Hip Joint (Part 2) .... (Muscle Group 11) 161
Movement of the Knee (& Hip Joint, Part 3) .... (Muscle Group 12) 169
Movement of the Ankle, Foot, and Toes .... (Muscle Group 13) 177
Intrinsic Muscles of the Foot 185

Chapter 7 – Summary Tables 187
Introduction 188
Summary of Actions – Upper Extremity 189
Summary of Actions – Axial Skeleton 191
Summary of Actions – Lower Extremity 193
Joint Summary 195
Ligament Summary 197
Innervation Summary 198

Chapter 8 – Study Tools 201
Introduction 202
How To Use the MusclePlus+ Flashcards 204
How to Use the Bony Landmark Practice Sheets 206
Samples of Other Study Tools 209

Bibliography 213

Index 214

Muscles – Alphabetic List 218

Muscles – List by Group .... inside back cover

Muscles – List by Group

Muscles are placed in groups based on the bones and joints they *move* as they contract.

----- (Chapter 4) -----

Group 1 – Scapula / Clavicle
Trapezius   p. 69-76
Levator scapula
Rhomboid major & minor
Serratus anterior
Pectoralis minor
Subclavius

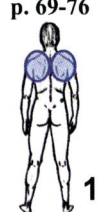

Group 2 – Shoulder Joint
Deltoid   p. 77-84
Supraspinatus
Infraspinatus
Teres minor
Subscapularis
Pectoralis major
Coracobrachialis
Latissimus dorsi
Teres major

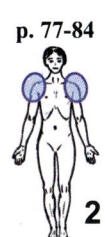

Group 3 – Elbow, Forearm
Biceps brachii   p. 85-92
Brachialis
Brachioradialis
Pronator teres
Pronator quadratus
Triceps brachii
Anconeus
Supinator

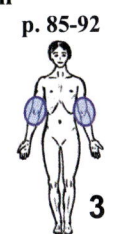

Group 4 – Wrist, Hand, Fingers
Flexor carpi radialis
Palmaris longus   p. 93-100
Flexor carpi ulnaris
Flexor digitorum superficialis
Flexor digitorum profundus
Extensor carpi radialis longus
Extensor carpi radialis brevis
Extensor carpi ulnaris
Extensor digitorum
Extensor indicis

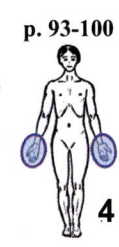

Group 5 – Thumb
Flexor pollicis longus   p. 101-108
Flexor pollicis brevis
Opponens pollicis
Adductor pollicis
Abductor pollicis longus
Abductor pollicis brevis
Extensor pollicis longus
Extensor pollicis brevis
Intrinsic muscles of the hand

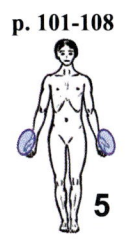

----- (Chapter 5) -----

Group 6 – Face, Jaw
Masseter   p. 113-120
Temporalis
Lateral pterygoid
Medial pterygoid
Occipitofrontalis
Platysma
Suprahyoids Group
    Geniohyoid, Mylohyoid,
    Stylohyoid, Digastric
Infrahyoids Group
    Sternohyoid, Sternothyroid,
    Omohyoid, Thyrohyoid
Muscles of facial expression

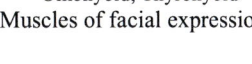

Group 7 – Neck, Head
Sternocleidomastoid   p. 121-128
Scalenes group
Longus capitis & longus colli
Suboccipital group
    Rectus Capitis Posterior Major
    Rectus Capitis Posterior Minor
    Oblique Capitis Superior
    Oblique Capitis Inferior
Splenius capitis
Splenius cervicis
Semispinalis capitis
Levator scapula*
Trapezius, upper fibers*
    *(revisited for reversed O/I actions)

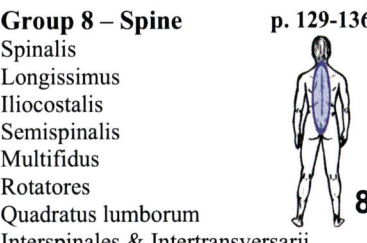

Group 8 – Spine   p. 129-136
Spinalis
Longissimus
Iliocostalis
Semispinalis
Multifidus
Rotatores
Quadratus lumborum
Interspinales & Intertransversarii

Group 9 – Thorax, Abdomen, Breathing
Rectus abdominis   p. 137-144
External oblique
Internal oblique
Transverse abdominis
Diaphragm
External intercostals
Internal intercostals
Serratus posterior superior
Serratus posterior inferior
Levator costae

----- (Chapter 6) -----

Group 10 – Hip Joint (Part 1)
Gluteus maximus   p. 153-160
Gluteus medius
Gluteus minimus
Piriformis (1st lateral rotator)
The other 5 lateral rotators
    Gemellus Superior
    Obturator Internus
    Gemellus Inferior
    Obturator Externus
    Quadratus Femoris
Iliopsoas
    (Iliacus & Psoas major)

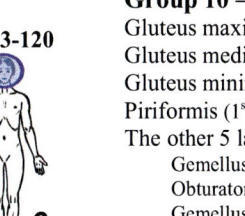

Group 11 – Hip Joint (Part 2)
Sartorius   p. 161-168
Tensor fascia latae
Pectineus
Adductor brevis
Adductor longus
Adductor magnus
Gracilis

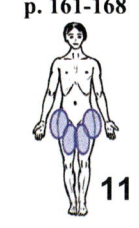

Group 12 – Knee (& Hip Joint, Part 3)
Rectus femoris   p. 169-176
Vastus lateralis
Vastus intermedius
Vastus medialis
Biceps femoris
Semitendinosus
Semimembranosus
Popliteus

Group 13 – Ankle, Foot, Toes
Gastrocnemius
Plantaris
Soleus   p. 177-186
Tibialis posterior
Flexor digitorum longus
Flexor hallucis longus
Peroneus longus
Peroneus brevis
Tibialis anterior
Extensor digitorum longus
Extensor hallucis longus
Intrinsic muscles of the foot

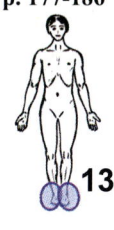

Muscles – Alphabetic List

A
Abductor digiti minimi of the foot 185
Abductor digiti minimi of the hand 108
Abductor hallucis 185
Abductor pollicis brevis 104
Abductor pollicis longus 104
Adductor brevis 164
Adductor hallucis 185
Adductor longus 164
Adductor magnus 164
Adductor pollicis 104
Anconeus 88

B
Biceps brachii 88
Biceps femoris 172
Brachialis 88
Brachioradialis 88
Buccinator 114

C
Coracobrachialis 80

D
Deltoid 80
Deep Six lateral rotators of the hip 155
Depressor anguli oris 114
Depressor labii inferioris 114
Diaphragm 140
Digastric 116
Dorsal interossei of the foot 185
Dorsal interossei of the hand 108

E
Erector spinae group 132
Extensor carpi radialis brevis 96
Extensor carpi radialis longus 96
Extensor carpi ulnaris 96
Extensor digiti minimi 99
Extensor digitorum (fingers) 96
Extensor digitorum brevis (toes) 185
Extensor digitorum longus (toes) 180
Extensor hallucis brevis 185
Extensor hallucis longus 180
Extensor indicis 96
Extensor pollicis brevis 104
Extensor pollicis longus 104
External abdominal oblique 140
External intercostals 140

F
Fibularis, see peroneus
Flexor carpi radialis 96
Flexor carpi ulnaris 96
Flexor digiti minimi of the foot 185

Flexor digiti minimi of the hand 108
Flexor digitorum brevis 185
Flexor digitorum longus 180
Flexor digitorum profundus 96
Flexor digitorum superficialis 96
Flexor hallucis brevis 185
Flexor hallucis longus 180
Flexor pollicis brevis 104
Flexor pollicis longus 104

G
Gastrocnemius 180
Gemellus inferior & superior 156
Geniohyoid 116
Gluteus maximus 156
Gluteus medius 156
Gluteus minimus 156
Gracilis 164

H
Hamstrings 172
Hypothenar eminence muscles 108

I
Iliacus 156
Iliocostalis 132
Iliopsoas 156
Infrahyoids group 116
Infraspinatus 80
Intercostals, external 140
Intercostals, internal 140
Internal abdominal oblique 140
Internal intercostals 140
Interspinales 130
Intertransversarii 130

L
Latissimus dorsi 80
Levator anguli oris 114
Levator costae 140
Levator labii superioris 114
Levator labii superioris alaeque nasi 114
Levator scapula 72
Levator scapula (reversed O/I action) 124
Longissimus 132
Longus capitis 124
Longus colli 124
Lumbricales of the foot 185
Lumbricales of the hand 108

M
Masseter 116
Mentalis 114
Multifidus 132
Mylohyoid 116

Muscles – Alphabetic List

N
Nasalis 140
O
Oblique, abdominal 140
Oblique capitis inferior 124
Oblique capitis superior 124
Obturator externus 156
Obturator internus 156
Occipitofrontalis 116
Omohyoid 116
Opponens digiti minimi 108
Opponens pollicis 104
Orbicularis oculi 114
Orbicularis oris 114
P
Palmar interossei 108
Palmaris longus 96
Pectineus 164
Pectoralis major 80
Pectoralis minor 72
Peroneus brevis 180
Peroneus longus 180
Peroneus tertius 183
Piriformis 156
Plantar interossei 185
Plantaris 180
Platysma 116
Popliteus 172
Procerus 114
Pronator quadratus 88
Pronator teres 88
Psoas major 156
Psoas minor 155
Pterygoid, lateral 116
Pterygoid, medial 116
Q
Quadratus femoris 156
Quadratus lumborum 132
Quadratus plantae 185
Quadriceps 172
R
Rectus abdominis 140
Rectus capitis anterior 122
Rectus capitis lateralis 122
Rectus capitis posterior major .. 124
Rectus capitis posterior minor .. 124
Rectus femoris 172
Rhomboid major 72
Rhomboid minor 72
Risorius 114
Rotator cuff muscles 80
Rotatores 132
S
Sartorius 164
Scalenes group 124
Semimembranosus 172
Semispinalis 132
Semispinalis capitis 124
Semitendinosus 172
Serratus anterior 72
Serratus posterior superior 140
Serratus posterior inferior 140
Soleus 180
Spinalis 132
Splenius capitis 124
Splenius cervicis 124
Sternocleidomastoid 124
Sternohyoid 116
Sternothyroid 116
Stylohyoid 116
Subclavius 72
Suboccipital group (4) 124
Subscapularis 80
Supinator 88
Suprahyoids group (4) 116
Supraspinatus 80
T
Temporalis 116
Tensor fascia latae 164
Teres major 80
Teres minor 80
Thenar eminence muscles 108
Thyrohyoid 116
Tibialis anterior 180
Tibialis posterior 180
Transverse abdominis 140
Transversospinal group 132
Trapezius 72
Trapezius, upper fibers only 124
Triceps brachii 88
V
Vastus intermedius 172
Vastus lateralis 172
Vastus medialis 172
Z
Zygomaticus major 114
Zygomaticus minor 114

About the Information in This Book

Books on muscles and movement are notoriously inconsistent in the details of the muscle attachments, actions and innervation assigned to individual muscles. Variations in artistic renderings of muscles and other structures present an additional challenge. There are many valid reasons for these differences, including human anatomical variation, measurement and analysis methods employed by anatomists, and editorial decision processes.

Suffice it to say, this book necessarily adds one more resource to the fray. As such, I will note the main resources and the process I used while making decisions about the muscle information I present in this book. The Bibliography on page 213 lists the main 25 resources I used while developing this book. Some resources were influential in my artistic choices, while others were given varying degrees of influence in my decisions about factual muscle information (origin, insertion, action, innervation).

For muscle details, my main sources were as follows. I started with Clemente's *Anatomy* [4], and Kendall's *Muscles, Testing and Function* [12]. I also considered two books commonly used in massage school kinesiology classes: Biel's *Trail Guide to the Body* [2], and Thompson's *Manual of Structural Kinesiology* [23]. When differences were not easily reconciled, I turned to the highly detailed analyses of anatomy and function in the Travell manuals [21] and [25]. In more difficult cases I made tables to compare sources and look for common ground, discussed the information with colleagues, consulted additional books, and studied cadaver dissections.

While drawing the origin/insertion (red/blue) pictures, I compared illustrations in Netter's *Atlas of Human Anatomy* [16], Clemente's *Anatomy* [4], Platzer's *Color Atlas: Locomotor System* [18], McMinn's *Color Atlas of Human Anatomy* [14], and *Gray's Anatomy* [6].

After weighing all of the above, I then "flavored" the presentation based on my specific approach, i.e., to be *brain-friendly*. Please read the Preface on page *i*, and read Chapter 3 – "How to Use the Muscle Chapters" to better understand this approach.

Acknowledgements

Over the six-plus years I have been developing this book there have been too many influences and contributors to name (or remember) them all individually. However, the following is an attempt to "name a few" of the people who graciously helped me along the way.

Those making direct contribution to the process and/or content of this book (in alphabetical order):
Ahwren Ayers, Jack Blackburn, Diane Brampton, Francis Brown, Barb Collins, Gwen Crowell, Pete Darcy, Elizabeth Fletcher Brown, Janae Fletcher-Almasi, George Gottlieb, Avilio Halme, Kinsey Jackson, Steve Johnson, Sara King, Carolyn Koehnline, Liz Lamm, Julie McKay, Brenda Mitchell, Lisa Nelson, Cynthia Price, Stacey Renando, Sharon Souders, Sari Spieler, Helen Thayer, Diana Thompson, Kristin Torok, Faith Van De Putte, and all my students and colleagues (2001-2007) at the Spectrum Center School of Massage.

Inspiration/influence past and present (though not directly involved with the writing of this book):
Bernard Siegfried Albinus, Emilie Conrad, Jen Cosgrove, Barbara D. Cummings, Eliot Goldfinger, Louise Gordon, John Hull Grundy, Lenore Jones, Deane Juhan, Frank H. Netter, Masahiro Takakura, Milton Trager, Edward Tufte, Gunther von Hagens.

And finally for patience, love, support, expert editorial advice, and a little fun in between the hard work:
Laura Pitts.

Chapter 1

Basic Information

Introduction 2

Anatomical Terms.. 2
    Anatomical Position 2
    Terms of Direction and Position 3
    Regions of the Body 4
    Planes 5
    Movement Definitions (Actions) 6
    A System for Describing Movement 9

Skeletal System – The Bones 11
    Bones 11
    Skeleton 12
    Classification of Bones by Shape 13
    Bony Landmarks 13

Articular System – The Joints 14
    Broad Classification of Joints – by Structure and Function 14
    Stability vs. Mobility Trade-off 14
    Synovial Joints – Structure 14
    Six Types of Synovial Joints 15
    Joints of the Upper Extremity 16
    Joints of the Body 17
    Range of Motion (ROM) 18

Muscular System – The Muscles 19
    Anatomy and Function 19
    Fiber Arrangements 20
    Tendons and Attachments 21
    Muscles and Opposing Movements 21

Nervous System – The Nerves 22
    Overall Organization 22
    Spinal Nerve Roots and Rami 23
    Nerve Plexuses and Naming of Major Nerves 24
    How Nerves Work With Muscles 25

Kinesiology Concepts 26
    Movement and Levers 26
    Types of Muscle Contractions 28
    Muscles Working Together 29
    Biomechanics 29
    Range of Motion Procedures 30

Introduction

By definition the term **kinesiology** means "the study of movement". To study movement of the human body one must learn the muscles, their functions, the joints they cross, and the bones they attach to. This book is designed to provide a rich and "brain-friendly" learning experience of these topics. It is organized as follows.

Chapter 1 – Basic Information provides foundational information for the study of muscles and movement. It gives definitions of terms, sets up a system for describing and analyzing body movement, and gives basic information about bones, joints, and muscles.

Chapter 2 – Bones, Joints and Bony Landmarks uses an atlas format to cover all the details about the individual bones of the skeleton. It also includes whole-body skeleton illustrations and joint information.

Chapter 3 – How To Use the Muscle Chapters is a must-read to set yourself up to fully utilize the main chapters that teach all the muscles of the body. Understanding how to proceed is an essential step to allow you to truly master the muscles and movements of the body.

Chapters 4, 5, and 6 provide the bulk of the muscles and movement information in a special format that emphasizes constantly comparing and contrasting, to help you understand and memorize the muscles of the body and their actions and innervations. This unique organization helps you build a rich neuronal network in your brain that will lead to true mastery of this subject.

Chapters 7 and 8 give useful summary information and describe the many available Study Tools that have been developed to specifically complement **Mastering Muscles & Movement**.

Anatomical Terms

Anatomical Position

Anatomical position is a standing posture in which the parts of the body are placed in specific ways. It provides a *reference position* that is used as the basis to name and describe body movements, positions, and directions. The components of anatomical position are:

- Erect posture
- Face forward
- Feet forward
- Arms at sides
- Palms of hands forward
- Fingers and thumbs in extension
  (straight, not closed in fists).

A vertical line called the **midline** divides the body into right and left symmetrical halves. Note that the body is not symmetrical from front to back, so the midline does not apply when viewing the body from the side.

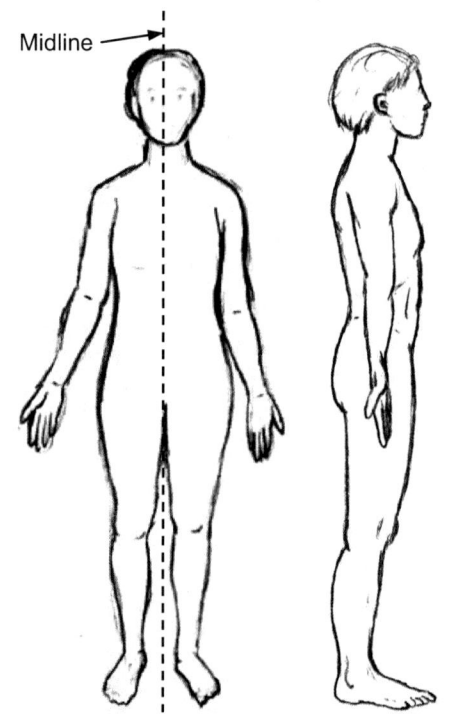

Anatomical Position

Terms of Direction and Position

The following terms are used to describe the relationships of one body structure to another, and to clarify body positions and movements. These terms are defined for a person standing in anatomical position. Therefore, it is easiest to learn the terms while visualizing the body in that position. Once learned, the terms can be used to precisely describe body positions and directions no matter what orientation the body is in.

Superior / Cranial / (also cephalad) – Closer to the head; situated above another structure. Example: The right lung is *superior* to the liver.
 	
Inferior / Caudal – Closer to the feet; situated below another structure. Example: The umbilicus (belly button) is *inferior* to the chin.
   The terms *cranial* and *caudal* may be used when referring to the trunk.
	Proximal – Nearer to the trunk or point of origin of a limb. Example: The knee is *proximal* to the foot.
	
	Distal – Further from the root of a limb. Example: The hand is *distal* to the elbow.
	   The terms *proximal* and *distal* are used when referring to the arms and legs.

Anterior / Ventral – Front of the body, or a structure closer to the front than another structure. Example: The abdomen is *anterior* to the spine. **Posterior / Dorsal** – Back of the body, or a structure closer to the back than another structure. Example: The spine is *posterior* to the sternum (breast bone). The terms *ventral* and *dorsal* may be used when referring to the trunk.	**Deep** – Beneath or inward from the surface of the body. Example: Muscles are *deep* to the skin, and bones are deep to the muscles. **Superficial** – Near the surface, or closer to the surface than another structure. Example: The muscles are *superficial* to the bones.
	Ipsilateral – Indicates that a structure is on the same side of the body as another structure. Example: The shoulder and ipsilateral hip moved toward each other. **Contralateral** – Refers to a structure being on the opposite side of the body from another structure.
Medial – Refers to a structure that is closer to the midline or median plane of the body. Example: The eyes are *medial* to the ears. **Lateral** – A structure that is further away from the midline. Example: The little toes are *lateral* to the big toes (in anatomical position).	**Supine** – Lying on the back (face up, belly exposed). **Prone** – Lying on the belly (face down, back exposed).

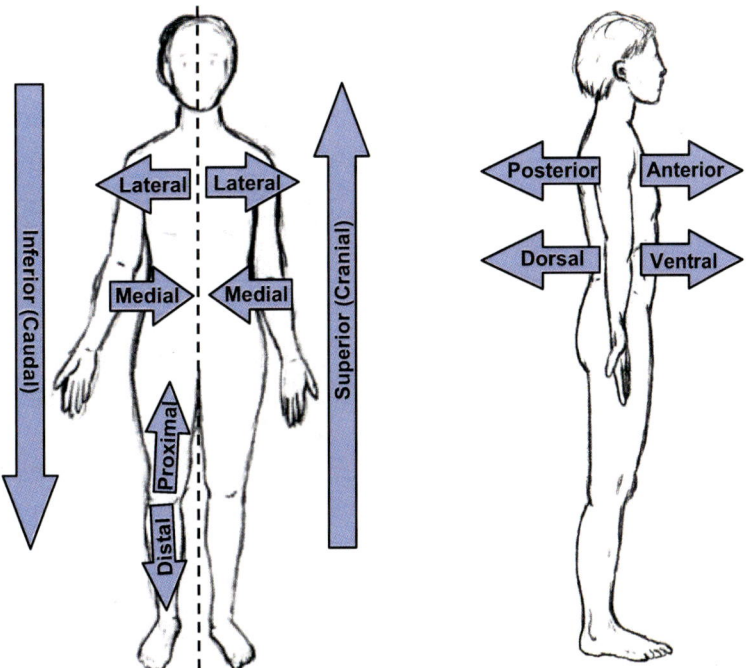

Terms of Direction and Position

Regions of the Body

The body is divided into many regions. Knowing the names of regions allows efficient and precise communication when talking about locations on the body. Here are some regional terms that are useful when studying kinesiology:

Cranial	- head	Lumbar	-	lower back
Cervical	- neck	Sacral	-	base of spine, tail bone
Thoracic	- upper trunk, ribcage	Inguinal	-	where lower abdomen meets thigh
Supraclavicular	- above the clavicle	Pubic	-	genital region
Axillary	- armpit	Gluteal	-	buttocks
Pectoral	- upper chest	Femoral	-	thigh
Brachial	- arm (upper arm)	Popliteal	-	behind knee
Antebrachial	- forearm	Crural	-	leg (below knee)
Cubital	- elbow	Sural	-	calf of leg
Antecubital	- front of elbow	Dorsum	-	top of foot, also back of hand
Palmar	- palm side of hand	Plantar	-	bottom (sole) of foot

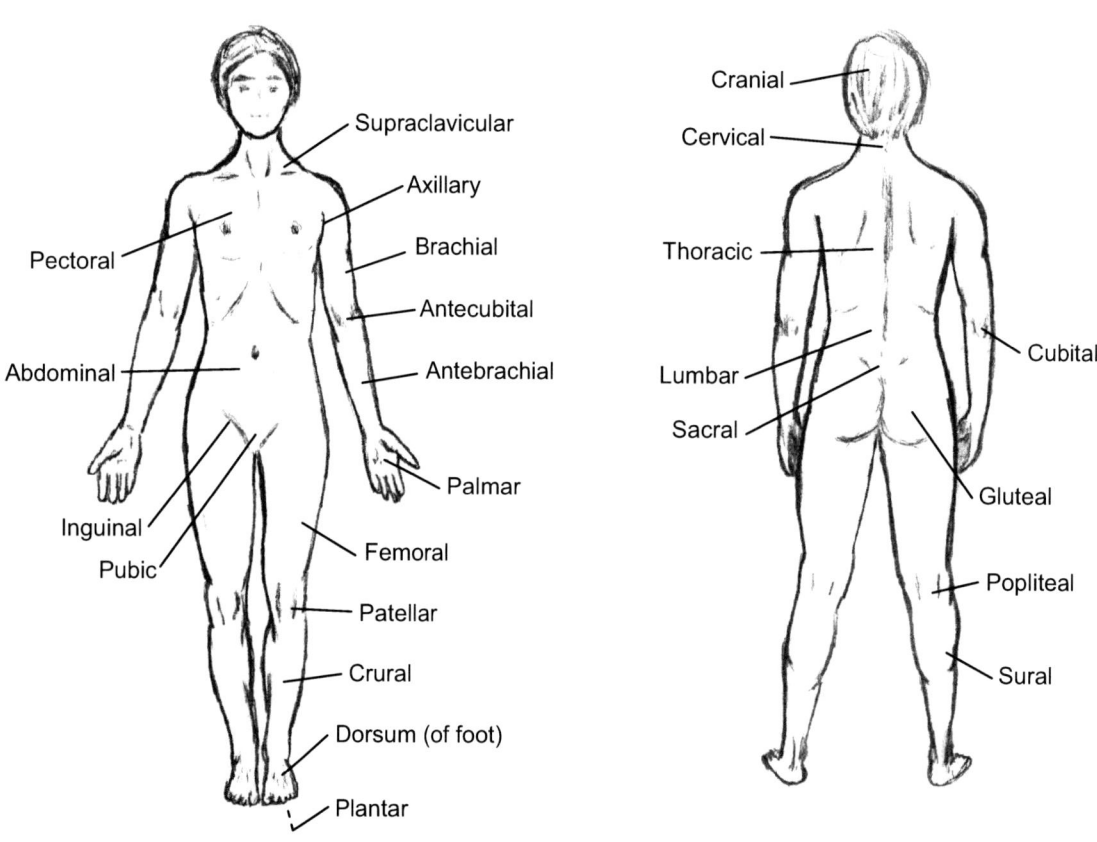

Regions of the Body

Planes

A plane is an imaginary flat surface (visualize a pane of glass or flat piece of cardboard). Basic body movements are defined to occur in one of three planes:

1. **Sagittal Plane** – A vertical plane passing from front to back, dividing the body into right and left portions. A special sagittal plane called the *median plane* passes through the midline and divides the body into equal halves. Forward and backward body movements occur in the sagittal plane.

2. **Coronal (Frontal) Plane** – A vertical plane extending from side to side, dividing the body into anterior and posterior portions. Side-to-side movements occur in the coronal plane.

3. **Transverse Plane** – A horizontal plane that divides the body into upper and lower portions. Rotational or twisting movements occur in the transverse plane.

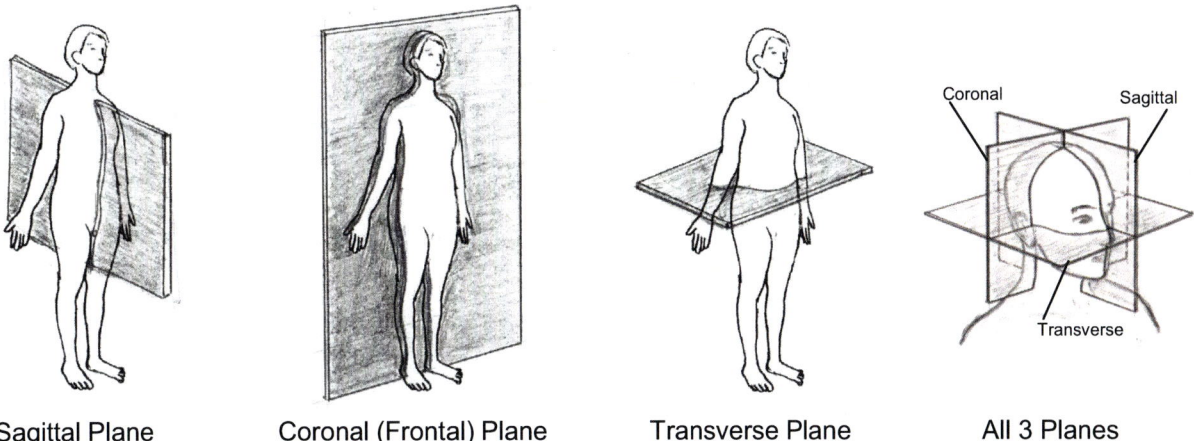

Sagittal Plane   Coronal (Frontal) Plane   Transverse Plane   All 3 Planes

Moving In a Plane: To visualize a body part moving in a plane, first place the plane so it passes through the joint that is moving. Pick a point on the body part that is doing a certain action, and that point will stay in contact with the plane throughout the movement. For example, if you are going to bend the neck and trunk forward (flexion of the spine) then place a sagittal plane at the midline of the body so it passes through the joints along the spine. Note that your nose lies in the plane, and as you flex forward notice how the tip of your nose travels (it stays *in* the sagittal plane).

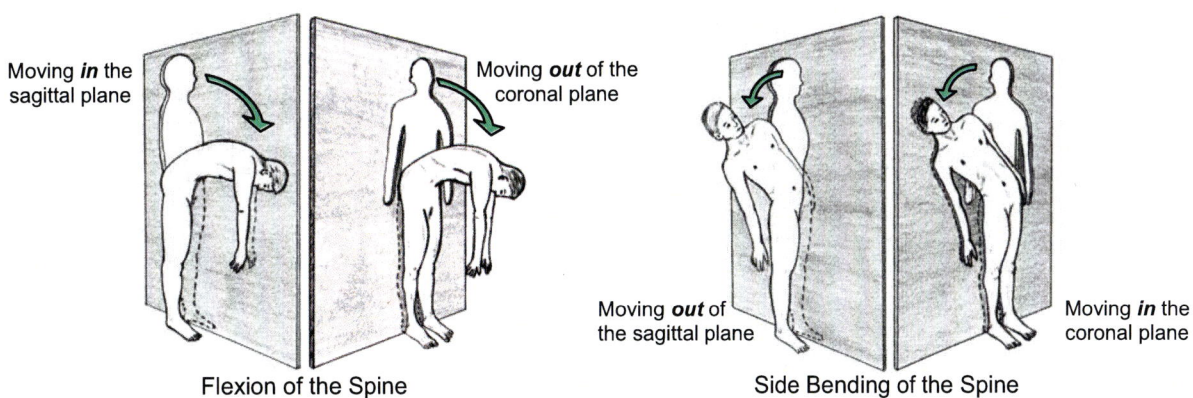

Flexion of the Spine   Side Bending of the Spine

Moving in one of the vertical planes constitutes moving *out* of the other vertical plane. For example, moving from anatomical position in the sagittal plane causes movement out of the coronal plane. In the above spine flexion example, your nose moves forward out of the coronal plane.

Movement Definitions (Actions)

The following set of terms enables clear and concise description of the movements (also called actions) of all parts of the body. These terms, like the terms of position and direction on page 3, are initially defined and most easily learned referring to a body in anatomical position. They may then be used to describe movements of the body in any orientation or position. The actions below are grouped in pairs of opposite actions.

Main Actions (apply to many parts of the body, see page 7):

<u>Movements in the Sagittal plane:</u>
1. Flexion   (limbs, axial skeleton)
2. Extension   (limbs, axial skeleton)

<u>Movements in the Coronal (Frontal) plane:</u>
3. **Ab**duction   (limbs)
4. Adduction   (limbs)

5. Lateral Flexion to the right   (axial skeleton)
6. Lateral Flexion to the left   (axial skeleton)

<u>Movements in the Transverse plane:</u>
7. Lateral Rotation (External Rotation)   (limbs)
8. Medial Rotation (Internal Rotation)   (limbs)

9. Right Rotation   (axial skeleton)
10. Left Rotation   (axial skeleton)

Special Actions (apply to specific structures, see page 8):

11. Pronation   (forearm)
12. Supination   (forearm)

13. Plantarflexion   (ankle)
14. Dorsiflexion   (ankle)

15. Inversion   (subtalar joint, below ankle)
16. Eversion   (subtalar joint, below ankle)

17. Protraction   (scapula, mandible)
18. Retraction   (scapula, mandible)

19. Elevation   (scapula, mandible)
20. Depression   (scapula, mandible)

21. Upward Rotation   (scapula)
22. Downward Rotation   (scapula)

23. Radial Deviation   (abduction)   (wrist)
24. Ulnar deviation   (adduction)   (wrist)

25. Horizontal Abduction   (shoulder joint)
26. Horizontal Adduction   (shoulder joint)

27. Circumduction (a combination action at ball & socket and ellipsoid joints)

Main Actions (shown as action pairs)

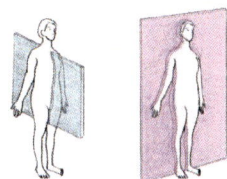

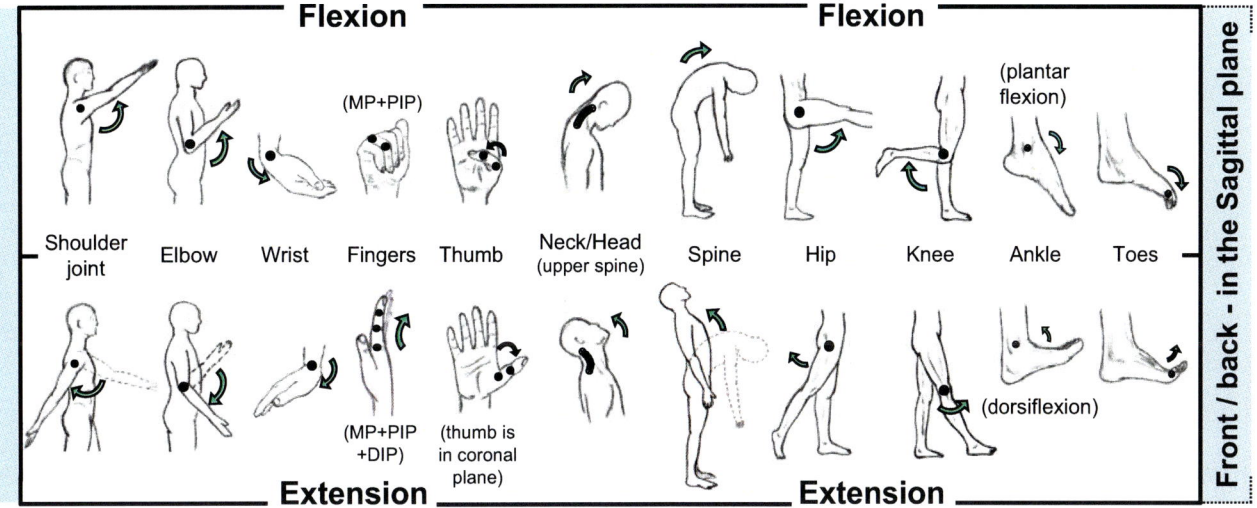

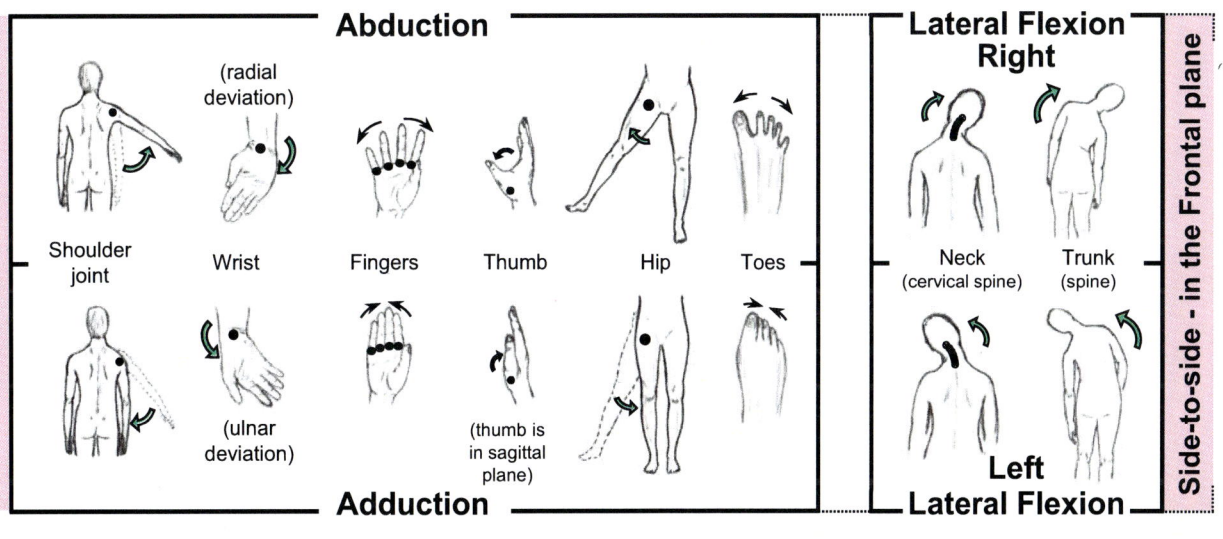

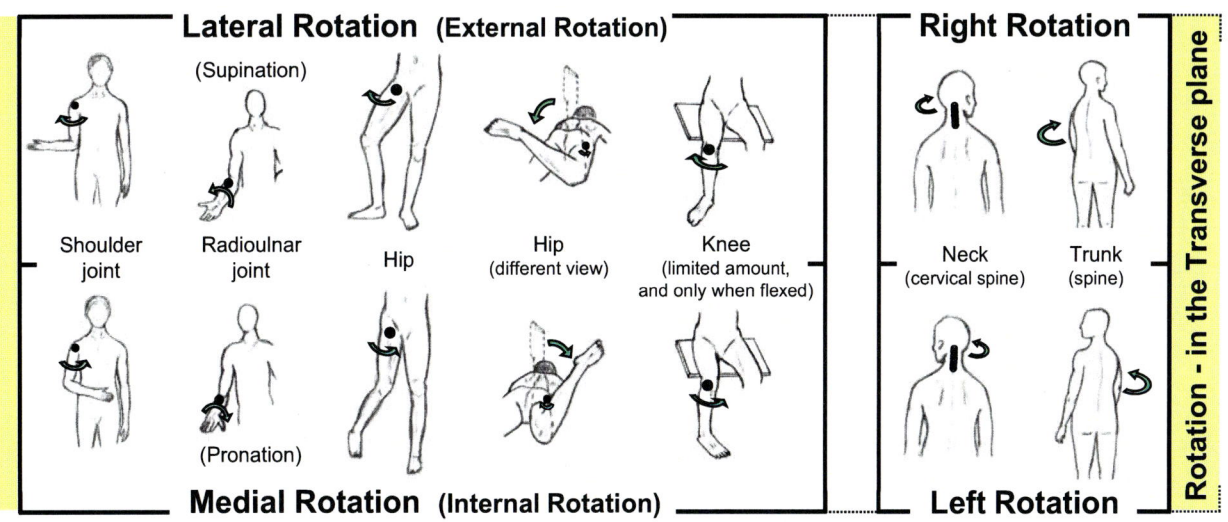

Special-Purpose Actions (shown as action pairs)

↱ = Direction of movement
● = Location of joint that is moving

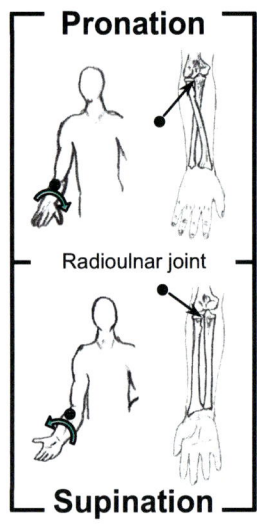
Pronation / Radioulnar joint / Supination

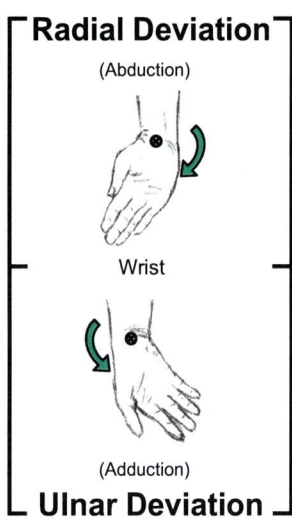
Radial Deviation (Abduction) / Wrist / Ulnar Deviation (Adduction)

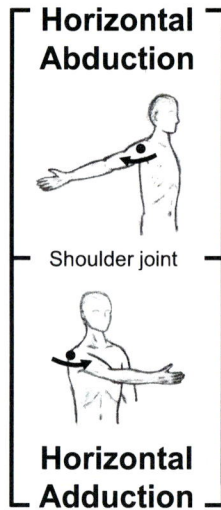
Horizontal Abduction / Shoulder joint / Horizontal Adduction

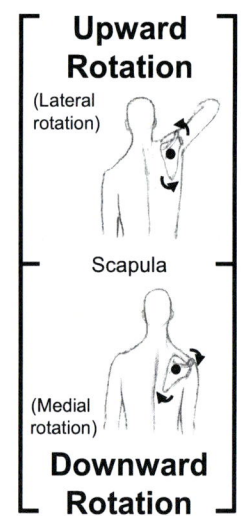
Upward Rotation (Lateral rotation) / Scapula / Downward Rotation (Medial rotation)

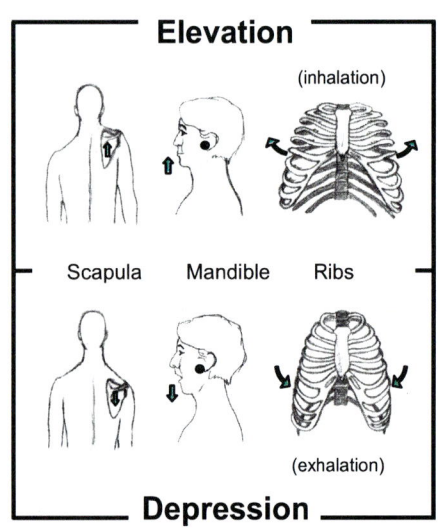
Elevation (inhalation) / Scapula — Mandible — Ribs / Depression (exhalation)

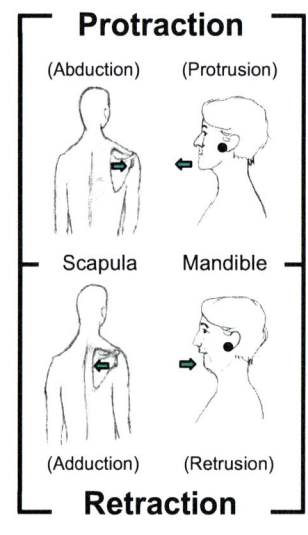
Protraction (Abduction) (Protrusion) / Scapula — Mandible / Retraction (Adduction) (Retrusion)

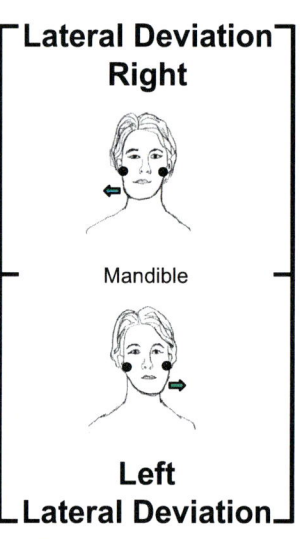
Lateral Deviation Right / Mandible / Left Lateral Deviation

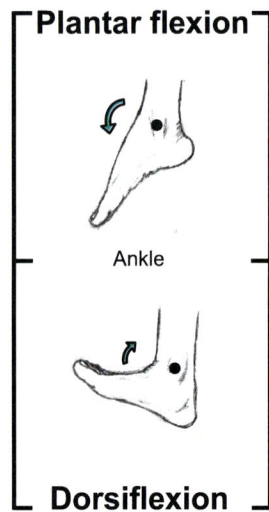
Plantar flexion / Ankle / Dorsiflexion

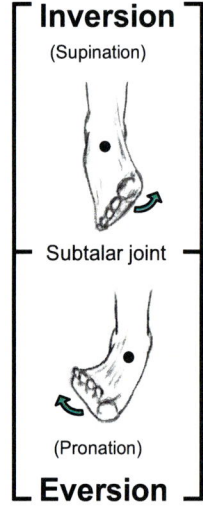
Inversion (Supination) / Subtalar joint / Eversion (Pronation)

A System for Describing Movement

Movements are defined to occur <u>in</u> a plane and <u>about</u> an axis. The plane and axis are positioned so that they both pass through the joint that is moving. The concept of moving in a plane was discussed earlier in this chapter on page 5. We will now define the axis, and discuss how to put a plane and axis together to set up a full system for describing movements.

Axes

An *axis* is a straight line (plural is axes). Visualize an arrow or the radio antenna on a car. Three axes at right angles establish an axis system for describing body movements in three dimensions.

1. **Sagittal Axis** – A horizontal front-to-back line, lying in the sagittal plane & at right angles to the coronal plane.

2. **Coronal (Frontal) Axis** – A horizontal side-to-side line, lying in the coronal plane & at right angles to the sagittal plane.

3. **Longitudinal Axis** – A vertical top-to-bottom line, lying at right angles to the transverse plane.

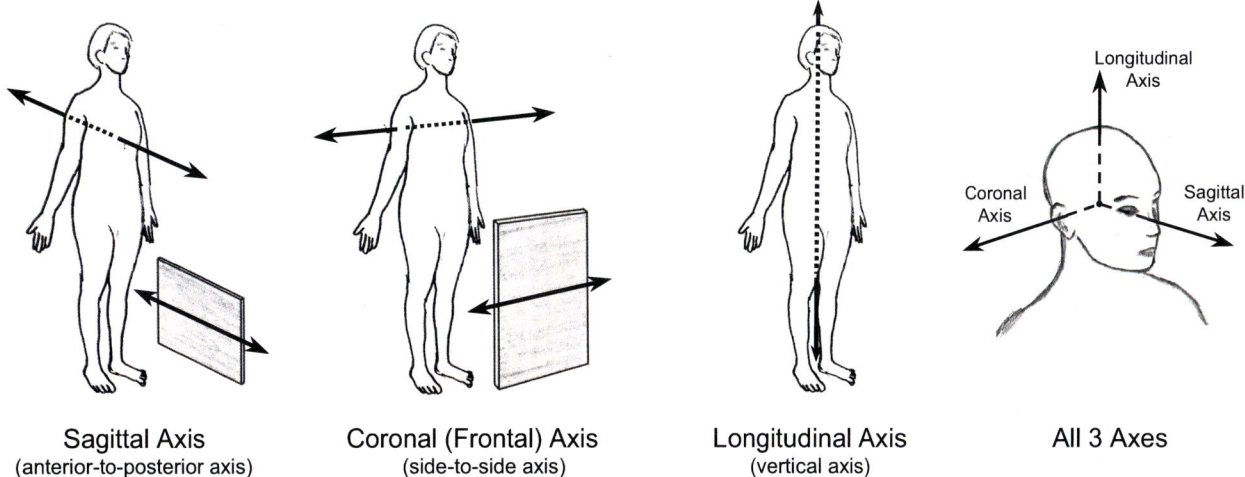

Sagittal Axis (anterior-to-posterior axis) | Coronal (Frontal) Axis (side-to-side axis) | Longitudinal Axis (vertical axis) | All 3 Axes

Moving About an Axis

To visualize moving about an axis, consider an axis skewering a block of wood. Assume the hole the axis is going through is loose enough to allow the block to spin on the axis. The block spinning on the axis is moving <u>in</u> a plane that is perpendicular (at right angles) to the axis.

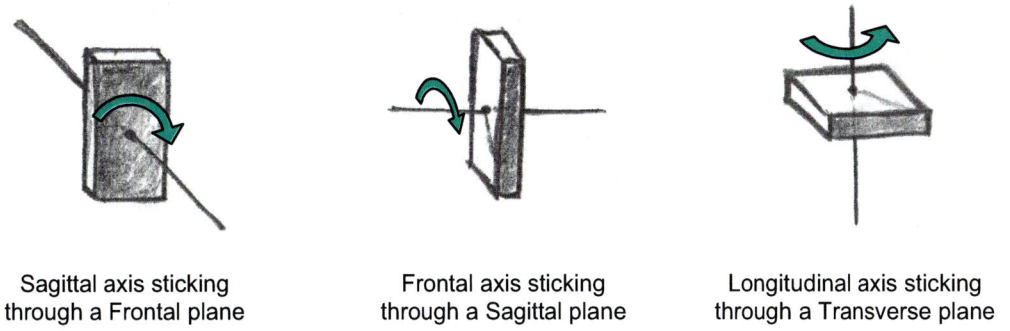

Sagittal axis sticking through a Frontal plane | Frontal axis sticking through a Sagittal plane | Longitudinal axis sticking through a Transverse plane

Putting It All Together

As a body part moves in a given plane, the joint turns about an axis. The axis is perpendicular (at right angles) to the plane, and the axis passes through the joint that is turning as the movement progresses. The following are the three planes and their associated right-angle axes:

Plane/Axis Pair	Movements in the Plane & about the Axis (see page 7)
1. Sagittal Plane & Coronal Axis	• Flexion, Extension
2. Coronal Plane & Sagittal Axis	• Abduction, Adduction, Lateral Flexion
3. Tranverse Plane & Longitudinal Axis	• Rotations

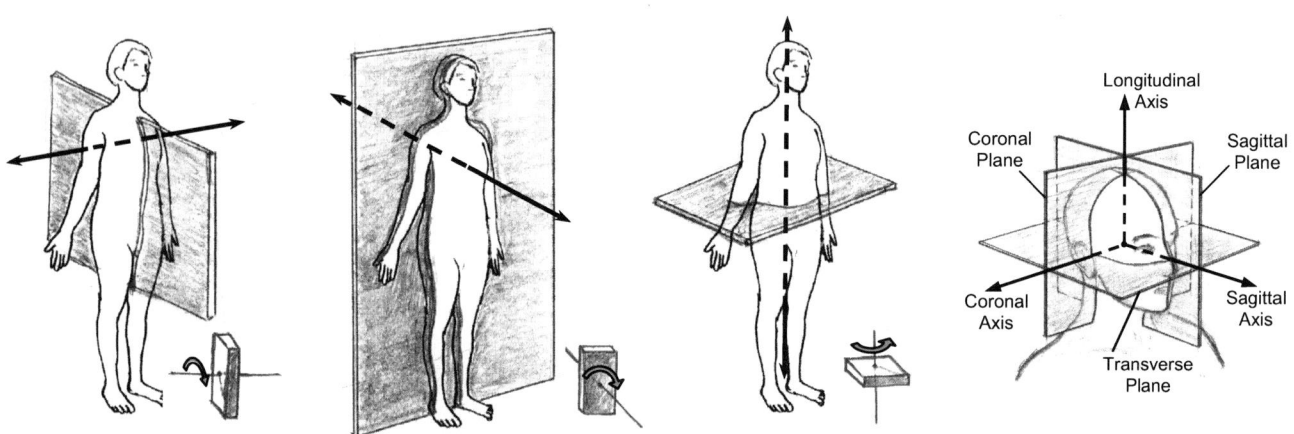

Sagittal Plane goes with Coronal Axis   Coronal Plane goes with Sagittal Axis   Transverse Plane goes with Longitudinal Axis   The Whole System

When talking about movement, the plane is usually named when describing a body part moving through space, and the axis is used when describing what the joint is doing at the point where the two bones articulate. For example, "When Brenda's arm swung forward in the sagittal plane, the shoulder joint was turning around a coronal axis."

Non-Planar Movements

Many movements a person does in real life are complex and not purely in one plane. The plane and axis system defined above, however, allows us to categorize and analyze body movement for the study of kinesiology.

For example, movements that do not fit into the simple 3-plane/3-axis system are:

- Oblique/diagonal/circular movements: They are analyzed as combinations and sequences of the basic planar movements.

- Specialty actions: Elevation, depression, lateral deviation (see page 8) – are sliding/gliding movements, so they don't turn about an axis. These movements are named and used as "exceptions" to the basic 3-axis system.

Skeletal System – The Bones

This section gives a brief overview of the skeletal system, which is made up of bones and cartilage. The human body has (at least) 206 bones. The bones are connected to each other at the joints to form the skeleton, which is the internal framework of the body. The joints, also called articulations, will be discussed in the next section. A complete skeleton is shown on page 12.

A primary function of the skeleton is to give support and shape to the rest of the body. It also provides protection for vital organs such as the heart, lungs, brain, and spinal cord. Finally, and most important for the study of kinesiology, bones and joints form a system of levers that muscles attach to and pull on to create body movement.

The skeleton is divided into two major parts (see page 12):

1. Axial Skeleton - The central structure of the body: Head, spine, ribcage
2. Appendicular Skeleton - The extremities: Arms & shoulder girdles, legs & hip bones

Basic List of Bones

Here is a basic list of bones to get started. As we study each section of the body we'll add more details about the bones in that part of the body. A complete list of bones, and details about each individual bone of the body are presented in **Chapter 2 – Bones, Joints and Bony Landmarks**.

Axial Skeleton	Appendicular Skeleton

Skull:
    Cranial bones:
        Occiput, Parietal, Temporal
        Frontal, Sphenoid, Ethmoid
 	
    Facial bones:
        Zygomatic, Maxilla, Nasal, Lacrimal,
        Palatine, Vomer, Inferior Nasal Concha
 	
    Mandible (technically also a facial bone)
 	
Hyoid
 	
Spine:
    Cervical Vertebrae
    Thoracic Vertebrae
    Lumbar Vertebrae
    Sacrum
    Coccyx
 	
Ribs
Sternum
	Upper Extremity:
	    Clavicle
	    Scapula
	    Humerus
	    Ulna
	    Radius
	    Carpals
	    Metacarpals
	    Phalanges of the hand
	
	Lower Extremity:
	    Coxal (hip) bone:
	        Ilium
	        Ischium
	        Pubis
	    Femur
	    Patella
	    Tibia
	    Fibula
	    Tarsals
	    Metatarsals
	    Phalanges of the foot

The Skeletal System

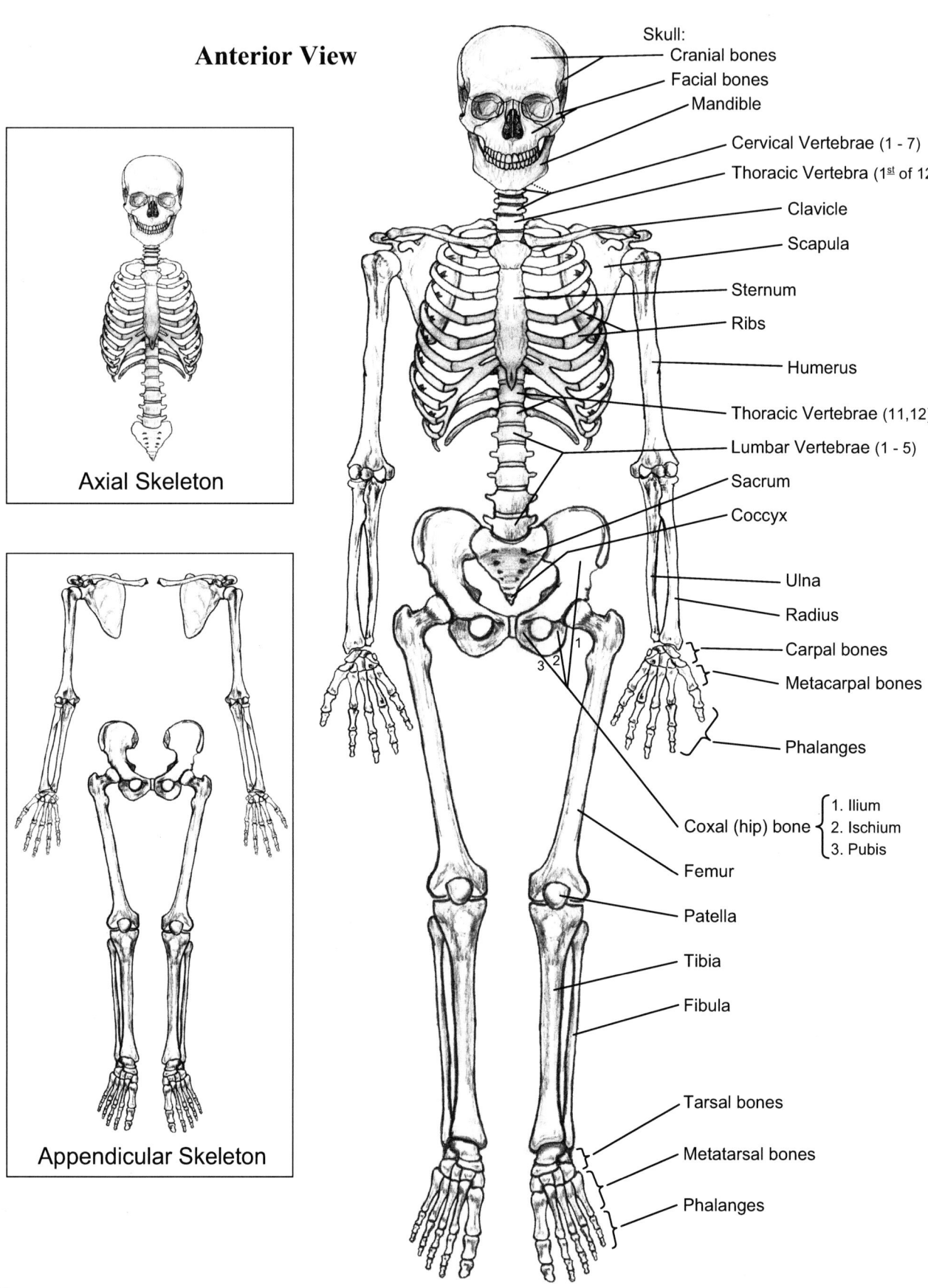

Classification of Bones by Shape

One way bones are classified is by their general shapes. Many individual bones, however, do not easily fit into one of these categories

1. Long Bones - Shaft with widened articulating ends (humerus, fibula, phalanges)
2. Short Bones - More or less cube-shaped (carpal bones, most tarsal bones)
3. Flat Bones - Have flat broad surfaces (scapula, ribs, ilium, parietal)
4. Irregular Bones - "Other" varied shapes (vertebrae, sphenoid, calcaneus)
5. Sesamoid Bones - Oval, small, suspended in tendon (patella, pair under base of big toe)

Bony Landmarks

Bony landmarks are words that name specific locations and features on bones. These words are used when specifying origins and insertions of muscles, i.e., the places where muscles attach to bones. **Chapter 2 – Bones, Joints and Bony Landmarks** has detailed drawings showing all bony landmarks used in this book. Here are a few sample bony landmark drawings (see Chapter 2, pages 40, 41 and 44 for full size versions).

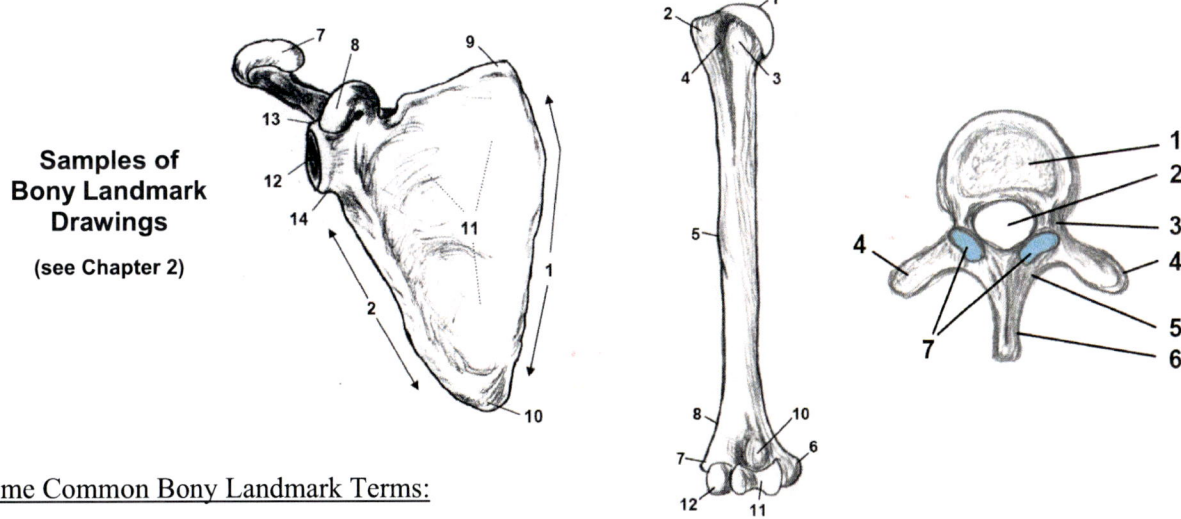

Samples of Bony Landmark Drawings
(see Chapter 2)

 Some Common Bony Landmark Terms:

Process – A part of a bone that "sticks out"

Tubercle, Tuberosity – A bump or bulging place on a bone

Fossa – A smooth, flat part of a bone (often slightly concave)

Head – Enlarged rounded end of a long bone

Condyle – Dual rounded protruding ends of a long bone that articulate with the next bone

Epicondyle - A place on a long bone just above the condyle

Foramen – A hole in a bone; vessels, nerves or other structures pass through the hole

Articular System – The Joints

A joint (also called an articulation) is the point of contact between two bones, between a cartilage structure and a bone, or between teeth and bones. Joints are the structures that allow the individual "rigid" bones of the skeleton to assemble into a freely moving body.

Throughout this book a special symbol ◄► is used to indicate the meeting point of the bones that make up a joint. For example, the tibiofemoral joint is the connection of the femur and the tibia at the knee. This could be represented by "femur ◄► tibia", or for greater detail, "condyles of femur ◄► condyles of tibia".

Broad Classifications of Joints - by Structure and Function

Joint structure determines function. The physical *structure* of a joint includes the shape of the articulating surfaces of the bones, how tightly they fit, and the types of tissue that hold the bones together. The *function* of a joint indicates how it moves (or doesn't move). There are three broad categories of joints:

Function	Structure	Examples
1. Synarthrotic (immovable)	Fibrous	Sutures, teeth in sockets, 1st rib ◄►sternum
2. Amphiarthrotic (slightly moveable)	Cartilagenous	Intervertebral discs, pubic symphysis, manubrium ◄► body of sternum
3. Diarthrotic (freely moveable)	Synovial	Most joints in the body; synovial joints are discussed in detail below

Stability vs. Mobility Trade-off

Stability is the ability of the body to maintain its integrity and form and to resist injury. *Mobility* is the ability of the body to move freely as required for the activities of life. A joint may allow a great deal of motion, as in the shoulder, or very little motion as in the tibiofibular joint. All joints function within an inherent trade-off: Joints that are more moveable provide less stability, while joints that are more stable tend to have less movement ability.

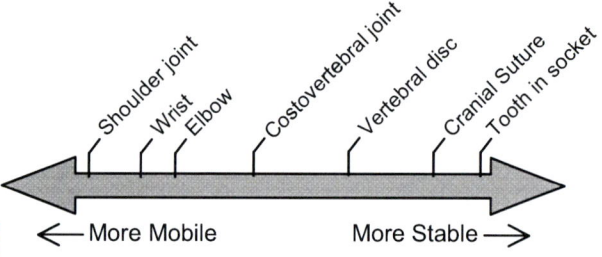

Synovial Joints – Structure

All synovial joints have the same structural components:

- Two bones that articulate
- Articular cartilage on each bone
- Joint capsule (fibrous outer shell, reinforced by ligaments)
- Synovial membrane (inner lining of capsule)
- Joint cavity
- Synovial fluid (in the cavity)

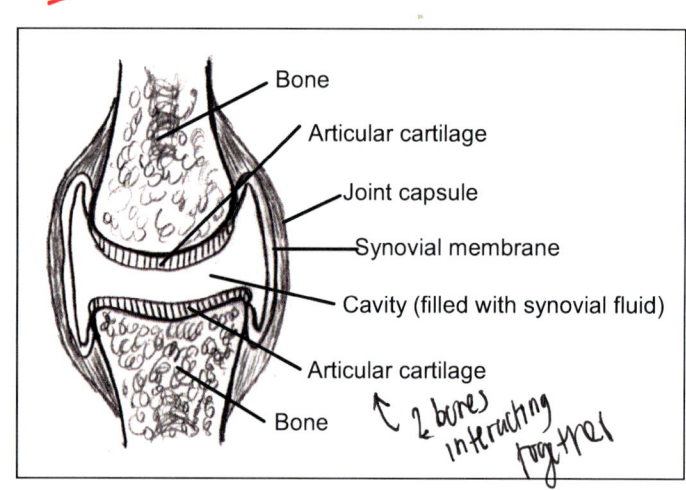

Structure of a Typical Synovial Joint

In addition to the basic joint capsule, synovial joints may be supported by one or more **accessory structures**. These structures include ligaments, cartilage pads, bursae, and fat pad "packing material".

Ligaments – Strong connective tissue bands that connect bone to bone and provide protection against the joint moving too far and becoming damaged. Most ligaments are outside the joint capsule (extracapsular ligaments) and span the bones that make up the joint. In addition, ligaments can be embedded within the fibrous material of the joint capsule itself, or can be completely inside the joint cavity (intracapsular ligaments).

Cartilage pads (small discs or menisci) – Extra padding, protection, shaping, and containment inside the synovial cavity.

Bursae – A bursa (plural is bursae) is a sac containing synovial fluid. The composition of the sac is similar to a joint capsule, i.e., fibrous connective tissue lined with a synovial membrane that secretes synovial fluid. These fluid-filled sacs serve as shock absorbers or reduce friction between moving structures, and are primarily where a tendon may rub on a bone.

Six Types of Synovial Joints

There are six types of synovial joints, based on commonality of bone shapes and supporting structures. Each type has a characteristic bone-shape/capsule/ligament arrangement that allows a certain set of actions.

A mnemonic that may help you remember the six types is BS-PHEGS.

	Joint Type	Description	# of Axes	Action Pairs	Examples
BS	Ball & Socket	A rounded end on one bone fits into a cupped socket on the other bone	Triaxial	Flexion, Extension Abduction, Adduction Medial & Lateral Rotation	Glenohumeral joint (shoulder joint), Coxal joint (hip joint)
P	Pivot	A rounded projection on one bone fits into a ring formed by bone and ligament	Uniaxial	Rotation (Medial and Lateral Rotation, or Right and Left Rotation)	Radioulnar joint, Atlantoaxial joint (dens part)
H	Hinge	Cylindrical surfaces fit together like a door hinge	Uniaxial	Flexion, Extension	Humeroulnar joint, Interphalangeal joint
E	Ellipsoid / Condyloid	A shallow rounded end on one bone meets an oval depression on another bone	Biaxial	Flexion, Extension Abduction, Adduction	Radiocarpal joint, Metacarpophalangeal joint
G	Gliding	Flat or slightly curved surfaces allow sliding in all directions	Non-axial	Gliding	Intervertebral facet joints, Intercarpal joints
S	Saddle	Surfaces resemble saddles – convex one way and concave the other	Biaxial+	Flexion, Extension Abduction, Adduction, Opposition (facilitated by a specialized type of rotation)	Carpometacarpal joint #1 (base of thumb near the wrist)

Joints of the Upper Extremity

To introduce the features of synovial joints, the following table presents joints of the upper extremity, beginning with the glenohumeral joint and moving distally. Note that many of the joint names are created by simply naming the bones that make up the joint.

The upper extremity contains at least one of each type of synovial joint. Compare the actions of each joint in this table with the six joint types (BS-PHEGS) listed on the previous page.

Joint Name	Common Name	Bones Involved (& bony landmarks)	Type of Joint	Actions
Glenohumeral Joint	Shoulder joint	Scapula ◄► Humerus (glenoid fossa) (head)	Ball & Socket	Flexion, Extension Abduction, Adduction Rotation (medial & lateral) Horizontal Abduction Horizontal Adduction
Scapulothoracic Joint	Shoulder blade	Scapula ◄► Ribs (subscapular fossa) (posterior surfaces) -with muscles sandwiched between-	"False" Joint	Elevation, Depression Abduction, Adduction Rotation (upward & downward)
Humeroulnar Joint	Elbow	Humerus ◄► Ulna (trochlea) (trochlear notch)	Hinge	Flexion, Extension
Radioulnar Joint (proximal)	Forearm	Radius ◄► Ulna (head) (radial notch)	Pivot	Supination, Pronation (i.e., rotations)
Radiocarpal Joint	Wrist	Radius ◄► Carpal bones (distal end) (proximal row)	Ellipsoid	Flexion, Extension Abduction, Adduction
Carpometacarpal Joints – digits #2-#5: fingers	Mid-palm of the hand	Carpal bones ◄► Metacarpals (distal row) (bases)	Gliding	Gliding
Carpometacarpal Joint – digit #1: thumb	Base of thumb at the wrist	Carpal bone ◄► Metacarpal #1 (trapezium) (base)	Saddle	Flexion, Extension Abduction, Adduction Opposition
Metacarpophalangeal Joints	Knuckles of the hand	Metacarpals ◄► Phalanges (heads) (bases)	Condyloid	Flexion, Extension Abduction, Adduction
Interphalangeal Joints (PIP-Proximal Interphalan..) (DIP-Distal Interphalangeal)	Fingers	Phalanges ◄► Phalanges (heads) (bases)	Hinge	Flexion, Extension

Joints of the Upper Extremity

The joints of the upper and lower extremities and the axial skeleton are described in detail later in this book, as they relate to the groups of muscles in Chapters 4 through 6. Also, a complete list of the joints of the body is included in Chapter 2 – Bones, Joints, and Bony Landmarks.

Joints of the Body

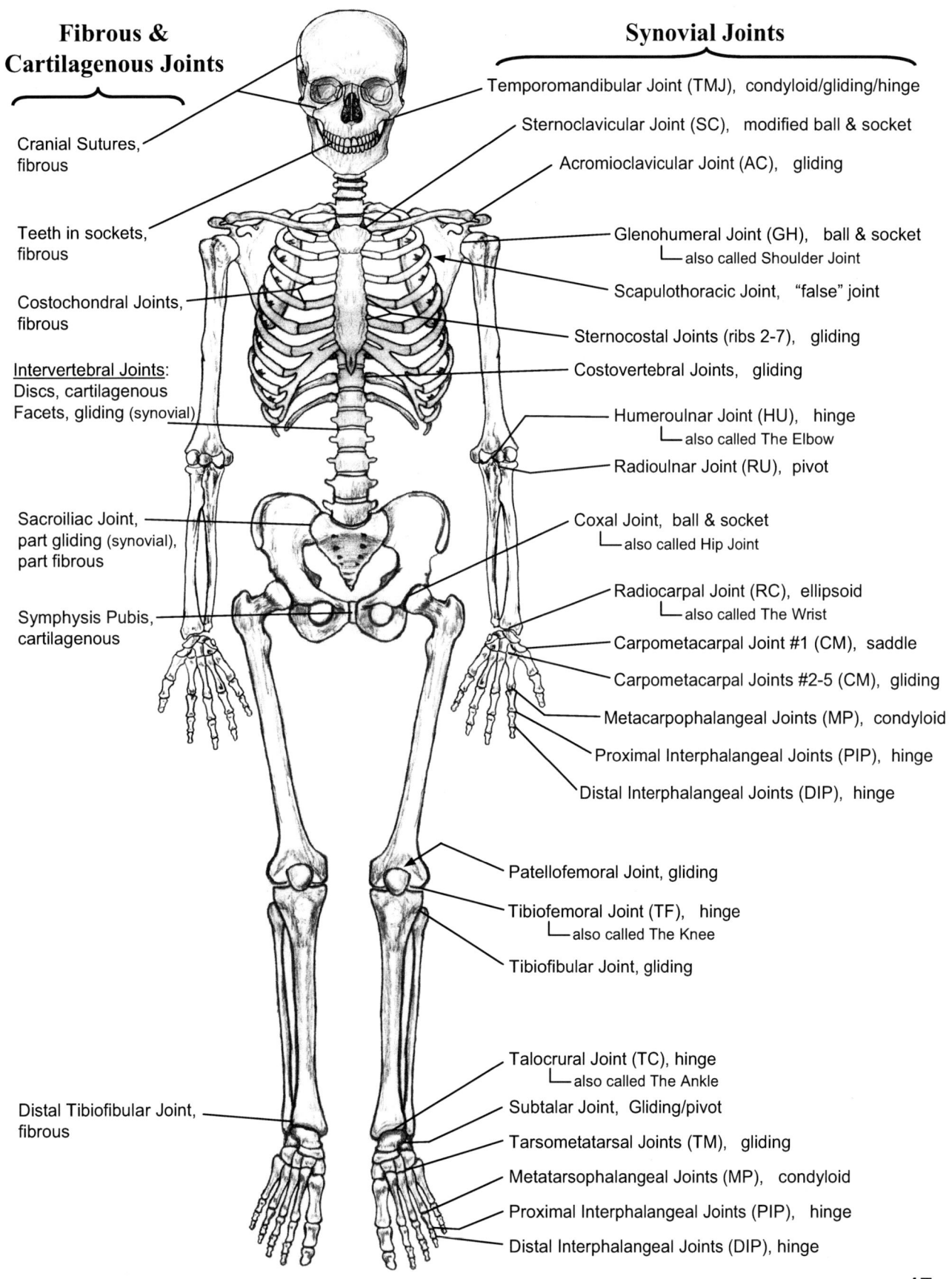

Range of Motion (ROM)

Range of motion (ROM) is a term that applies to each joint in the body. ROM is the amount of motion, usually expressed in degrees, allowed by the shape of the joint and the soft tissue surrounding it. The range is defined separately for each action the joint is capable of doing.

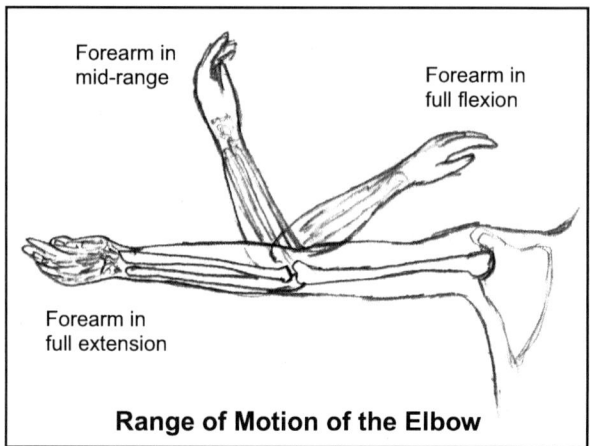

Range of Motion of the Elbow

Each joint has a "normal" range through which it can move, e.g., from full extension to full flexion. The range is dictated by bone shapes, angles of articulation, and any normal inhibiting structures. Normal inhibiting structures include ligaments, joint capsule, muscle length, tissues, other bones, etc.

Abnormal ROM can mean a pathology exists.
1. Less than normal range (restricted, limited, what is stopping it?)
2. More than normal range (hypermobile, loose, what is *not* stopping it that should?)

Recall the three axes of movement described earlier in this chapter (p. 9-10). ROM is considered <u>individually</u> for each axis of movement available for a joint. A *pair* of opposite actions take a joint through its full range of motion about an axis. For example, going from maximum flexion through to the limit of extension travels the full ROM (moving in the extension direction) for a joint moving around its frontal axis.

Most reference books list ROM as number of degrees moved from anatomical position for each half of an action pair. For example, a "normal" wrist joint allows 80° of flexion and 65° of extension from anatomical position.

Joint Degrees of Freedom

A joint's capability can be classified by how many of the three axes it can move around:

uniaxial = 1 axis, **biaxial** = 2 axes, **triaxial** = all 3 axes

More axes mean the joint has more movement options.

Flexion & Extension pair	= one axis
Abduction & Adduction pair	= a second axis
Rotation pair	= a third axis

So, for the six joint types….

1. A Ball and Socket is a triaxial joint (it does all three of the above action pairs)
2. A Pivot is a uniaxial joint (it does only rotation)
3. A Hinge is a uniaxial joint (it does only flexion/extension)
4. An Ellipsoid (Condyloid) is a biaxial joint (it does flexion/extension and abduction/adduction)
5. A Gliding is a non-axial joint (it does small sliding movements, not moving about an axis)
6. A Saddle is a biaxial+ (does flexion/extension, abduction/adduction, and a limited form of rotation)

Muscular System – The Muscles

Anatomy and Function

This section gives a brief overview of the anatomy and function of skeletal muscles. The figure below shows the components of a typical skeletal muscle.

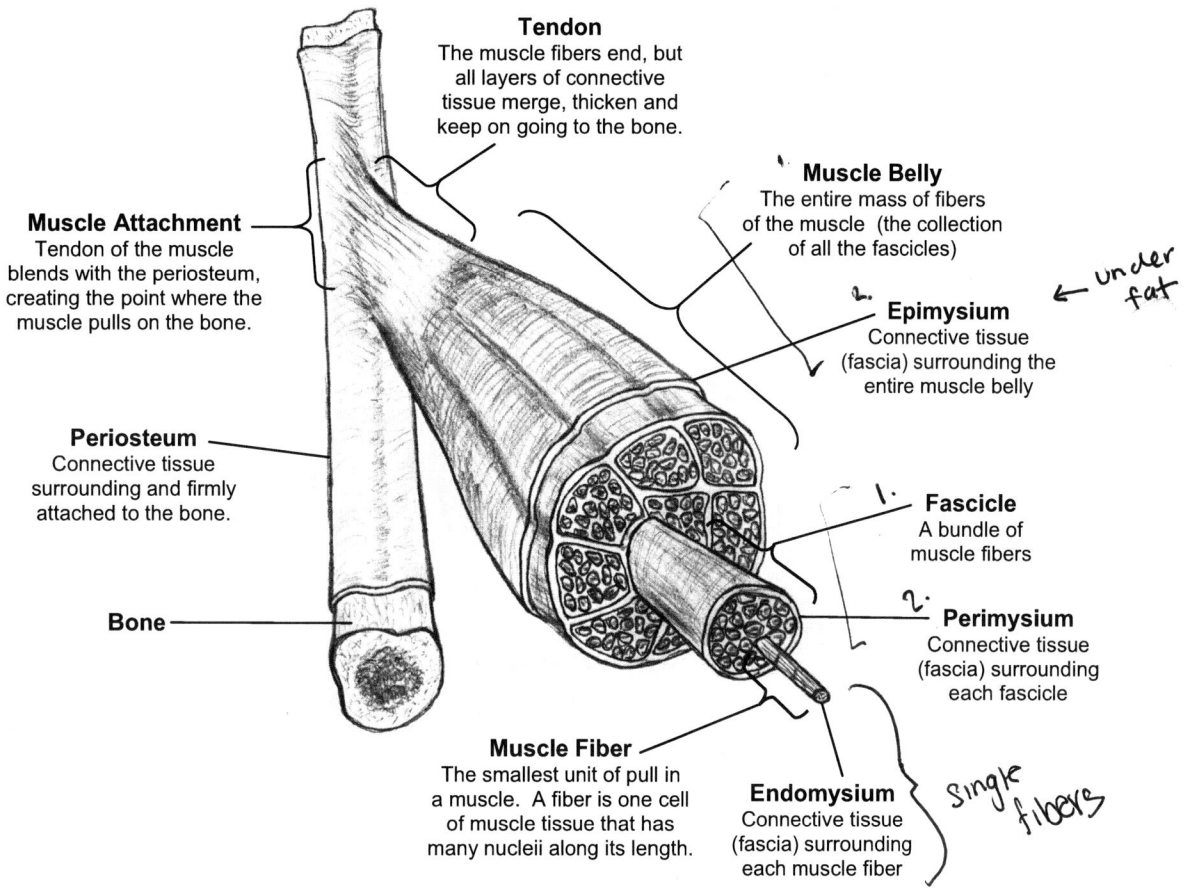

Anatomy of a Typical Skeletal Muscle

The muscle belly has a striated look and feel due to the muscle fibers being lined up with each other. Most muscle fibers run the entire length of the belly, from tendon to tendon. When relaxed, the muscle feels soft and has a neutral length and girth. When stimulated by the nervous system the muscle contracts, and the muscle belly feels firm/hard. If the bones are moveable, the muscle becomes shorter and thicker as it moves the bones. Conversely, when stretched beyond the relaxed/neutral state by an outside force, the muscle gets longer and thinner and develops a taut feel.

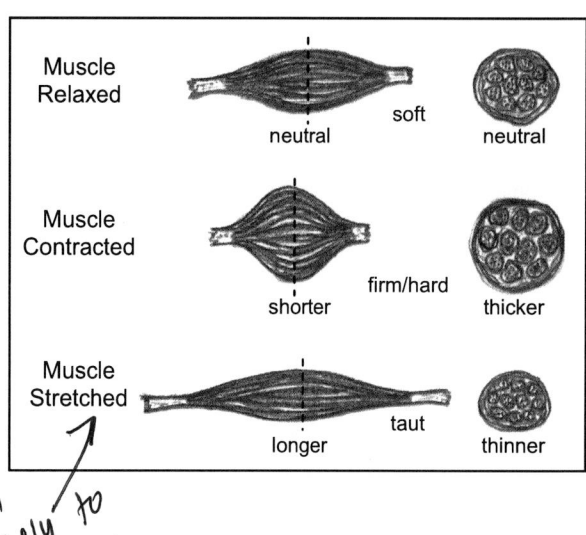

Line of Pull

Generally, a muscle attaches to two bones and crosses the joint between those two bones (note that there are many exceptions to this). The direction of the muscle fibers delineates a *line of pull* between the bones. When the muscle contracts, the bone that is more moveable moves toward the bone that is more stable and the joint rotates through an angle.

It is usually possible to visualize what body movements (actions) a muscle will create. First, observe the shape and fiber direction of the muscle and the points where its tendons attach to the bones. This creates a line of pull on the bones. Next, look where that line of pull is positioned relative to the joint between the bones. Finally, visualize how the bones will move around the joint when the line of pull shortens. Knowing the type of joint (see page 15) helps you know what actions are allowed.

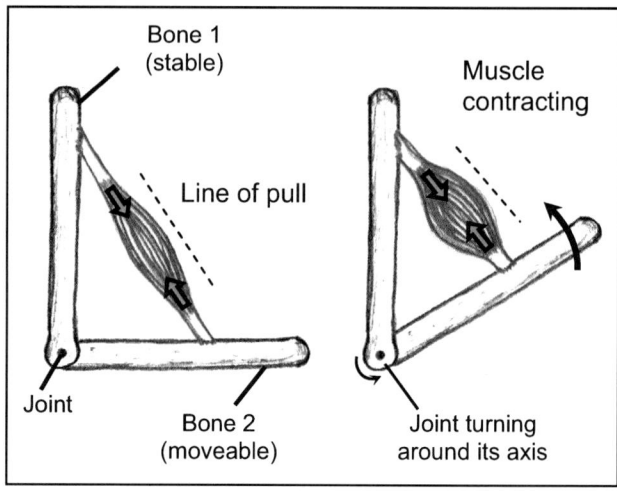

Schematic of a Muscle Contracting

Fiber Arrangements

The arrangement of the muscle fibers has a direct influence on how a muscle pulls on its attached bones. Speed of movement, range of movement, strength of pull, and direction(s) of pull are all related to the fiber lengths and angles. Here are the major categories:

1. **Fusiform** – Long fibers closing to a narrow tendon at each end. Can create long range and high speed of movement.

2. **Pennate** – Many short fibers connecting at oblique angle to a central tendon (creating a feather-like look). Has lots of fibers so is good for strength, but fibers are short so range is smaller.

3. **Convergent** – Narrow attachment one end, broad attachment other end. This creates a varying angle of pull, so different fiber regions can cause different actions.

4. **Parallel** – Wide with all parallel fibers; all fibers pull in the same direction. These can be thin with fewer fibers or thick with more fibers (and therefore more strength).

5. **Circular** – Fibers squeeze in to close an opening.

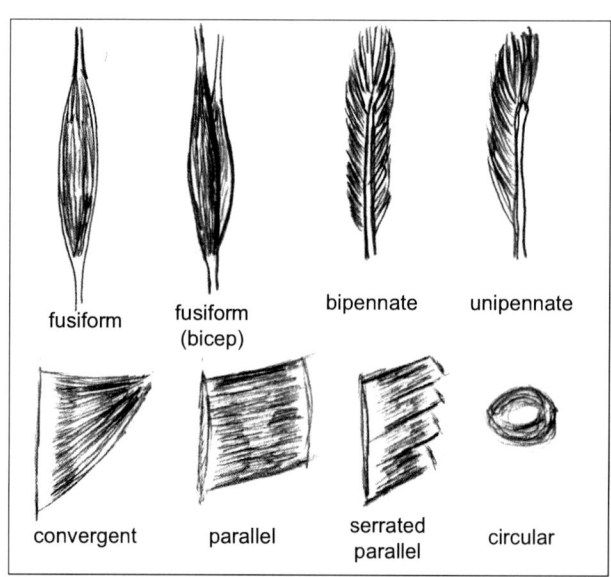

Muscle Fiber Arrangements

Tendons and Attachments

The tendon portion of a muscle can have many shapes: narrow, broad, thick, thin, long, short, nonexistent. Here are some possibilities:

1. Tendons at the two ends of a fusiform muscle attaching to a single point on each bone it pulls.

2. Long tendon where space is limited. The muscle belly is far away from the bone attachment where the pulling force occurs. Long tendons going through narrow spaces usually have synovial sheaths to facilitate easy sliding.

3. Tendon goes around a bend to reach the bone where it attaches, often gaining advantage by changing the direction from which it pulls on the bone.

4. Broad area on a flat bone. Little or no tendon portion; the fascia of the muscle fascicles blends right into the periosteum.

5. Elongated, oblong, odd shaped area on a bone, or multiple bones.

6. Attachment to interosseus membrane between two bones.

7. Attachment to another muscle via fascia or aponeurosis. An *aponeurosis* is a broad, flat tendon that covers a relatively large area on the body.

With few exceptions, each muscle attaches directly to the bones it moves. When a bone has several muscles attaching to it, the attachment sites fit together like puzzle pieces so each muscle has its own point where it pulls on the bone. See page 79 for an example showing the scapula and humerus attachments.

Muscles and Opposing Movements

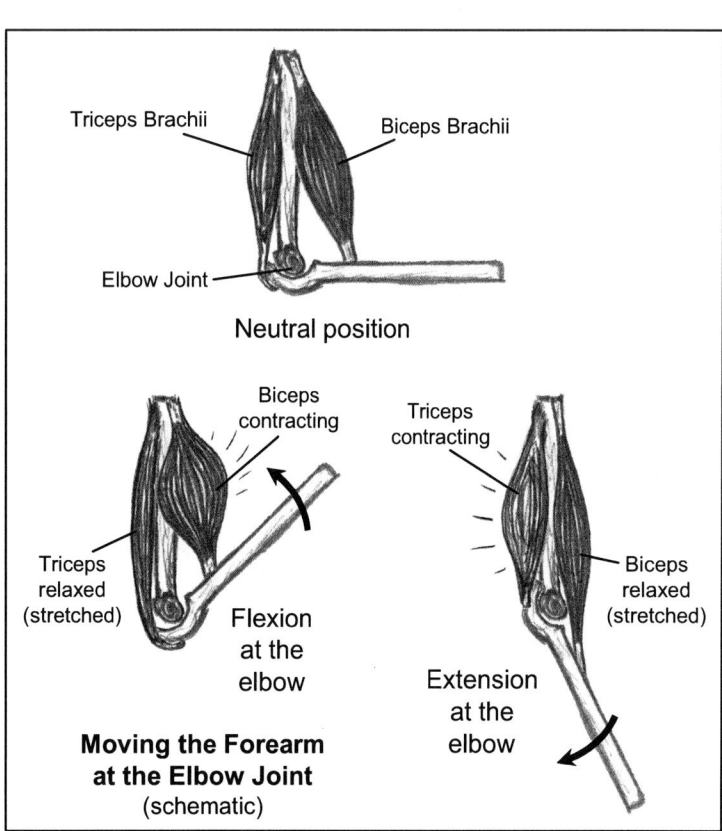

For each action created by a muscle acting on a joint, there is an opposite action created by a muscle pulling on the joint from the opposite side of the body. The figure at the right uses the actions at the elbow joint of the biceps brachii and the opposing triceps brachii to illustrate this concept.

When the biceps brachii contracts it moves the forearm in flexion at the elbow. Meanwhile the triceps must relax and become longer to accommodate the action.

Conversely, when the triceps brachii contracts it creates extension at the elbow (the opposite action to flexion). To allow this, the biceps must relax and become longer.

The nervous system naturally helps this process by reflexively relaxing a muscle if its opponent is being stimulated to contract (this is called *reciprocal inhibition*).

Nervous System – The Nerves

This section gives a brief description of the nervous system as it relates to muscles and movement. The nervous system in its entirety is quite complex, and it is beyond the scope of this book to cover most of its features. Please refer to any good Anatomy & Physiology textbook for more complete information.

The goal of this section is to give you an overview of how the nervous system is organized as it relates to the workings of skeletal muscles and the movements they create. Additionally, this section will define the terminology used in the "Innervation" portions of the "B" tables in Chapters 4, 5, and 6 (see page 82 for a sample of a B table). To help understand this section, please preview the bony landmarks on pages 43-45.

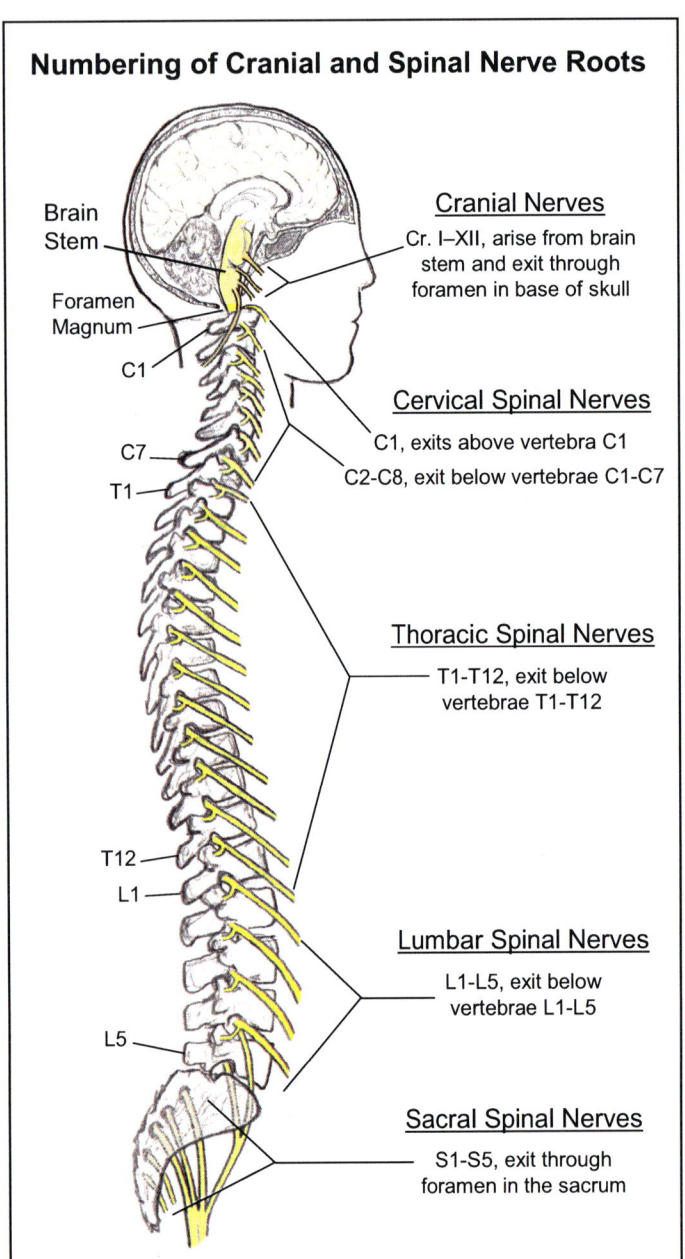

Numbering of Cranial and Spinal Nerve Roots

Brain Stem
Foramen Magnum
C1
C7
T1
T12
L1
L5

Cranial Nerves
Cr. I–XII, arise from brain stem and exit through foramen in base of skull

Cervical Spinal Nerves
C1, exits above vertebra C1
C2-C8, exit below vertebrae C1-C7

Thoracic Spinal Nerves
T1-T12, exit below vertebrae T1-T12

Lumbar Spinal Nerves
L1-L5, exit below vertebrae L1-L5

Sacral Spinal Nerves
S1-S5, exit through foramen in the sacrum

Overall Organization

The human nervous system is divided into the *central nervous system*, which includes the brain and spinal cord, and the *peripheral nervous system*, which includes the nerves that carry signals to and from the parts of the body. For the study of kinesiology, we are primarily interested in the nerves that activate muscles and cause movement (motor nerves). However, sensory nerves also play an important role, both in giving reasons for the body to want to move, and in providing feedback as a movement is in progress.

To understand where the nerves come from that activate muscles we are interested in the brain stem and the spinal cord. The *brain stem* resides inside the cranial vault along with the rest of the brain. At its inferior end it becomes the *spinal cord* which passes through the foramen magnum at the base of the skull and travels down through the vertebral column.

Level by level down the vertebral column, groups of nerves emerge from each side of the spinal cord. At each space between two vertebrae, the nerves bundle together into *spinal nerve roots*, exit bilaterally from the vertebral column, and go out into the body. At the point of exit the nerves become part of the peripheral nervous system.

The figure at left illustrates the naming scheme used to identify nerve roots. The cranial nerves are numbered with roman numerals I-XII, and the spinal nerves are numbered using vertebra-related identifiers C1-C8, T1-T12, L1-L5, and S1-S5.

Spinal Nerve Roots and Rami

The spinal cord passes down the vertebral column inside the *spinal canal*, which is created by the lined-up vertebral foramen of the vertebrae. At each place where two adjacent vertebrae meet, a side channel is created which allows nerves near that level to emerge from the spinal canal and go out into the body.

The side-exit opening is called the *intervertebral foramen*, and it is formed by the meeting of the archways of the pedicle portions of the vertebrae above and below. Note that movements of the spine, such as flexion/extension, right/left rotation and lateral flexion, cause the shape of the intervertebral foramen to change as the pedicles of adjacent vertebrae move relative to each other. This creates a dynamic environment of pressures and strains on the nerve roots.

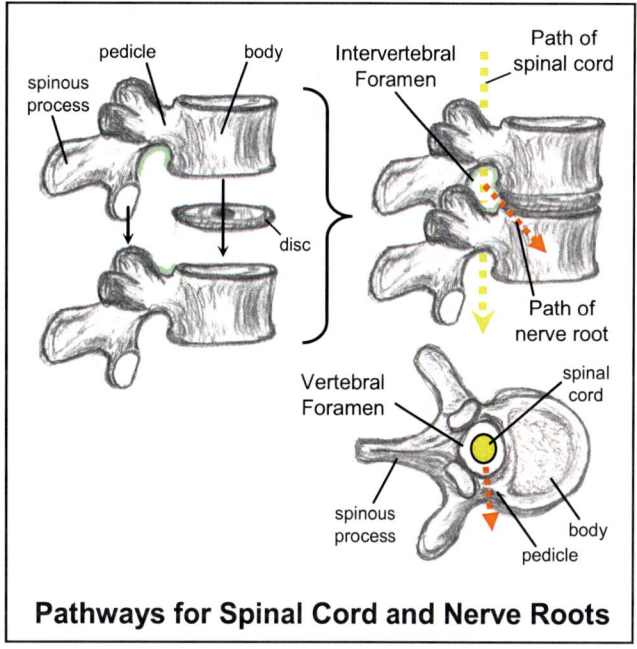

Pathways for Spinal Cord and Nerve Roots

At each segmental level several nerve roots arise from the anterior and posterior aspects of the spinal cord, wrap around laterally, and bundle into a single spinal nerve root on each side. The anterior roots are motor nerves sending impulses out to the peripheral body, and the posterior roots are sensory nerves bringing sensory signals from the body into the spinal cord and up to the brain.

Shortly after a spinal nerve root exits from the vertebral column, it begins branching to take nerves to different places in the nearby body area. The initial, or *primary*, split divides the root into two *rami* (branches). The *dorsal ramus* is smaller and carries nerves directly to the skin and muscles of the back. The *ventral ramus* is much larger because it contains more nerves to supply the relatively larger portion of the body that is anterior to the spine.

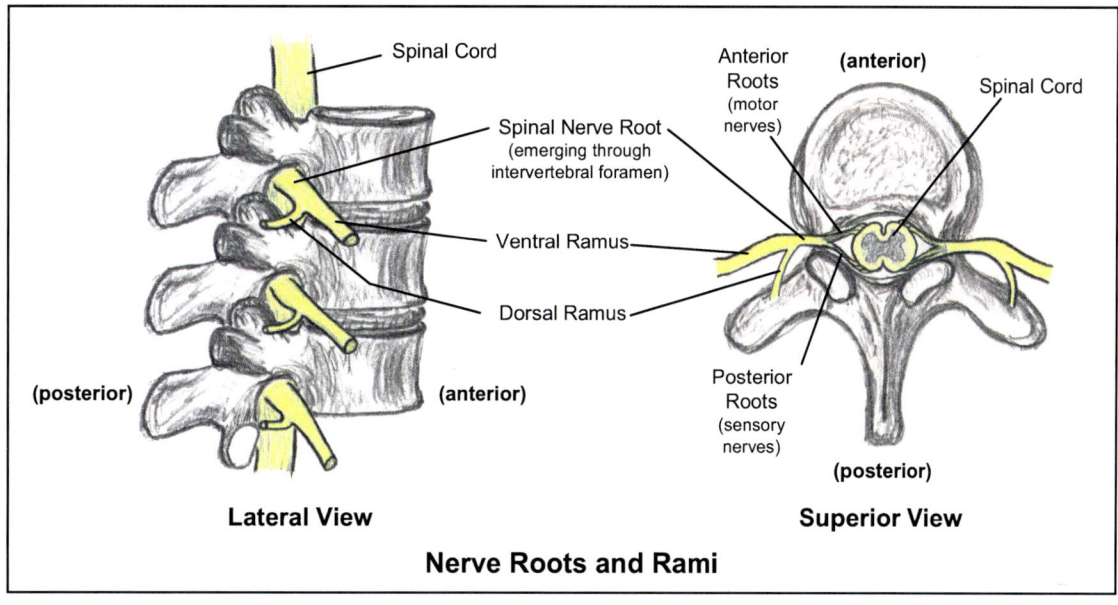

Nerve Roots and Rami

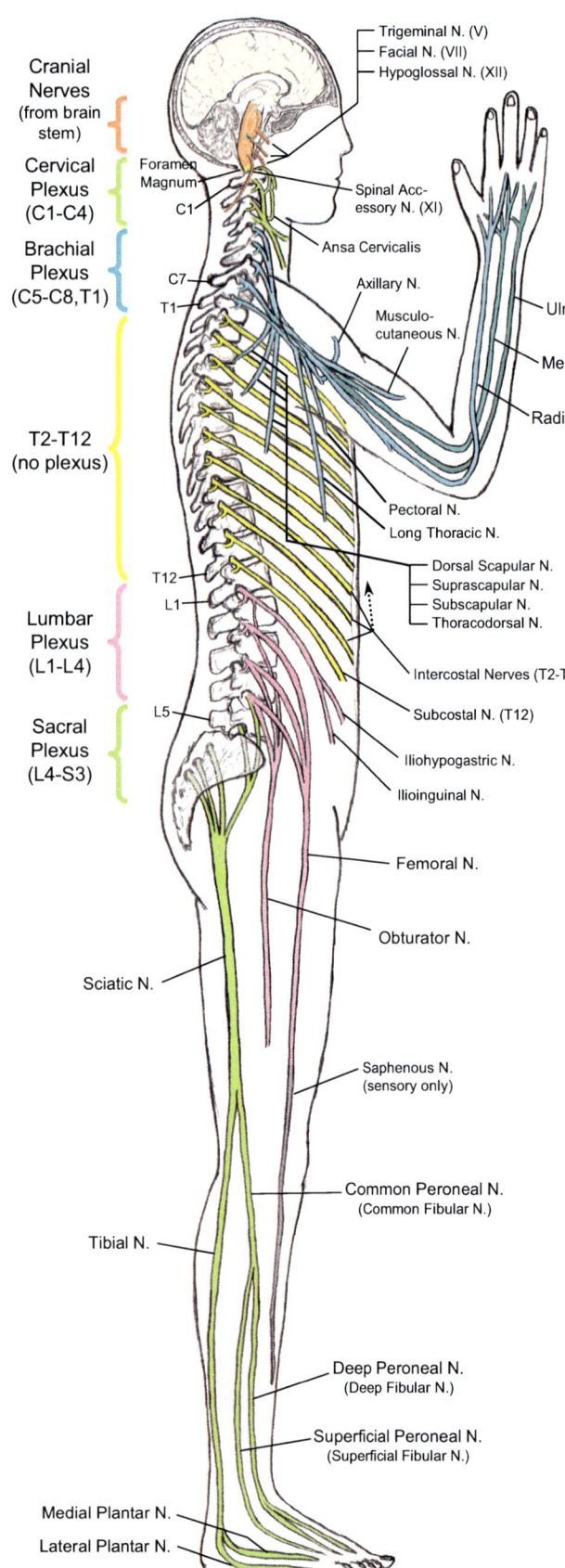

Nerve Plexuses and Naming of Major Nerves

This section completes the description of how the nervous system is organized as it relates to muscles and movement.

The ventral rami of spinal nerves (except for thoracic nerves T2-T12) cross-connect and mingle within their regions to create braid-like collections of nerves called *plexuses*. Each plexus supplies specific regions of the body, as seen in the illustration at left. There are four major plexuses:

Cervical Plexus – originates from segments C1-C4 and supplies muscles of the neck, upper shoulders, and diaphragm.

Brachial Plexus – originates from segments C5-C8 and T1 and supplies muscles that move the shoulder girdle and upper limb.

Lumbar Plexus – originates from segments L1-L4 and supplies muscles of the lower abdomen and anterior and medial aspects of the thigh.

Sacral Plexus – originates from segments L4-L5 and S1-S3 and supplies the buttocks, perineum, posterior thighs, legs, and feet.

Note that the term *lumbosacral plexus* is often used to refer to the lumbar and sacral plexuses combined. Also note that nerve roots T2-T12 do not form a plexus. They individually travel out along the line of each rib.

Names of Major Nerves

Major nerves that carry multiple nerve bundles to specific body areas have names that help identify where they reside in the body. These major nerves are labeled in the illustration at left (Radial N., Femoral N., etc.). The names are used when specifying the innervation of muscles. Many of the major nerves contain nerve bundles that arise from multiple spinal segments due to the cross linking that occurs in plexuses.

How Nerves Work With Muscles

The main function of skeletal muscles is to contract and create movement by pulling on bones or other structures. Muscles only contract when they are stimulated by the *motor nerves* that supply them. Nerve impulses can be sent by the willful control of the brain to accomplish a desired movement (for example, to pick up a book), or by reflex actions that loop back directly from the spinal cord (for example, to jerk away from a flame when sensory receptors sense skin about to get burned).

Each muscle of the body is activated by impulses from one or more specific motor nerves. The set of nerves that control a given muscle are called its *innervation*. For example, the rectus femoris muscle is supplied by branches of the femoral nerve (see page 24).

Motor nerves come from nerve roots that exit from the spine at one or more spinal segments. Also, some motor nerves come from cranial nerves. Therefore a more complete way to specify the innervation of a muscle is to give both its major nerve name and the spinal segments where the nerve originates. To complete the above example: The rectus femoris muscle's innervation is the femoral nerve originating from segments L2, L3, and L4.

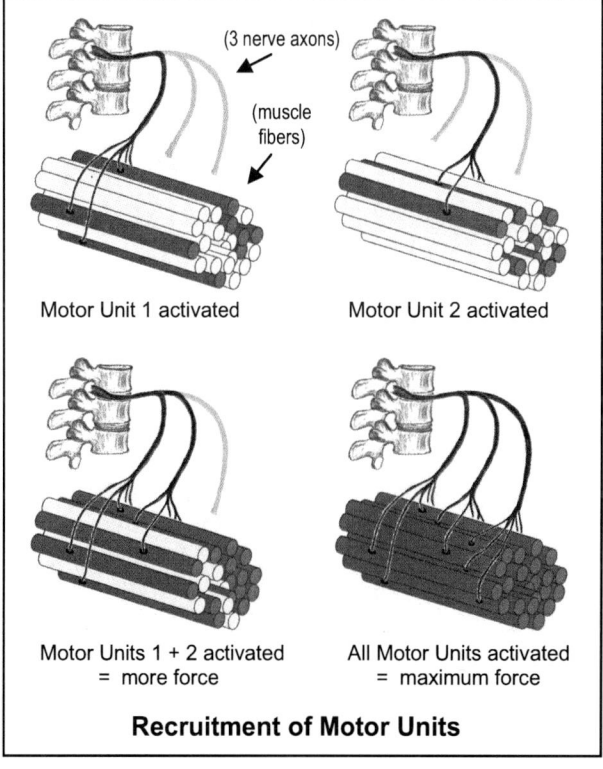

Recruitment of Motor Units

Each muscle fiber within a muscle has a branch of a motor axon from the nervous system connected to it. When the brain sends a nerve impulse down the motor axon, the muscle fiber reacts by making itself shorter in length. This shortening is called a *contraction*, and a muscle fiber contracts fully whenever it is stimulated. Each contraction is "all-or-none", meaning there are no gradations of strength of pull in any one muscle fiber.

A single motor axon divides into many small branches when it arrives at the muscle belly, and each small branch then connects to a single muscle fiber. Thus, the firing of one axon causes a number of fibers to contract. The fibers are typically *not* adjacent to each other. One motor nerve axon, its branches, and the set of connected muscle fibers is called a *motor unit*.

Each motor unit creates a specific amount of pulling force based on the number of fibers contained in the motor unit (more fibers = more force). To gain more pulling force, the brain recruits more muscle fibers by firing the motor axons that connect to additional motor units.

For the study of kinesiology, we are primarily interested in the nerves that activate muscles and cause movement (motor nerves). However, sensory nerves also play an important role, both in giving reasons for the body to want to move, and in providing feedback as a movement is in progress. Sensory nerves include the nerves that sense body movement and position in space (proprioceptive nerves), nerves that sense such things as heat, pressure and pain (sensory nerves), and special senses such as sight and sound that help inform the brain how to orchestrate a series of muscle contractions.

Kinesiology Concepts

When muscles contract they, simply stated, shorten in the line of the fibers. This creates a pulling force at both ends of the muscle where it attaches to bone. Usually one end is attached to a more stable bone, so the other (more moveable) bone moves. Sometimes both bones are free, so they move toward each other.

Movement and Levers

1. Muscles pull bones around a joint (an axis for the movement passes through the joint).

2. Movement occurs by the opening, closing, rotating, and gliding of bones at the joints.

3. The bones act as **Levers,** and the muscles attach to and pull on these levers.

    A leverage unit is made up of a lever with an associated fulcrum, resistance, and effort.

    F   Fulcrum = The joint connecting the two bones

    R   Resistance = The load to be lifted or moved

    E   Effort = The muscle pulling at attachment point

- First-class lever –
    Joint (fulcrum) is between muscle and load.
    Examples: Seesaw = balance, pry bar = power
    For balance, as used in the body.

- Second-class lever –
    Load (resistance) is between joint and muscle.
    Example: Wheelbarrow
    For power, with small range of motion.

- Third-class lever –
    Muscle (effort) is between joint and load.
    Examples: Shoveling, paddling a canoe
    For speed and large range of motion (in the body, the point where the muscle pulls is near the joint).

    Third-class are by far the most common levers in the body.

A mnemonic that may help you remember the three levers is to name what is in the middle:

    Class 1, 2, 3 = F R E

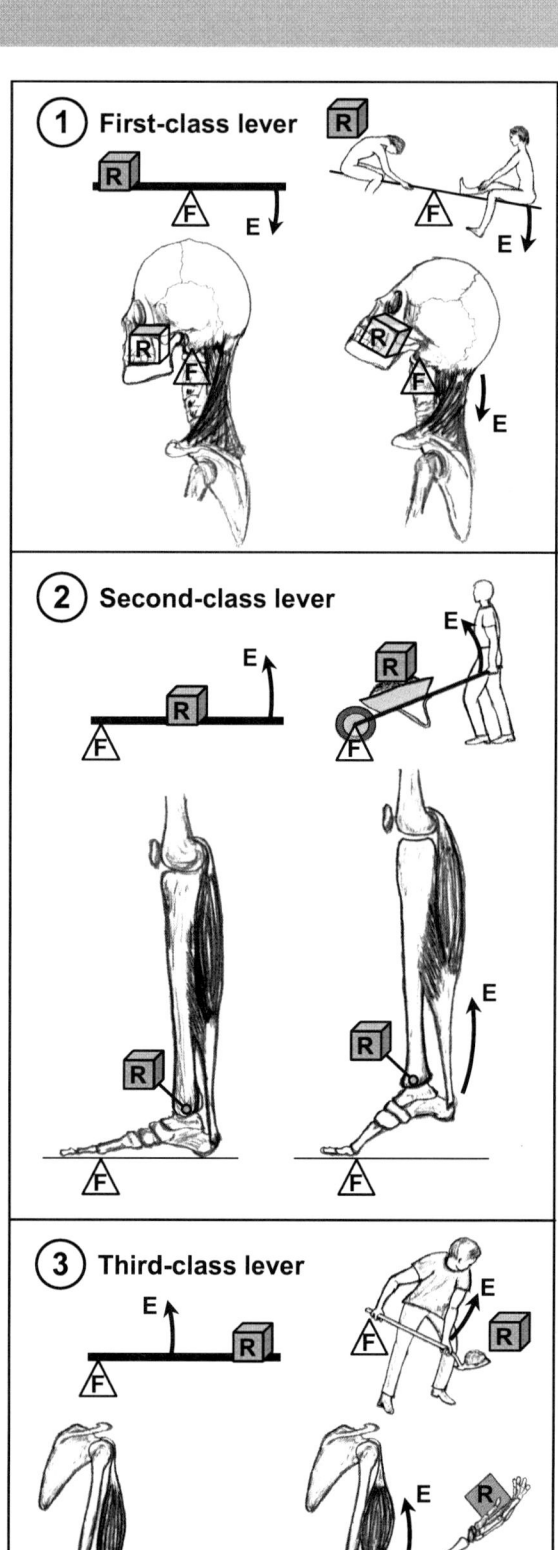

26 Chapter 1 – Basic Information

Roles of Muscle Attachments

Muscle **Origin**

The muscle attachment to the more stable bone is called the *origin*. The origin is usually the more proximal attachment on a limb. On the axial skeleton, the origin is often the more inferior or more medial attachment, but there are many exceptions to this.

Muscle **Insertion**

The attachment to the more moveable bone is called the *insertion*. The insertion is usually the more distal attachment on a limb.

Origins and insertions are shown in red and blue on bone drawings in Chapters 4-6. Knowing the origin vs. insertion can help you visualize the action of a muscle when it contracts.

Muscle **Action**

The movements that the muscle creates when it contracts are called its *actions*. Actions occur at the joint (or joints) that exist between the bone of origin and the bone of insertion.

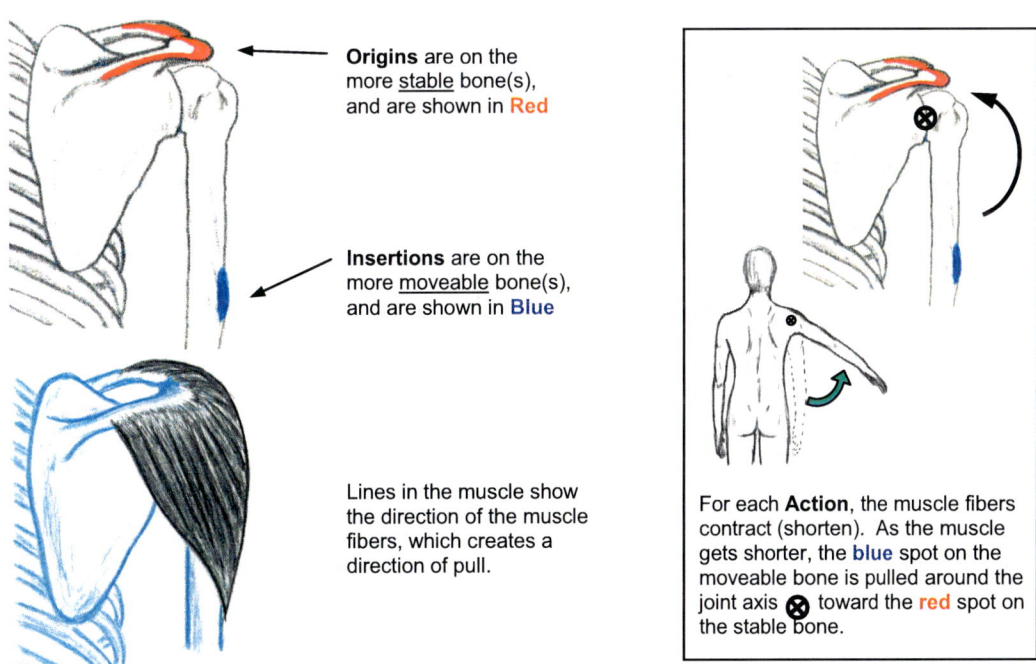

Origins are on the more <u>stable</u> bone(s), and are shown in Red

Insertions are on the more <u>moveable</u> bone(s), and are shown in Blue

Lines in the muscle show the direction of the muscle fibers, which creates a direction of pull.

For each **Action**, the muscle fibers contract (shorten). As the muscle gets shorter, the blue spot on the moveable bone is pulled around the joint axis ⊗ toward the red spot on the stable bone.

Origin, Insertion and Action of the Deltoid Muscle

"Reversed O/I" Muscle Action

The origin, insertion, and action for a muscle are defined based on which bone moves when doing "normal" activities. In many cases, there are ways that the so-called "more moveable" bone can be stabilized, which makes the so-called "more stable" bone then become the moveable one. In that case we do *not* rename the origin and insertion, but instead simply refer to the action as a **reversed O/I action**.

Types of Muscle Contractions

There are three categories of muscle contractions, with an important sub-division for isotonic contractions.

1. **Isotonic** – The muscle is contracting (actively working) and the bone is moving.

    Two types of isotonic contractions:

    Concentric contraction – *Creates* a movement, which is the action of the muscle. The muscle is working and is getting shorter.

    (also called: active shortening)

    Eccentric contraction – *Controls* or slows down a movement that is going in the opposite direction of the action of the muscle. The muscle is working, but is getting longer while working.

    (also called: lengthening contraction)

2. **Isometric** – The muscle is contracting but the bones are *not* moving. No movement occurs either because the muscle force exactly matches the resistance (e.g., when holding a book out in front of you), or because the muscle cannot overcome a stable object (e.g., pushing against a wall).

3. **Tonic** – Sustained small sequencing contractions when the muscle is at rest, or is holding posture.

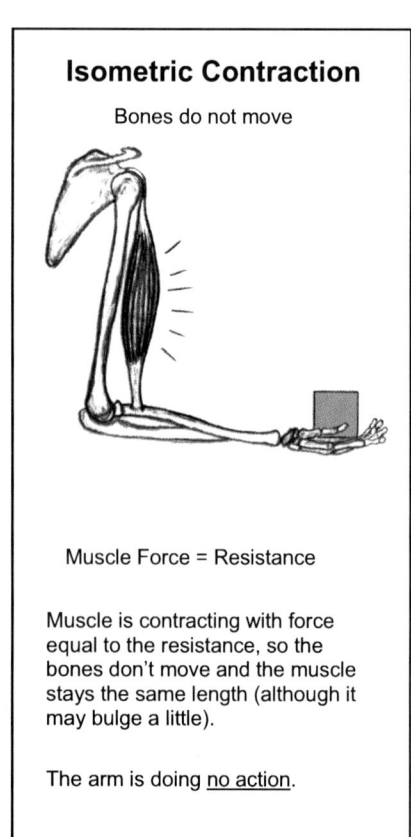

Isometric Contraction

Bones do not move

Muscle Force = Resistance

Muscle is contracting with force equal to the resistance, so the bones don't move and the muscle stays the same length (although it may bulge a little).

The arm is doing no action.

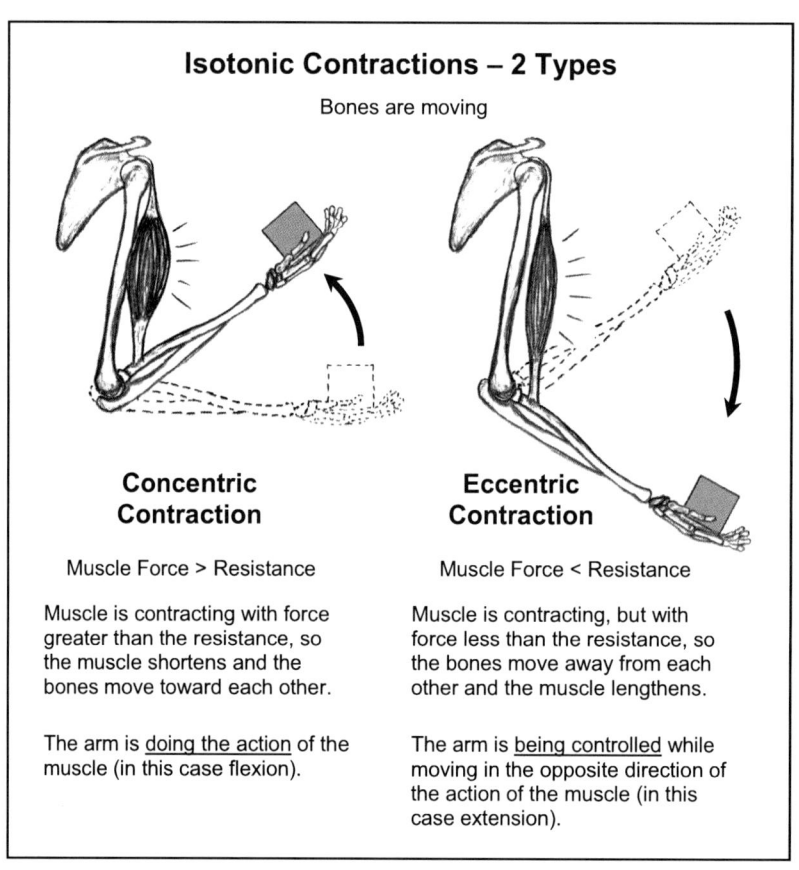

Isotonic Contractions – 2 Types

Bones are moving

Concentric Contraction

Muscle Force > Resistance

Muscle is contracting with force greater than the resistance, so the muscle shortens and the bones move toward each other.

The arm is doing the action of the muscle (in this case flexion).

Eccentric Contraction

Muscle Force < Resistance

Muscle is contracting, but with force less than the resistance, so the bones move away from each other and the muscle lengthens.

The arm is being controlled while moving in the opposite direction of the action of the muscle (in this case extension).

Muscles Working Together

There are several roles a muscle can play when moving a part of the body in a given direction. Sometimes the muscle can be the main driving force of the action, and sometimes the muscle can play a supporting role such as stabilizing another part of the body, helping the action, or slowing down or controlling the smoothness of the action.

1. Agonist

    - The main muscle(s) creating a given action. Sometimes called "prime mover".
    - There can be more than one agonist for a given action at a given joint

2. Synergist

    - "Assist": Helps the agonist with the movement, but with a weaker contribution than the agonist.
    - "May assist": Even weaker, only active when really needed, or when bone angles make it useful.
    - Note: The word *synergist* is also used to indicate <u>all</u> the muscles that contribute directly to a given action at a given joint, as in, "Name all the synergists that create flexion at the elbow."

3. Antagonist

    - Opposes the action of the agonist, thus restricting or controlling the speed and smoothness of the movement.

4. Stabilizer

    - Stabilizes against unwanted movements (involved in coordination, controlling the movement)

Naming the synergists and antagonists for a muscle can *only* be done in the context of a *specific* action of that muscle. For example, biceps brachii is a synergist with brachialis in the action of *flexion*, but is a synergist with supinator in the action of *supination* (see page 90).

Biomechanics

Many mechanical factors are involved when a muscle contracts, affecting the way the bones move, the strength of the pulling force, the speed of the movement, and effects and stresses on other parts of the body. Analyzing these many factors is called the study of *biomechanics*. Biomechanical analysis is beyond the scope of this basic text, but as you learn the muscles and actions in this book you will be naturally applying many of the concepts that are formally organized in the field of biomechanics.

A few of the concepts involved are:

- Angle of pull of a muscle on a bone, or combined angles of pull of multiple muscles on a bone.

- Mechanical advantage created by position of muscle attachment in relation to the joint axis.

- Relationships changing as a joint moves through its range of motion.

- Working with gravity: Stability, equilibrium, balance, lifting objects or body parts.

Range of Motion Procedures

There are many ways to use range of motion (ROM) while working with patients/clients. The following examples require that you have selected a specific muscle you wish to assess, treat, strengthen, or stretch.

Shortening a muscle, i.e., moving the bones in a way that brings the bony attachments of the muscle closer together.

Active Shortening – You instruct your client and they perform the movement using the muscle of interest.

Passive Shortening – The client relaxes while you move their body parts.

Resisted Shortening – The client actively contracts their muscle while you provide resistance. For the muscle to shorten, they "win" the tug-of-war (creating an isotonic, concentric contraction). If their bones stay in place, i.e., their pull matches your resistance, then it is an isometric contraction.

Lengthening a muscle, i.e., moving the bones to pull its bony attachments further apart from each other.

Active Lengthening – You instruct your client to do a movement that is the opposite of the muscle's action (they will be using the antagonist muscle), so the muscle you are interested in gets longer.

Passive Lengthening – You move the client's body parts in a way that makes the muscle belly longer. Sometimes the movement is well within their range of motion, and sometimes you want to lengthen the muscle to the end of its range (stretch it).

Resisted Lengthening – The client actively contracts against your resistance, and your resistance is greater than their pull (you "win" the tug-of-war). This creates an isotonic, eccentric contraction of the muscle.

Uses of ROM

Moving parts of the body while considering range of motion is a useful tool in many ways.

For identification – Use movement while palpating to locate bony landmarks, specific muscles, and other body structures.

For assessment – Use movement to test the client's body for capabilities and/or pathologies.

For treatment – Move the client's body parts as a component of a treatment technique.

Chapter 2
Bones, Joints and Bony Landmarks

Introduction 32

Bones 33
    Bone List 33
    Skeleton – Anterior 34
    Skeletons – Lateral and Posterior 35

Joints 36
    Joint List 36
    Skeleton – Anterior - Joints 37
    Skeleton – Upper Body – Comparison of Bone and Joint Names 38
    Skeleton – Lower Body – Comparison of Bone and Joint Names 39

Bony Landmarks 40

    <u>Upper Extremity</u>
    Scapula 40
    Shoulder Girdle (Scapula & Clavicle) 40
    Upper Arm (Humerus) 41
    Forearm (Radius, Ulna) 41
    Hand 42

    <u>Axial Skeleton</u>
    Skull and Hyoid 43
    Spine (Vertebral Column) 44
    Vertebrae – Special Features by Section 45
    Thorax (Ribs, Sternum) 46
    Rib Articulations 46

    <u>Lower Extremity</u>
    Pelvis 47
    More About Pelvis 47
    Thigh (Femur) 48
    Leg (Tibia, Fibula) 48
    Foot 49

Introduction

Chapter 2 – Bones, Joints and Bony Landmarks provides a central location in this book where all terminology on bones is collected. This information is centralized in one location because you will need to refer to it frequently as you study the muscles in Chapters 4-6. Chapter 2 is primarily an atlas of the **bones** of the body with **bony landmarks** labeled on them. Also included are full skeleton illustrations and some information about **joints**.

The basic anatomy of bones and joints was described in Chapter 1 – Basic Information, pages 11-18.

How This Chapter is Organized

This chapter contains four parts:

1. A list of bones and drawings showing the bones of the body
2. A list of joints and drawings showing the joints of the body
3. Drawings to show a comparison between the names of bones and the names of the associated joints
4. Drawings of the individual bones of the body with landmarks and other features labeled

How To Use This Chapter

To fully learn the bones and bony landmarks, you should be able to recall the information from both visual and verbal directions. That is, when you *read* the name of a landmark you can visualize where it is on the bone, and conversely, when you *see* a place on a bone you can recall its bony landmark name.

Each page of bone drawings is arranged with the bones on one side of the page and a list of bone names, bony landmarks, and joints on the other side of the page. This arrangement allows you to cover the list of names to hide it, and then use the labels on the drawings to test yourself as you memorize the names. This facilitates learning the landmarks from a <u>visual</u> direction.

The Mastering Muscles & Movement **Study Tools** packet includes identical bone drawings, but with the opposite labeling arrangement. That is, the bones have the labels removed from the drawings and the list of bony landmark names is left intact. With this opposite arrangement you can practice learning the bony landmarks by starting with the names and then marking places on the drawings. This helps you memorize the information from a <u>verbal</u> direction. (See Chapter 8 for more information on the MMM Study Tools packet.)

To make your study time even more effective: Add a kinesthetic dimension, i.e., add touching and moving your own body or that of a partner while naming bones, joints, and bony landmarks. Touching body structures for the purpose of identification and assessment is called *palpation*.

Bone List

Axial Skeleton

Skull
    Cranial Bones   (8 total)
        Occiput   1
        Parietal   2
        Temporal   2
        Frontal   1
        Sphenoid   1
        Ethmoid   1

    Facial Bones   (14 total)
        Zygomatic   2
        Nasal   2
        Vomer   1
        Lacrimal   2
        Palatine   2
        Maxilla   2
        Inferior Nasal Concha   2

        Mandible   1

Hyoid

Spine
    Cervical Vertebrae (7)
    Thoracic Vertebrae (12)
    Lumbar Vertebrae (5)
    Sacrum (5 V. fused)
    Coccyx (2-4 V. fused)

Ribs (12 each side = 24 total)
    7 true ribs (each has its own cartilage to sternum)
    5 false ribs
        └── 3 connected to cartilage of rib 7
        └── 2 floating

Sternum

Appendicular Skeleton

Upper Extremity

Clavicle
Scapula
Humerus
Ulna
Radius
Carpals (8) –
    Scaphoid, Lunate, Triquetrum, Pisiform,
    Trapezium, Trapezoid, Capitate, Hamate
Metacarpals (5)
Phalanges of the Hand (14) –
    Digit #1 – Thumb (Pollux) has 2 phalanges
    Digits #2-5 – Fingers have 3 phalanges each
Sesamoids of the Hand

Lower Extremity

Coxal (Hip) Bone
    Ilium
    Ischium
    Pubis
Femur
Patella
Tibia
Fibula
Tarsals (7) –
    Talus, Calcaneus, Cuboid,
    Navicular, 1st, 2nd and 3rd Cuneiforms
Metatarsals (5)
Phalanges of the Foot (14) –
    Digit #1 – Big toe (Hallux), has 2 phalanges
    Digits #2-5 – Toes, have 3 phalanges
Sesamoids of the Foot

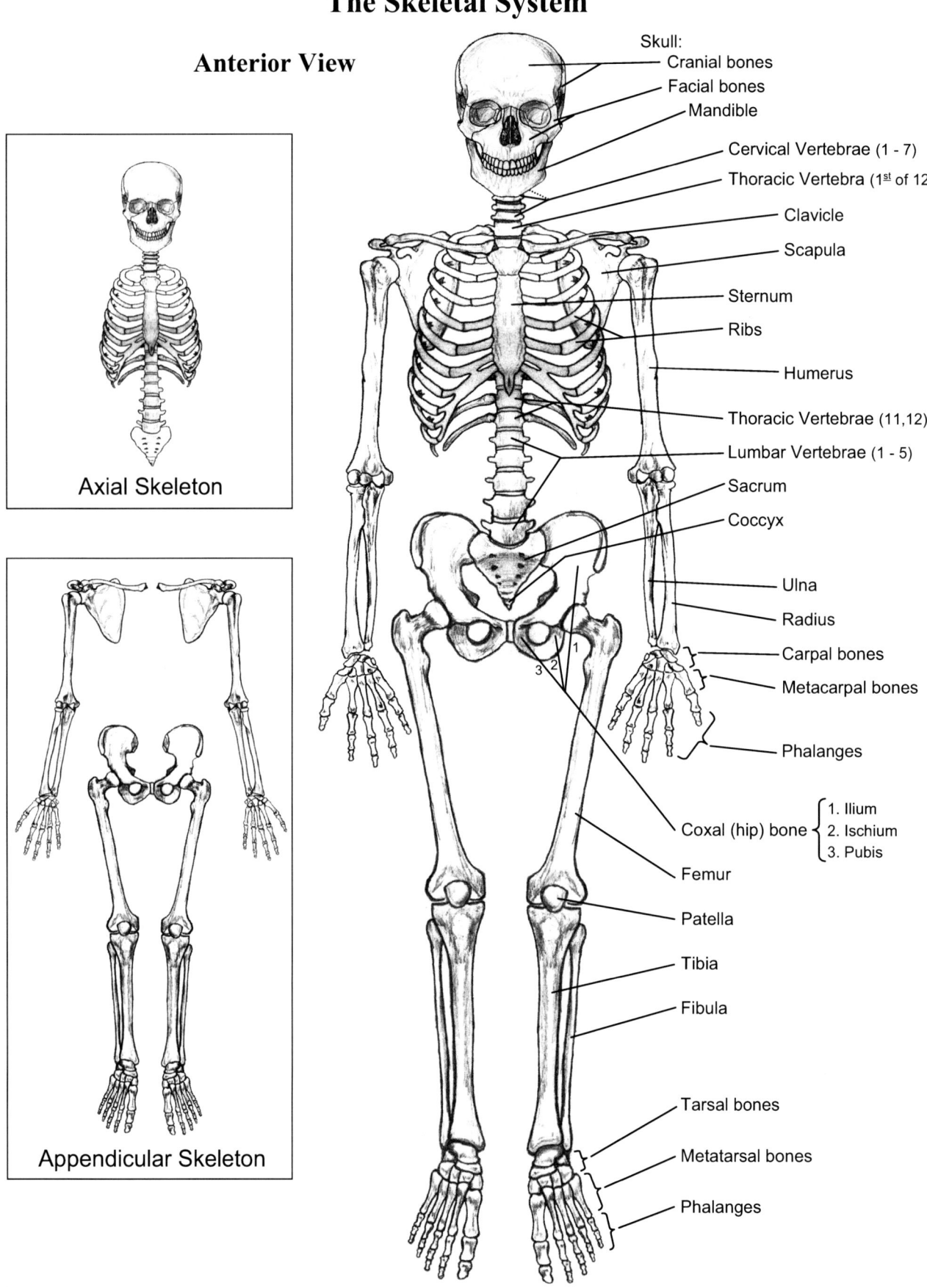

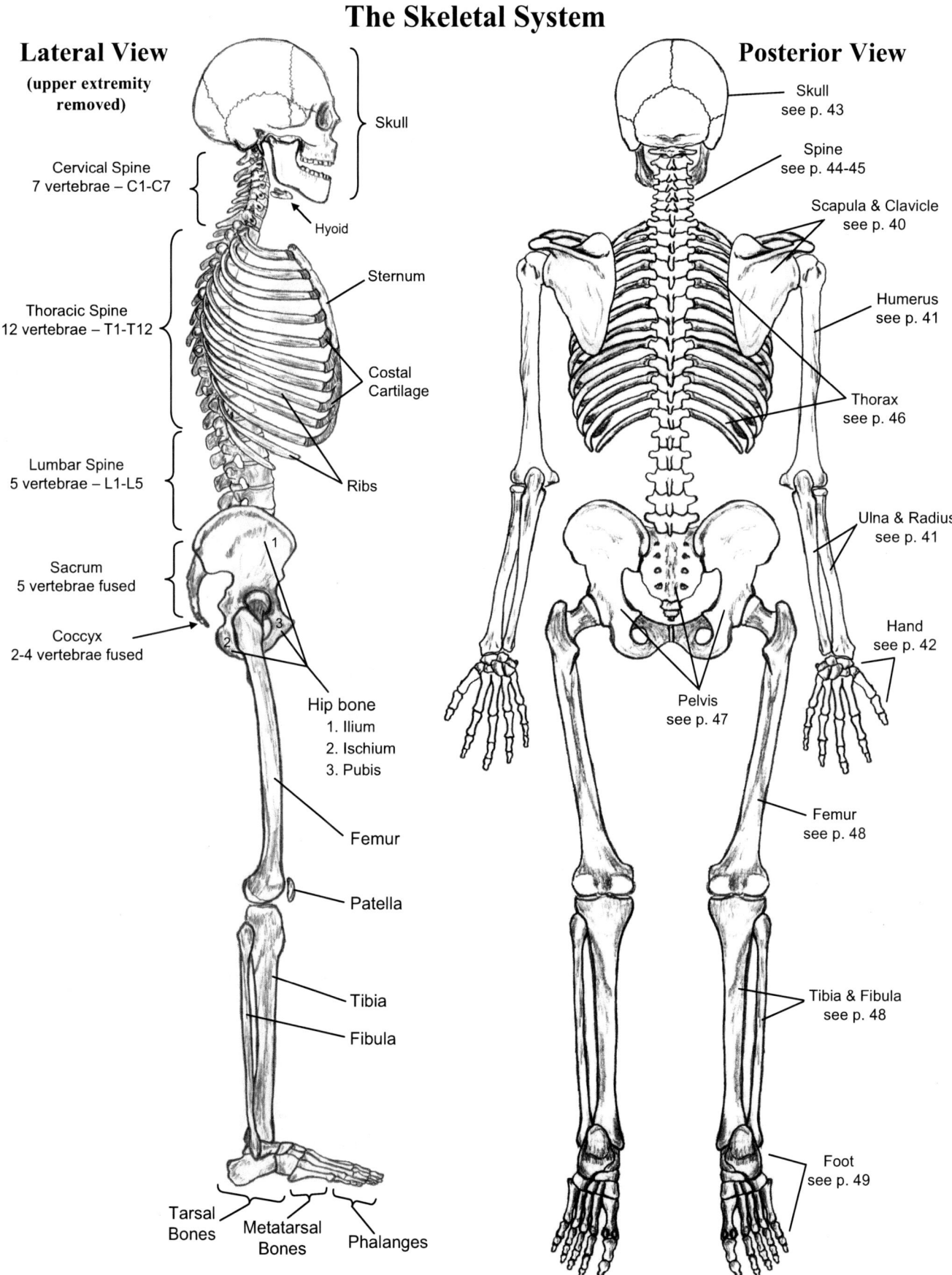

Joints of the Body

SYNOVIAL JOINT	JOINT TYPE	COMMENTS	("◄►"=articulation)

Head
   Temporomandibular — condyloid/gliding/hinge — 6 actions: elevation, depression, protraction, retraction, R.&L. lateral deviation

Spine
   Atlantooccipital (Occipitoatlantal) — ellipsoid — 2 points of contact (at facets) = 1 ellipsoid joint
   Atlantoodontoid — pivot — C1/C2: anterior arch of atlas ◄► dens of axis
   Atlantoaxial — gliding — C1/C2: inferior facets C1 ◄► superior facets C2
   Intervertebral Facets — gliding
   Intervertebral Discs — (amphiarthrotic) — Cartilaginous joint (not synovial)
   Sacroiliac — part gliding — Part fibrous (synarthrotic), part synovial

Trunk
   Sternocostal (rib 1) — (synarthrotic) — Fibrous joint (not synovial)
   Sternocostal (ribs 2-7) — gliding — Ribs 2-7, sternum ◄► costal cartilage
   Costochondral — (synarthrotic-fibrous) — Ribs 1-10, "junctions" of rib ◄► costal cartilage
   Costovertebral — gliding — Head of rib ◄► costal facet on vertebral body
   Costotransverse — gliding — Tubercle of rib ◄► costal facet on transverse process

Shoulder Complex
   Glenohumeral — ball and socket — This is the true shoulder joint, humerus ◄► glenoid fossa. Does 6 B&S actions + Horiz. abduction, Horiz. adduction
   Sternoclavicular (SC) — modif. ball & socket/hinge — SC, AC, & ST joints move together, with 6 actions:
   Acromioclavicular (AC) — gliding — elevation, depression, protraction, retraction,
   Scapulothoracic (ST) — false — upward rotation, downward rotation

Elbow
   Humeroulnar — hinge — The elbow joint
   Radioulnar (Proximal) — pivot — Does supination & pronation (i.e., rotations)
   Radioulnar (Distal) — pivot (but not much) — Pivots slightly to help accommodate supin. & pronation

Wrist, Hand, Fingers
   Radiocarpal — ellipsoid — Radius ◄► scaphoid, lunate, & triquetrum
   Intercarpal, (aka mid-carpal) — gliding — Articulations of carpal bones to other carpal bones
   Carpometacarpal (CM) #1 (thumb) — saddle — Trapezium ◄► 1st metacarpal
   Carpometacarpal (CM) #2-5 — gliding
   Metacarpophalangeal (MP) — condyloid
   Interphalangeal (PIP & DIP) — hinge — Fingers 1-4 have DIP & PIP, thumb has DIP only

Hip
   Coxal — ball and socket — Head of femur ◄► acetabulum of hip bone

Knee
   Tibiofemoral — modified hinge — The knee – "modified" because also rotates when flexed
   Patellofemoral — gliding — Kneecap
   Tibiofibular (Proximal) — gliding
   Tibiofibular (Distal) — (synarthrotic-fibrous) — Where tib./fib. connect to form upper part of ankle joint

Ankle, Foot, Toes
   Talocrural — hinge — Ankle joint (plantarflex=flexion, dorsiflex=extension)
   Subtalar (=Talocalcaneal joint) — gliding/pivot — Inversion & eversion occur at the subtalar joint
   Talocalcaneonavicular (TCN) — gliding, rotation
   Transverse Tarsal — gliding — } These create the springy arches of the foot
   Tarsometatarsal (TM) — gliding
   Metatarsophalangeal (MP) — condyloid — Ball of foot
   Interphalangeal (PIP & DIP) — hinge — Toes 1-4 have DIP & PIP, big toe has DIP only

SYNOVIAL JOINTS (DIARTHROTIC) – TYPES & MOVEMENTS

Ball & Socket: flexion, extension, abduction, adduction, lateral rotation, medial rotation
Pivot : rotation   **Hinge:** flexion, extension   **Ellipsoid/Condyloid:** flexion, extension, abduction, adduction
Gliding: gliding   **Saddle:** same actions as ellipsoid, + opposition

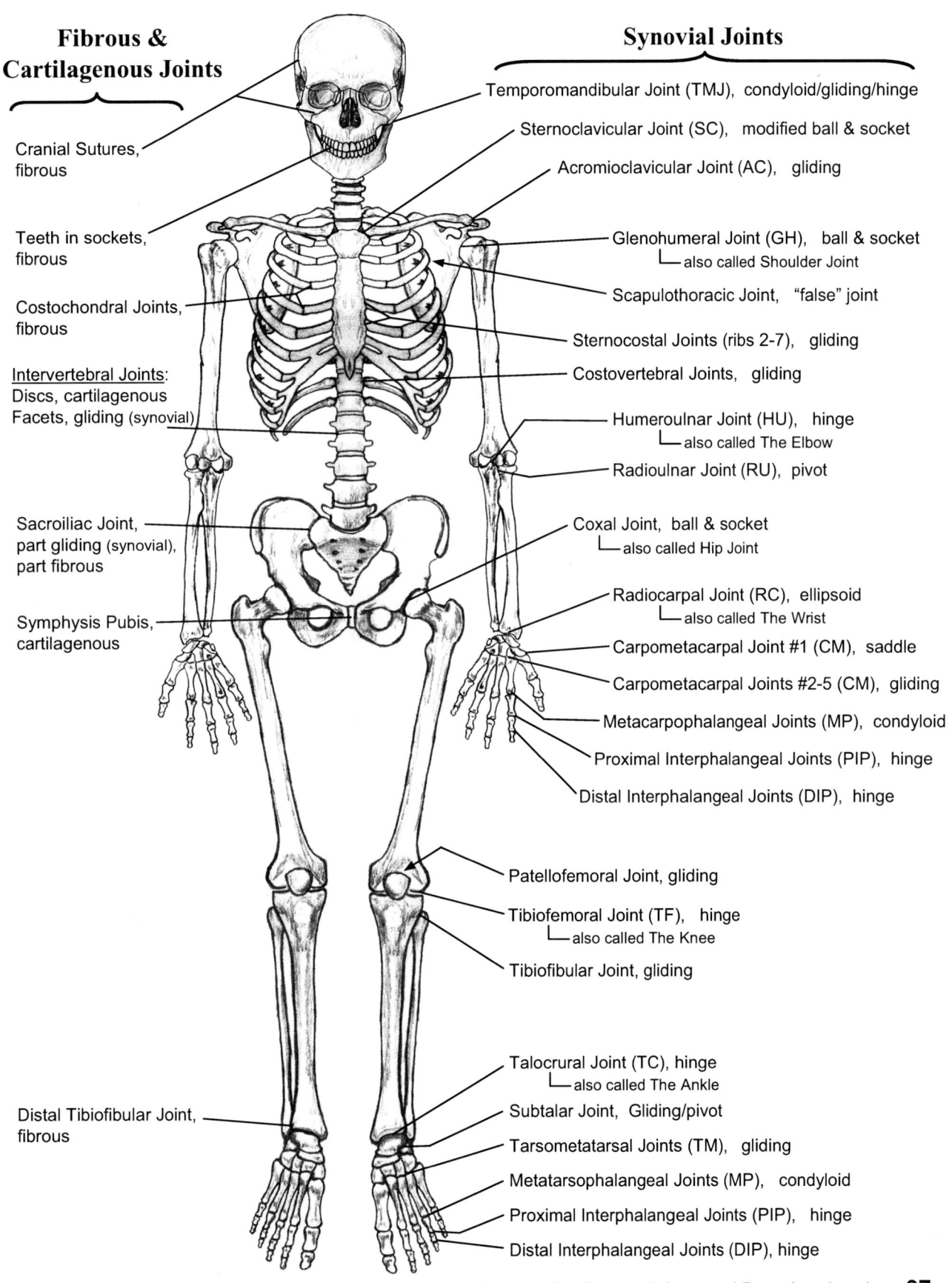

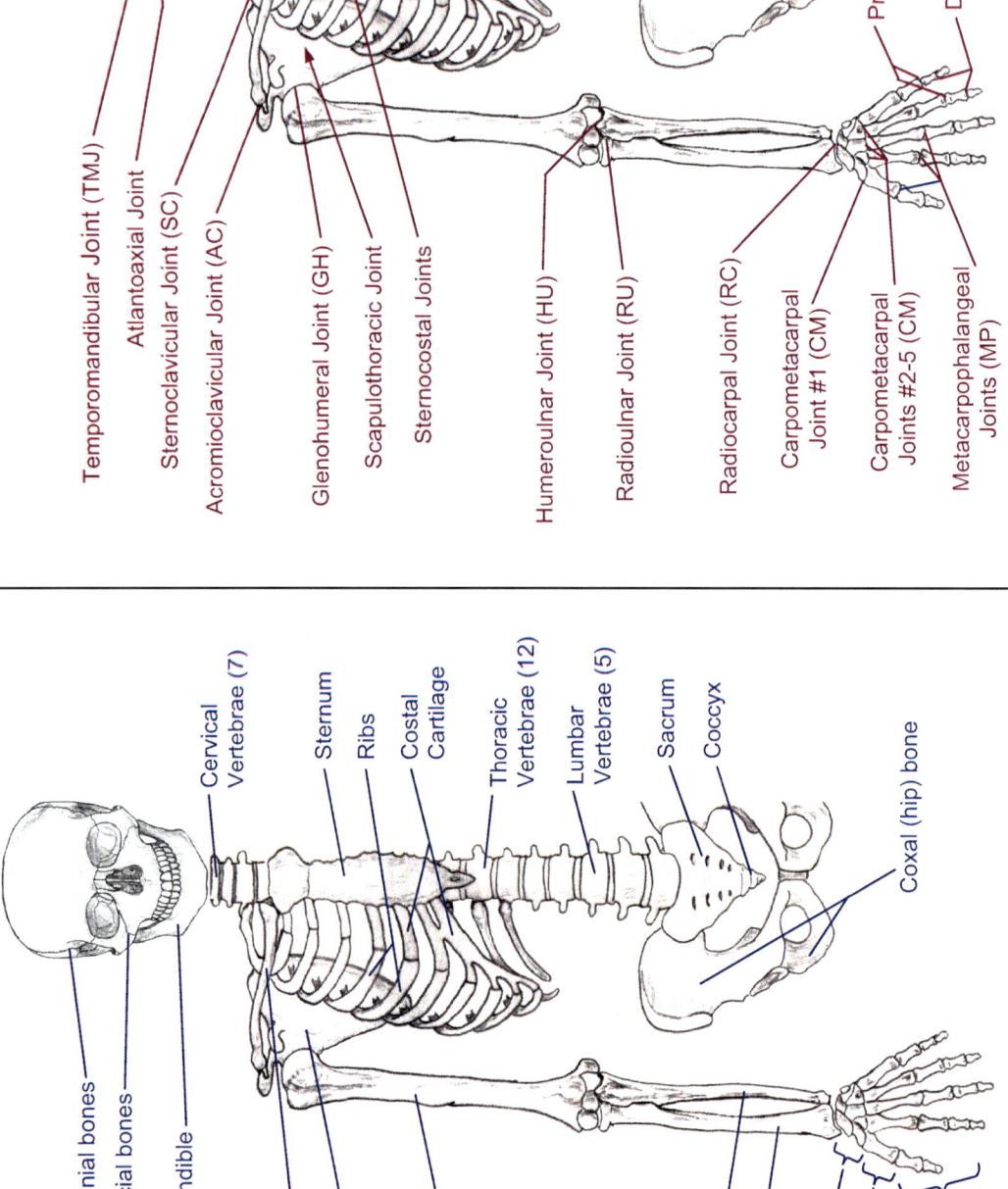

Comparison of Bone and Joint Names

38 Chapter 2 - Bones, Joints and Bony Landmarks

Lower Body – Joints

- Costovertebral Joint
- Intervertebral Joint
- Sacroiliac Joint
- Coxal Joint (or Hip Joint)
- Symphysis Pubis
- Patellofemoral Joint
- Tibiofemoral Joint (TF)
- Tibiofibular Joint
- Talocrural Joint (TC)
- Subtalar Joint
- Tarsometatarsal Joint (TM)
- Metatarsophalangeal Joint (MP)
- Proximal Interphalangeal Joint (PIP)
- Distal Interphalangeal Joint (DIP)

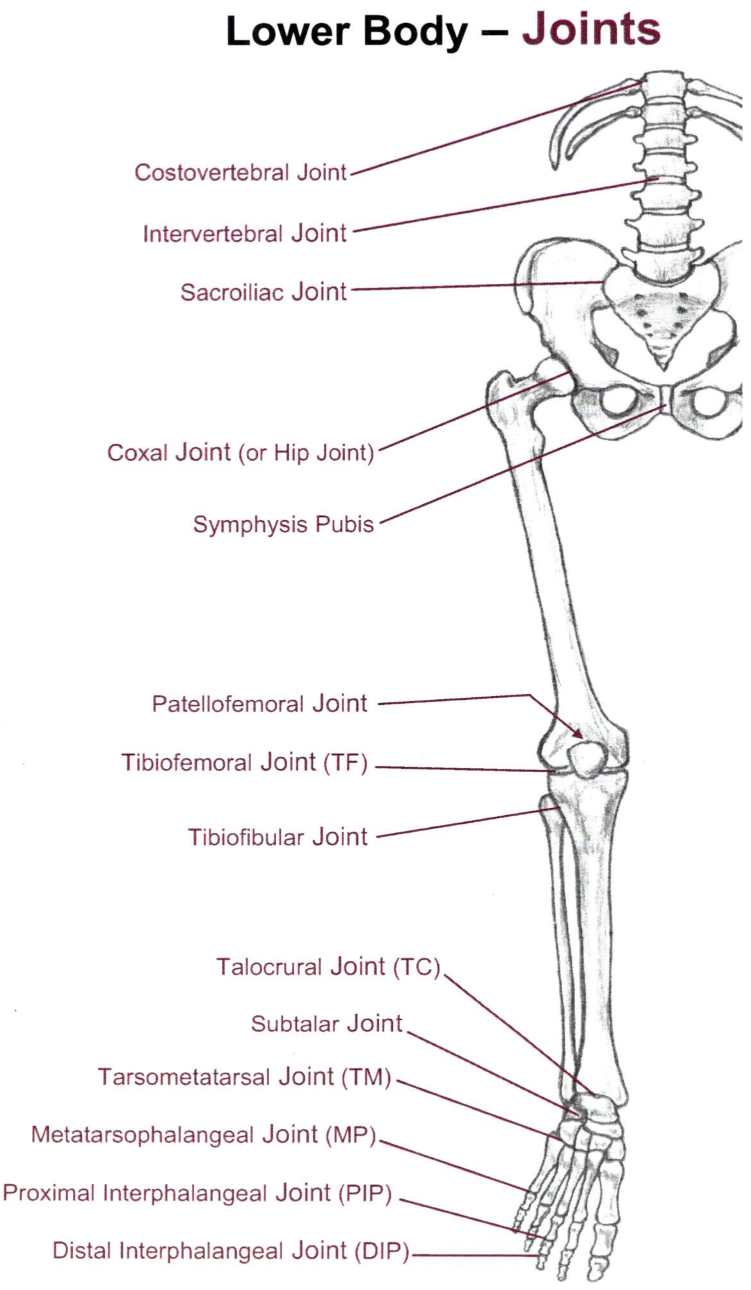

Comparison of Bone and Joint Names

Lower Body – Bones

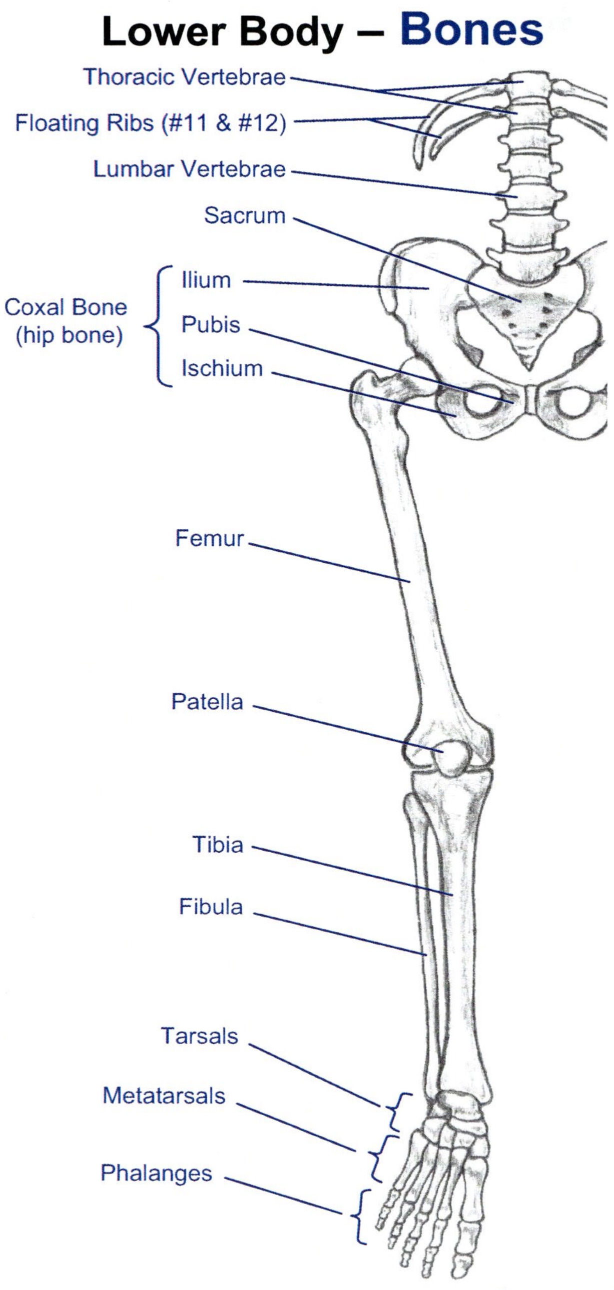

- Thoracic Vertebrae
- Floating Ribs (#11 & #12)
- Lumbar Vertebrae
- Sacrum
- Coxal Bone (hip bone)
  - Ilium
  - Pubis
  - Ischium
- Femur
- Patella
- Tibia
- Fibula
- Tarsals
- Metatarsals
- Phalanges

Mastering Muscles & Movement © 2007   Chapter 2 - Bones, Joints and Bony Landmarks

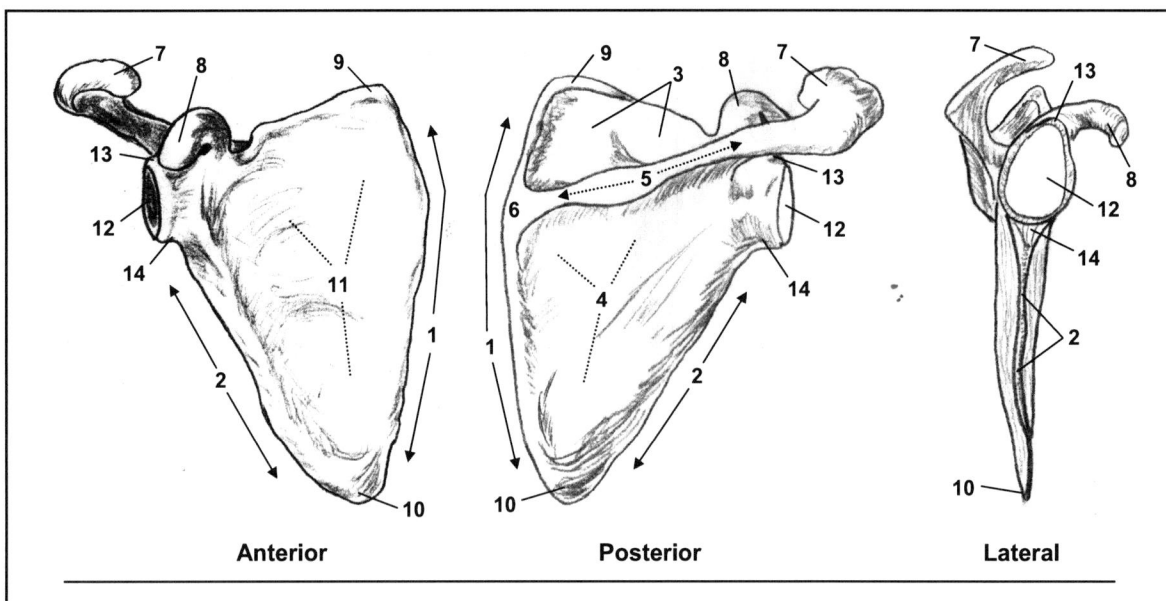

Right Scapula

1. Medial (vertebral) border
2. Lateral (axillary) border
3. Supraspinous fossa
4. Infraspinous fossa
5. Spine
6. Root of spine
7. Acromion
8. Coracoid process
9. Superior angle
10. Inferior angle
11. Subscapular fossa
12. Glenoid fossa
13. Supraglenoid tubercle
14. Infraglenoid tubercle

Scapula

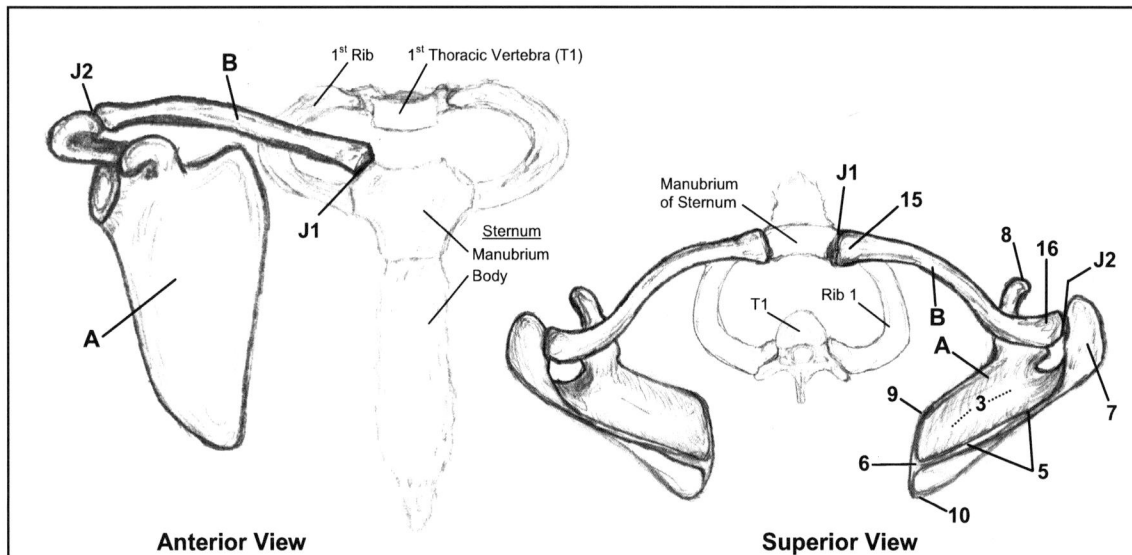

Shoulder Girdle (= Scapula + Clavicle)

A. Scapula
3. Supraspinous fossa
5. Spine of scapula
6. Root of spine
7. Acromion
8. Coracoid process
9. Superior angle
10. Inferior angle

B. Clavicle
15. Sternal (medial) end
16. Acromial (lateral) end

Joints
J1. Sternoclavicular joint
J2. Acromioclavicular joint

Shoulder Girdle

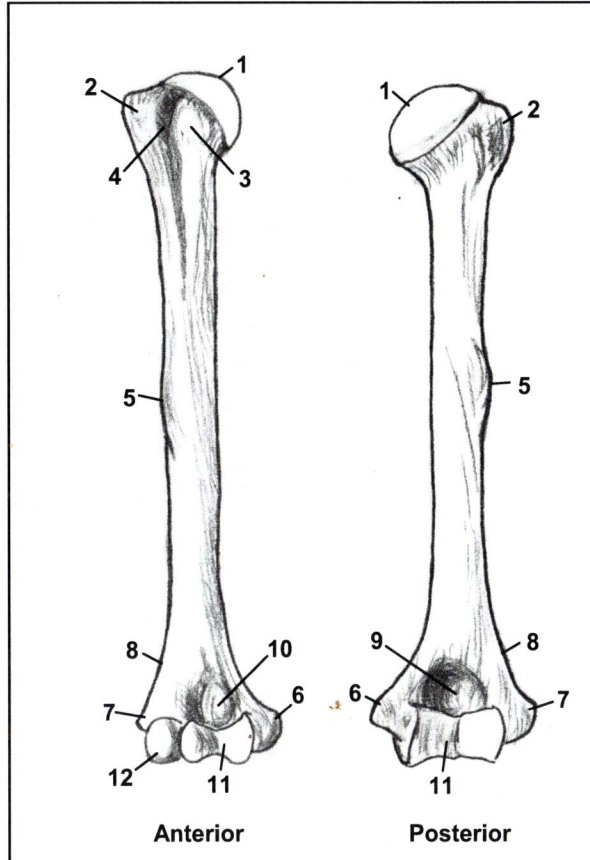

Right Humerus

1. Head
2. Greater tubercle
3. Lesser tubercle
4. Intertubercular groove (Bicipital groove)
5. Deltoid tuberosity
6. Medial epicondyle
7. Lateral epicondyle
8. Lateral supracondylar ridge
9. Olecranon fossa
10. Coronoid fossa
11. Trochlea
12. Capitulum

Upper Arm

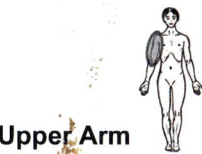

Anterior   Posterior

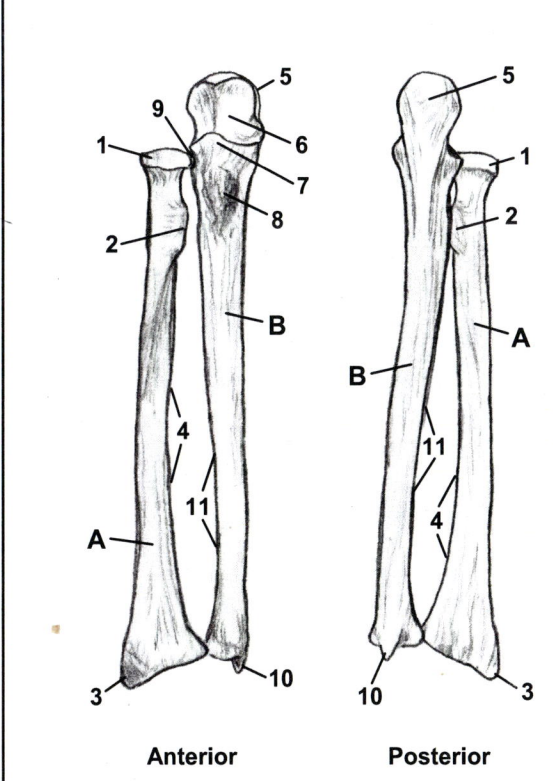

Right Radius & Ulna

A. Radius
1. Head of radius
2. Radial tuberosity
3. Styloid process
4. Interosseus border

B. Ulna
5. Olecranon process
6. Trochlear notch
7. Coronoid process
8. Ulnar tuberosity
9. Radial notch
10. Styloid process
11. Interosseus border

Forearm

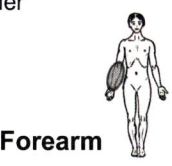

Anterior   Posterior

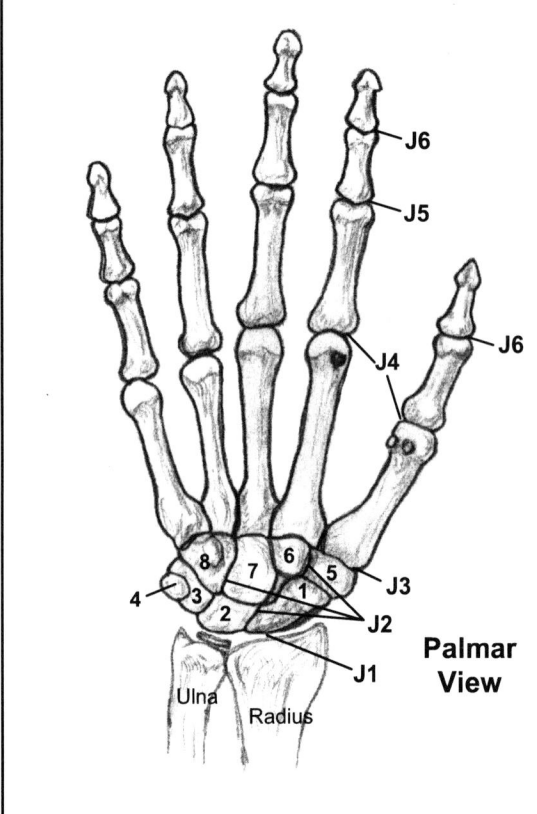

Palmar View

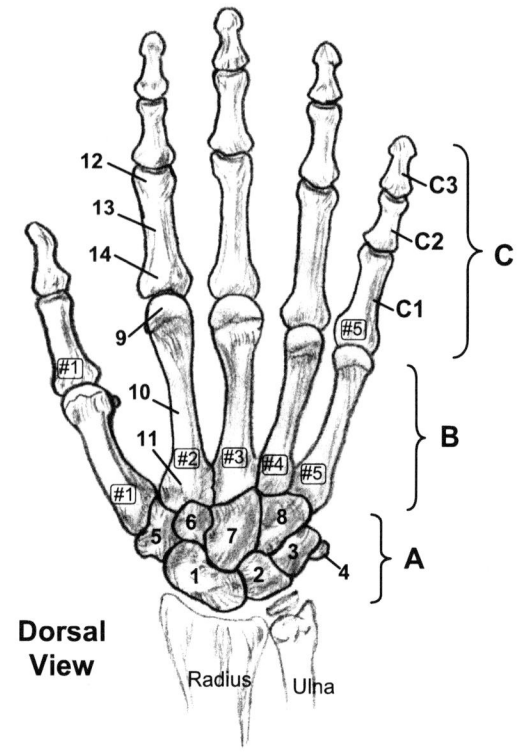

Dorsal View

Right Hand

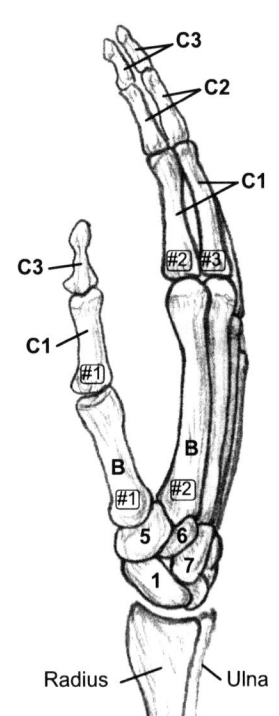

Radial View

A. Carpal Bones

Proximal Row	Distal Row
1. Scaphoid	5. Trapezium
2. Lunate	6. Trapezoid
3. Triquetrum	7. Capitate
4. Pisiform	8. Hamate

B. Metacarpal Bones

#1 radial side to #5 ulnar side

9. Head
10. Shaft
11. Base

C. Phalanges

12. Head
13. Shaft
14. Base

C1 – Proximal phalanx
C2 – Middle phalanx
C3 – Distal phalanx

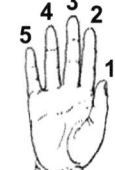

Numbering of Digits

Digit #1 = Thumb (Pollux)

Digit #2 - #5 = Fingers

Joints

J1.	Radiocarpal (RC)	
J2.	Intercarpal	(many)
J3.	Carpometacarpal (CM)	(5)
J4.	Metacarpophalangeal (MP)	(5)
J5.	Proximal Interphalangeal (PIP)	(4)
J6.	Distal Interphalangeal (DIP)	(5)

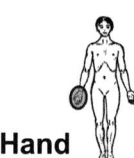

Hand

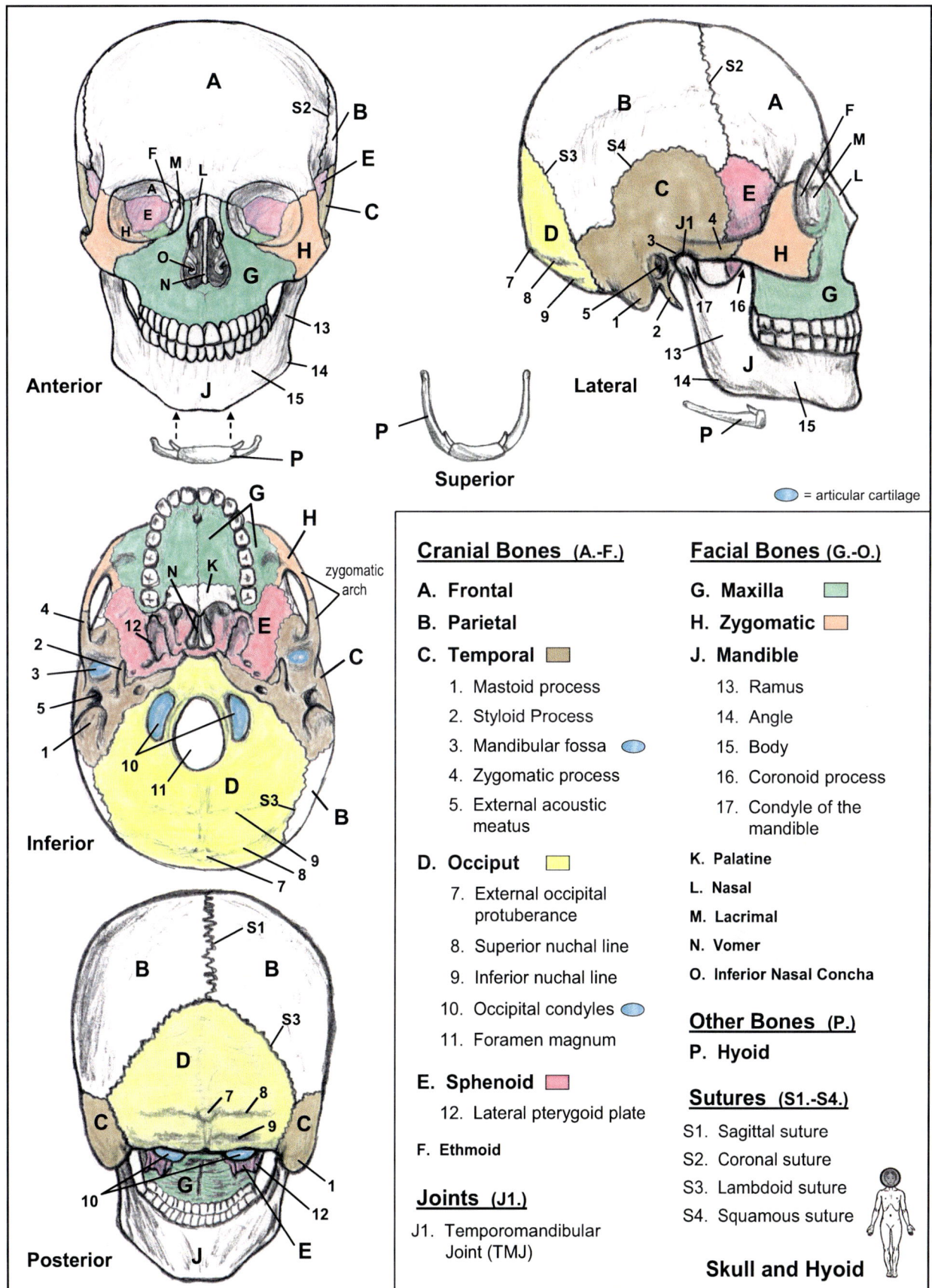

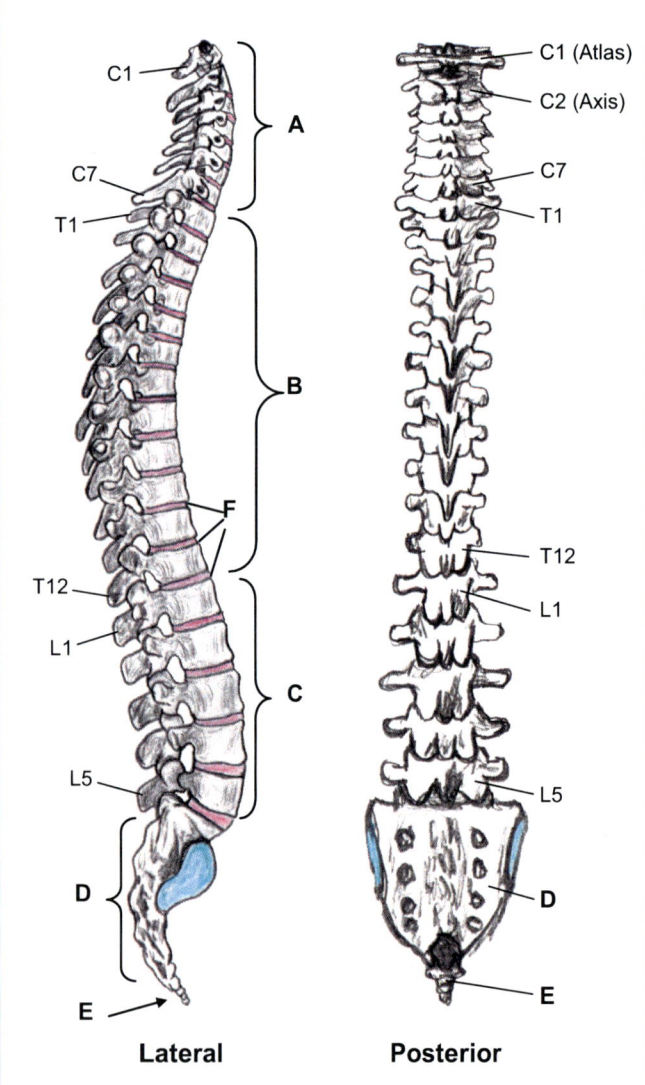

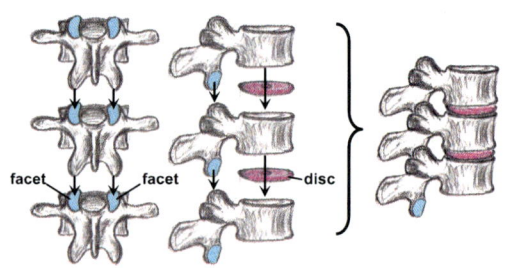

Intervertebral Joints
(3 articulations per joint: 1 disc + 2 facet joints)

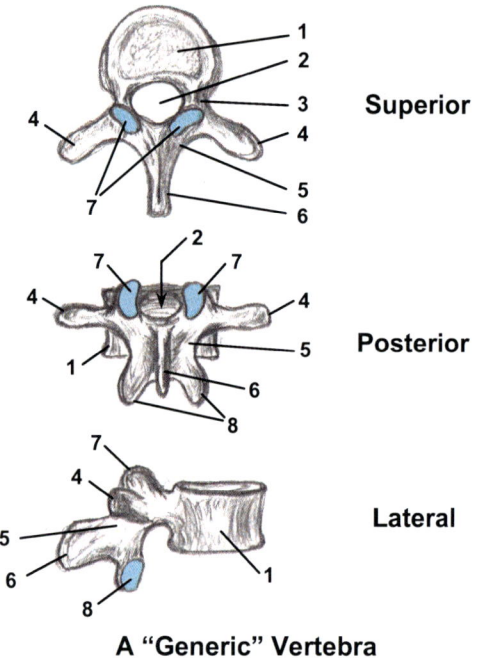

A "Generic" Vertebra

Vertebral Column

 = articular cartilage

A. Cervical Spine
  7 vertebrae: C1-C7
  Lordotic curve (secondary curve)

B. Thoracic Spine
  12 vertebrae: T1-T12 – ribs attach
  Kyphotic curve (primary curve)

C. Lumbar Spine
  5 vertebrae: L1-L5
  Lordotic curve (secondary curve)

D. Sacrum
  5 vertebrae: S1-S5 (fused)
  Kyphotic curve (primary curve)

E. Coccyx
  2 - 4 vertebrae (fused)

F. Intervertebral Discs (23)

Landmarks Common to All Vertebrae (except C1)

(see page 45 for landmarks unique to each spinal section)

1. Body
2. Vertebral foramen
    (spinal cord passes through it)
3. Pedicle
4. Transverse process (TVP)
5. Lamina
6. Spinous process (SP)
7. Superior facet (articular process)
8. Inferior facet (articular process)

Spine

◯ = articular cartilage

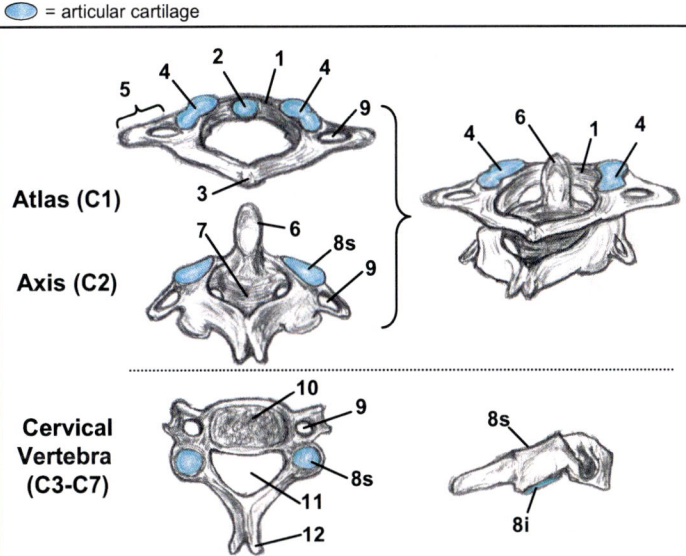

Special Features of Cervical Vertebrae
(see page 44 for features common to all vertebrae)

Atlas (C1):
1. Anterior arch (C1 has no body)
2. Articular facet for dens
3. Posterior tubercle, no spinous process (SP)
4. Superior facets – match occipital condyles
5. Wide transverse processes (TVPs)

Axis (C2):
6. Odontoid process (Dens)
7. Body – has inferior face only

All Cervicals:
8. Facets – nearly horizontal — 8s. Superior facets / 8i. Inferior facets
9. Transverse foramen in TVPs
10. Small bodies
11. Large vertebral foramen
12. Bifid SPs (except C7)

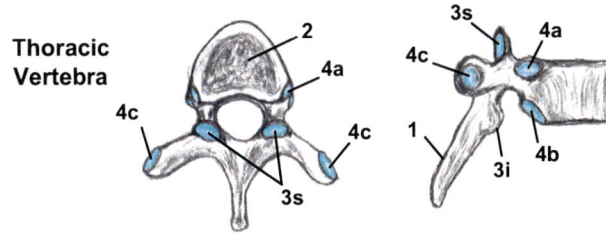

Special Features of Thoracic Vertebrae
(see page 44 for features common to all vertebrae)

1. Long, sloped SPs
2. Heart shaped body
3. Facets – nearly vertical, & facing front-to-back
   - 3s. Superior facets
   - 3i. Inferior facets
4. Extra facets – for ribs:
   - 4a. Superior costal facet
   - 4b. Inferior costal facet
   - 4c. Transverse costal facet

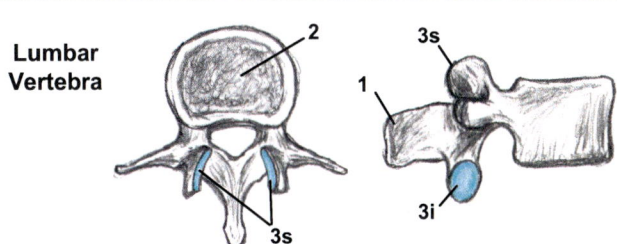

Special Features of Lumbar Vertebrae
(see page 44 for features common to all vertebrae)

1. Thick, short SPs
2. Large body
3. Facets – nearly vertical, & facing side-to-side
   - 3s. Superior facets
   - 3i. Inferior facets

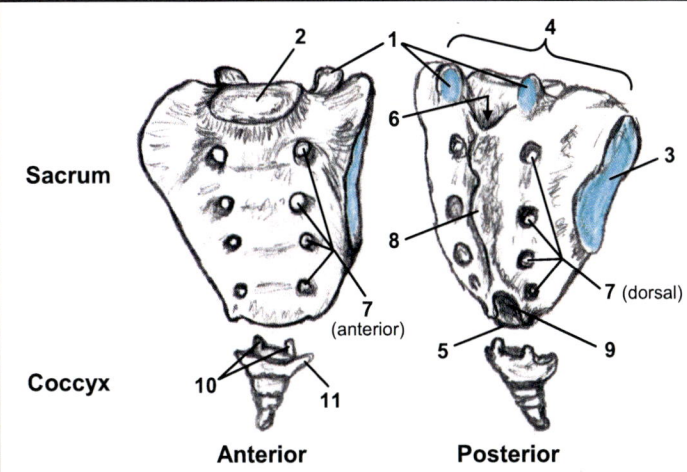

Sacrum
1. Facets (articulate with L5 inferior facets)
2. Lumbosacral articular surface (to disk with L5)
3. Auricular surface (to sacroiliac joint)
4. Base of sacrum
5. Apex of sacrum
6. Sacral canal
7. Sacral foramina
8. Sacral crest
9. Sacral hiatus

Coccyx
10. Coccygeal horns
11. Transverse processes

Vertebrae & Sacrum

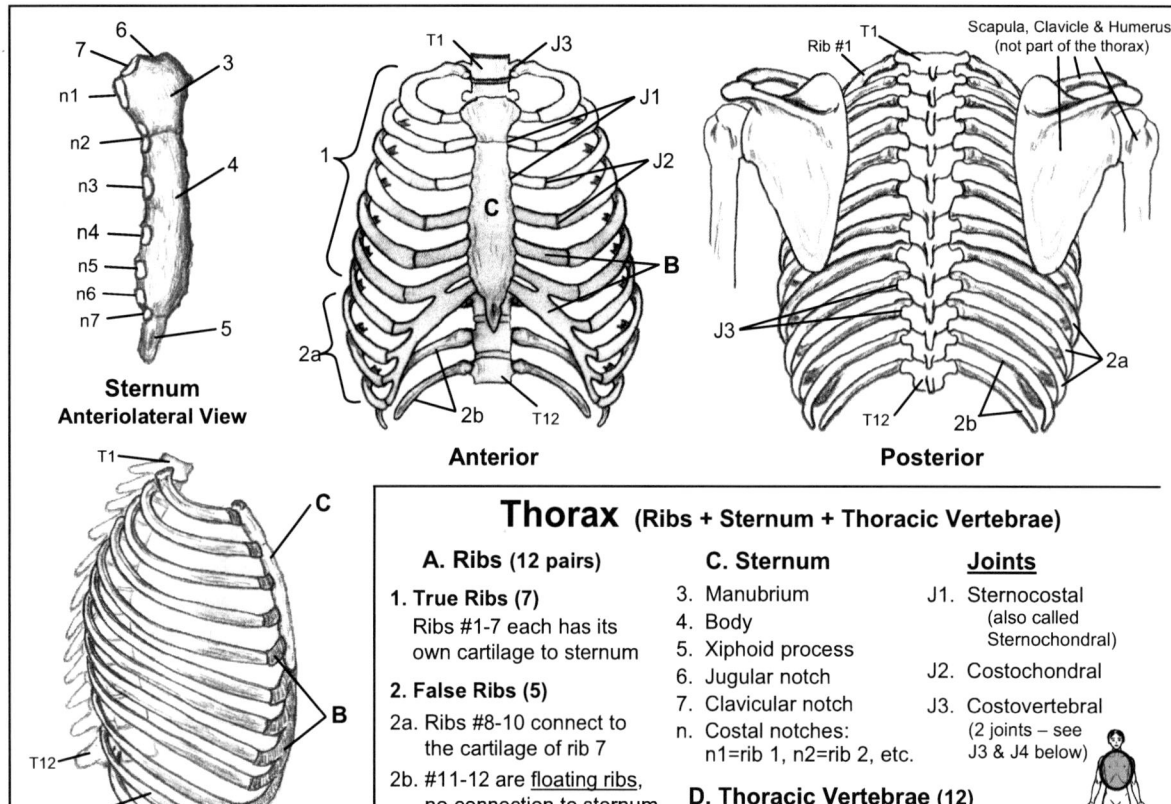

Thorax (Ribs + Sternum + Thoracic Vertebrae)

A. Ribs (12 pairs)

1. True Ribs (7)
   Ribs #1-7 each has its own cartilage to sternum
2. False Ribs (5)
   2a. Ribs #8-10 connect to the cartilage of rib 7
   2b. #11-12 are <u>floating ribs</u>, no connection to sternum

B. Costal Cartilage

C. Sternum

3. Manubrium
4. Body
5. Xiphoid process
6. Jugular notch
7. Clavicular notch
n. Costal notches: n1=rib 1, n2=rib 2, etc.

D. Thoracic Vertebrae (12)

T1. Thoracic vertebra #1
T12. Thoracic vertebra #12

Joints

J1. Sternocostal (also called Sternochondral)
J2. Costochondral
J3. Costovertebral (2 joints – see J3 & J4 below)

Thorax

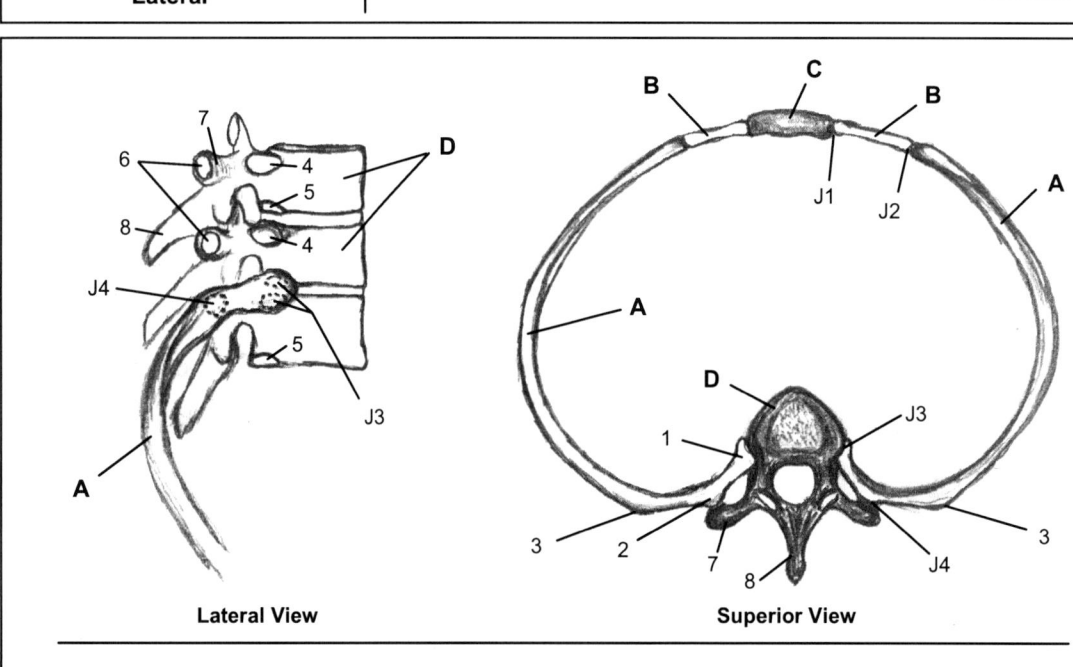

Lateral View / Superior View

A. Rib
1. Head of rib
2. Tubercle of rib
3. Angle of rib (costal angle)

B. Costal Cartilage

C. Sternum

D. Thoracic Vertebra
4. Superior costal facet
5. Inferior costal facet
6. Transverse costal facet
7. Transverse process (TVP)
8. Spinous process (SP)

Joints
J1. Sternocostal joint
J2. Costochondral joint
J3. Costovertebral joint
J4. Costotransverse joint

Rib Articulations

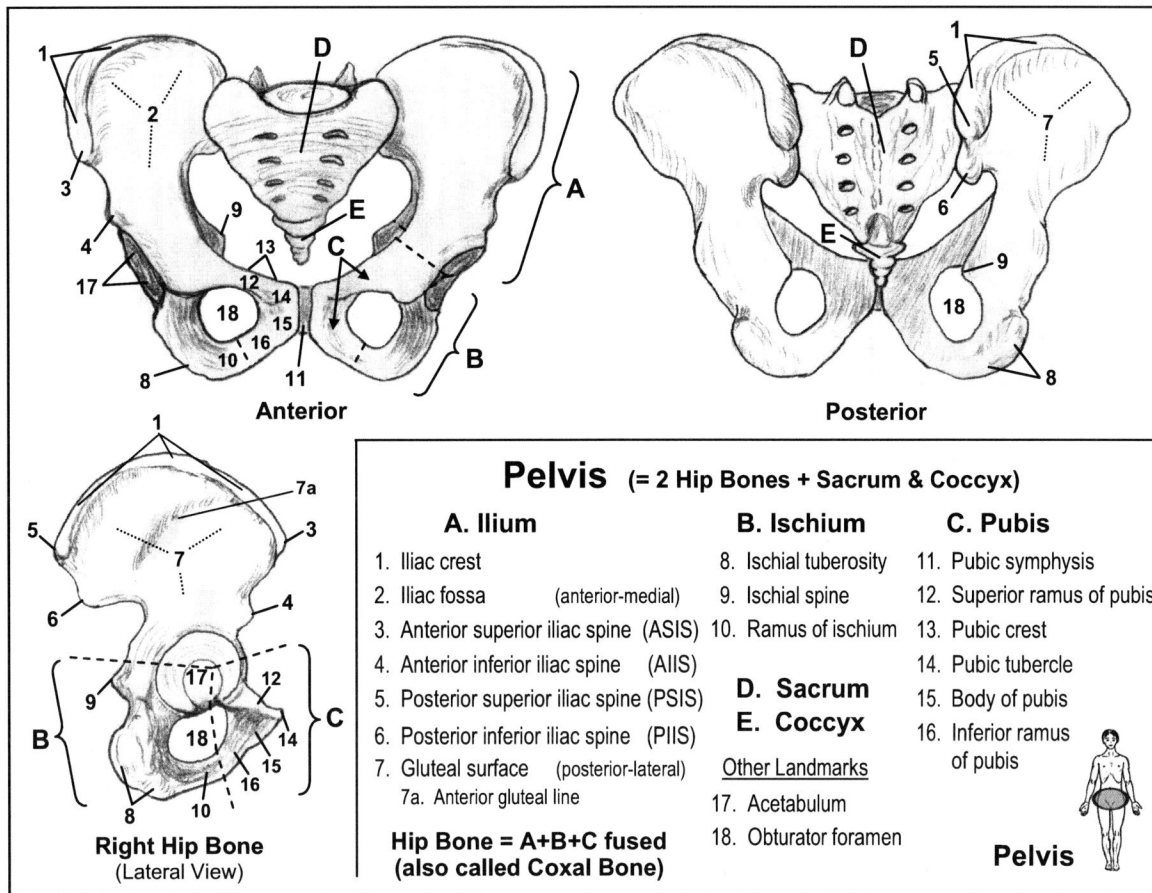

Pelvis (= 2 Hip Bones + Sacrum & Coccyx)

A. Ilium
1. Iliac crest
2. Iliac fossa (anterior-medial)
3. Anterior superior iliac spine (ASIS)
4. Anterior inferior iliac spine (AIIS)
5. Posterior superior iliac spine (PSIS)
6. Posterior inferior iliac spine (PIIS)
7. Gluteal surface (posterior-lateral)
   7a. Anterior gluteal line

B. Ischium
8. Ischial tuberosity
9. Ischial spine
10. Ramus of ischium

D. Sacrum
E. Coccyx

Other Landmarks
17. Acetabulum
18. Obturator foramen

C. Pubis
11. Pubic symphysis
12. Superior ramus of pubis
13. Pubic crest
14. Pubic tubercle
15. Body of pubis
16. Inferior ramus of pubis

Hip Bone = A+B+C fused (also called Coxal Bone)

Pelvis

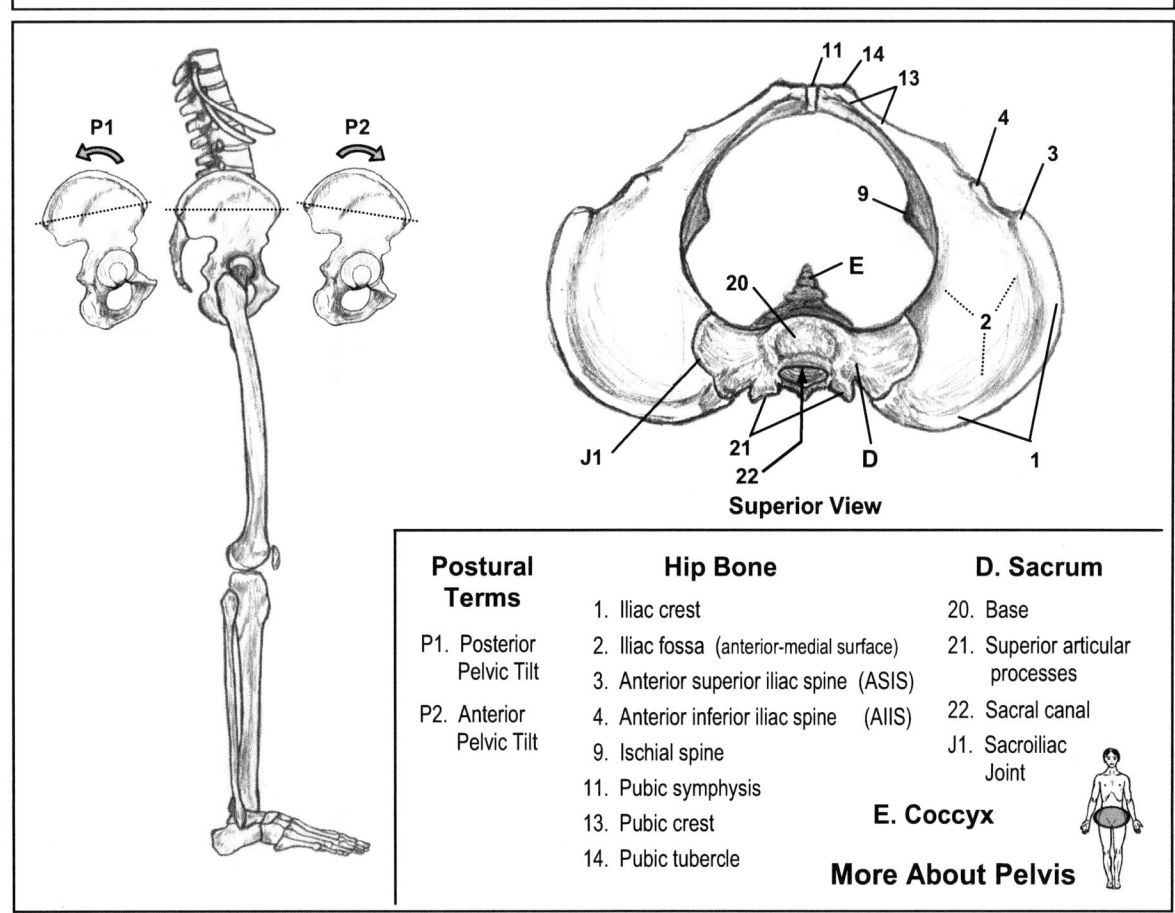

Postural Terms
P1. Posterior Pelvic Tilt
P2. Anterior Pelvic Tilt

Hip Bone
1. Iliac crest
2. Iliac fossa (anterior-medial surface)
3. Anterior superior iliac spine (ASIS)
4. Anterior inferior iliac spine (AIIS)
9. Ischial spine
11. Pubic symphysis
13. Pubic crest
14. Pubic tubercle

D. Sacrum
20. Base
21. Superior articular processes
22. Sacral canal
J1. Sacroiliac Joint

E. Coccyx

More About Pelvis

Mastering Muscles & Movement © 2007   Chapter 2 – Bones, Joints and Bony Landmarks   47

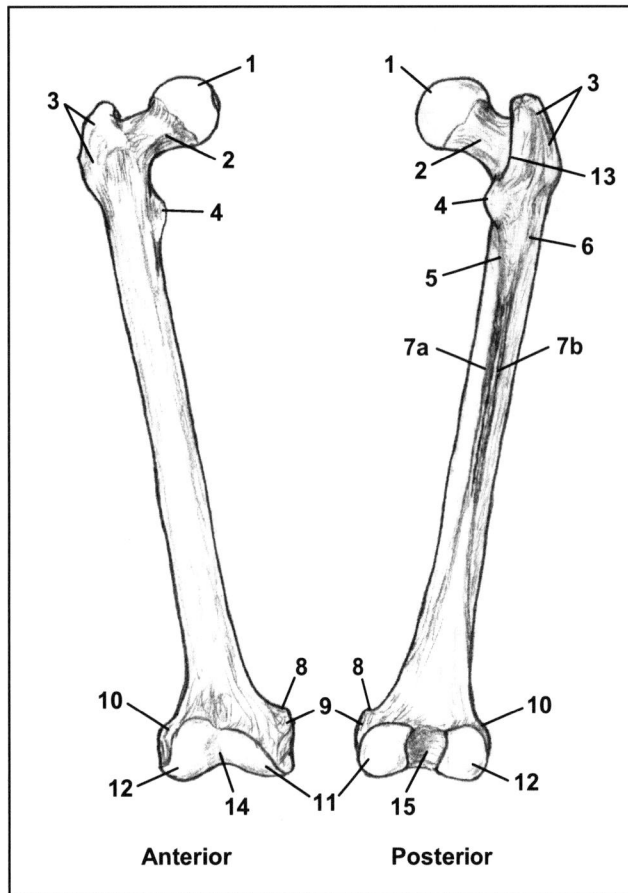

Right Femur

1. Head
2. Neck
3. Greater trochanter
4. Lesser trochanter
5. Pectineal line
6. Gluteal tuberosity
7. Linea aspera { 7a: Medial lip, 7b: Lateral lip }
8. Adductor tubercle
9. Medial epicondyle
10. Lateral epicondyle
11. Medial condyle
12. Lateral condyle
13. Intertrochanteric crest
14. Patellar surface
15. Intercondylar fossa

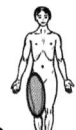

Thigh

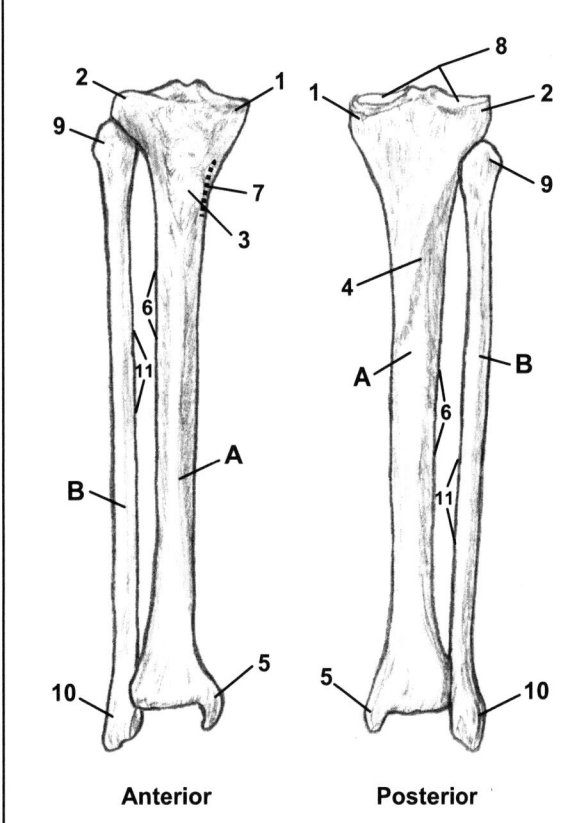

Right Tibia & Fibula

A. Tibia

1. Medial condyle
2. Lateral condyle
3. Tibial tuberosity
4. Soleal line
5. Medial malleolus
6. Interosseus border
7. Proximal Medial Shaft (Pes anserinus attachment)
8. Superior articular surfaces (Tibial plateau)

B. Fibula

9. Head
10. Lateral malleolus
11. Interosseus border

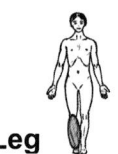

Leg

Right Foot

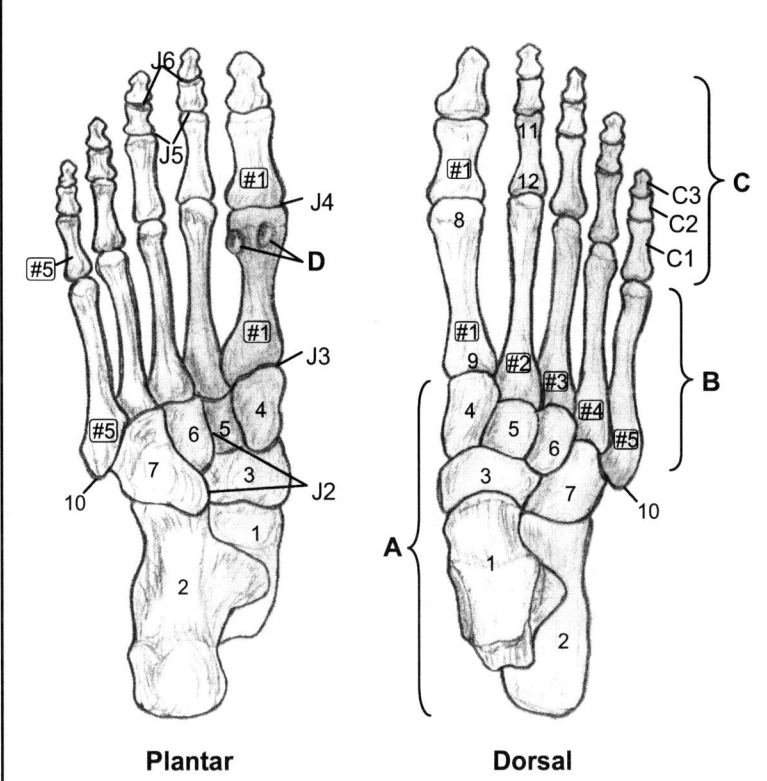

Plantar   **Dorsal**

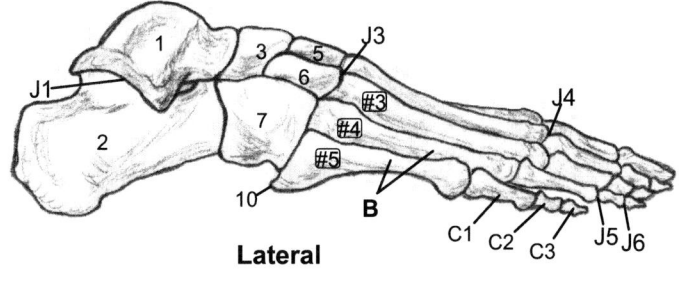

Lateral

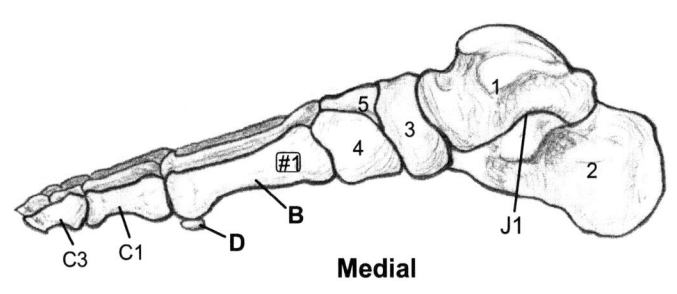

Medial

A. Tarsal Bones (7)
1. Talus
2. Calcaneus
3. Navicular
4. Medial Cuneiform (1st cuneiform)
5. Middle Cuneiform (2nd cuneiform)
6. Lateral Cuneiform (3rd cuneiform)
7. Cuboid

B. Metatarsal Bones (5)
8. Head
9. Base
10. Tuberosity of 5th metatarsal

[#1] medial side to [#5] lateral side

C. Phalanges (14)
11. Head
12. Base

C1 – Proximal phalanx
C2 – Middle phalanx
C3 – Distal phalanx

Digit [#1] = Big toe (Hallux)
Digits [#2] - [#5] = toes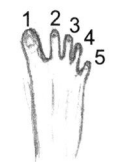

D. Sesamoid bones (2)

Joints
J1. Subtalar (talocalcaneal)
J2. Intertarsals
J3. Tarsometatarsal (TM)
J4. Metatarsophalangeal (MP)
J5. Proximal Interphalangeal (PIP)
J6. Distal Interphalangeal (DIP)

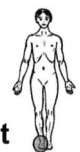
Foot

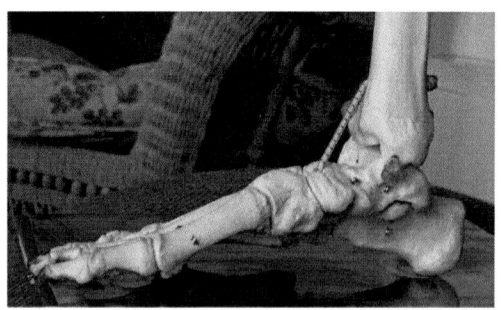

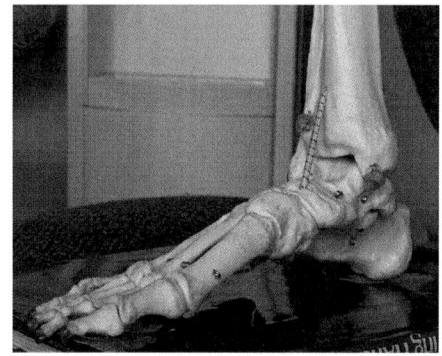

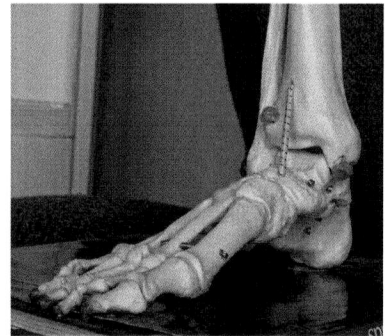

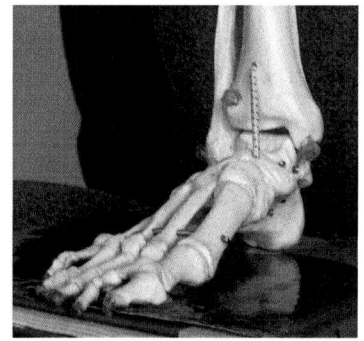

Chapter 3

How to Use the Muscle Chapters

Introduction	**52**
8-Page Format for Each <u>Group</u> of Muscles	53
How To Use the General Information Pages	**54**
Joints	54
Action Drawings	54
Bones and Bony Landmarks	55
Muscle Overview Drawings	55
How To Use the Tables and Figures	**56**
The "A" Table and Figure	56
The "B" Table and Figure	58
An Example: Learning the Deltoid Muscle	60
About Mastering the Muscles	**62**
What To Learn About Each Muscle	62
How to Use the Information You've Learned	63
Palpation Techniques	**64**
Summary and Generalizations	**65**

Introduction

Chapter 3 – How to Use the Muscle Chapters provides an orientation before you begin learning all the muscles described in the following Chapters (4, 5, and 6). You will learn how to use the tables and figures in Chapters 4-6 and how to get the most out of studying the muscles.

Because this book is for the study of Kinesiology (i.e., movement of the body), the muscles are organized into groups based on the bones and joints they move as they contract. For example, all muscles whose primary action is to move the humerus around an axis at the shoulder joint are given in "Muscle Group 2 – Movement of the Shoulder Joint". This arrangement makes it easier to recall the information when you are looking at the body from a movement viewpoint.

Thirteen muscle groups are organized in Chapters 4, 5 and 6 based on major body divisions:

>Chapter 4 – Muscles That Move the Upper Extremity (5 groups)
>Chapter 5 – Muscles That Move the Axial Skeleton (4 groups)
>Chapter 6 – Muscles That Move the Lower Extremity (4 groups)

For each of these 13 groups, there is a consistent 8-page format that gives all the information for the group of muscles. A sample of the 8-page format is shown on the facing page (p. 53).

8-Page Format

- **General Information for the Muscle Group** – the first 3 pages (see p. 54-55 for details)

    These three pages include a list of the muscles, a description of the joints and actions involved, a list of the associated bones and bony landmarks, and overview drawings showing all the muscles and their attachments in place on the skeleton.

- **"A" Table and Figure** – the next 2 pages (see p. 56-57 for details)

    Table (A) – Origin, Insertion, Action is a table of the basic verbal data for the muscles.

    Figure (A) – Muscle Attachments has drawings of bones with red and blue areas showing the origins and insertions of the muscles. These drawings provide visual cues to go with the verbal information in Table (A).

- **"B" Table and Figure** – the next 2 pages (see p. 58-59 for details)

    Table (B) – Synergists & Antagonists is a special table to study and compare muscle *actions*. This table also includes the *innervation* for each muscle.

    Figure (B) – Muscle Pictures has pictures of the muscles, lined up to match the Figure (A) muscle attachment drawings two pages prior. This allows the reader to simply lift the page to directly compare the physical muscle shape with the red and blue muscle origin and insertion locations.

- **Notes Page** – the final (8th) page

    This final page has small pictures of the muscles with blank areas to allow writing additional notes about each muscle. See page 84 for an example.

8-Page Format For Each Group of Muscles

General info about the Group (3 pages)

See page 54 for details

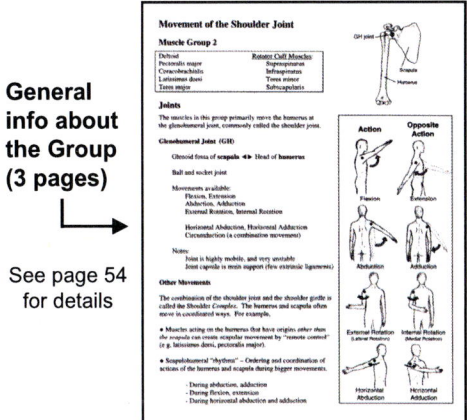
1. Descriptions of Joint(s), Actions & General Information

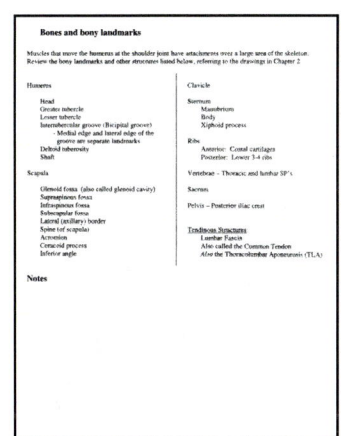
2. Bones, Landmarks, Structures & Note-Taking Area

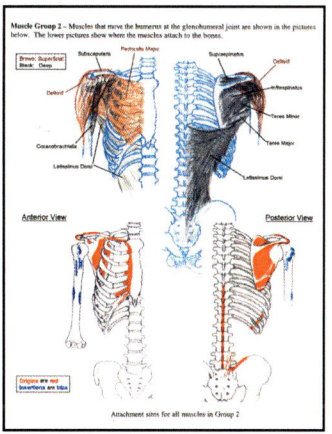
3. All muscles shown together & All origins & insertions together

"A" Table

See page 56 for details

"A" Figure

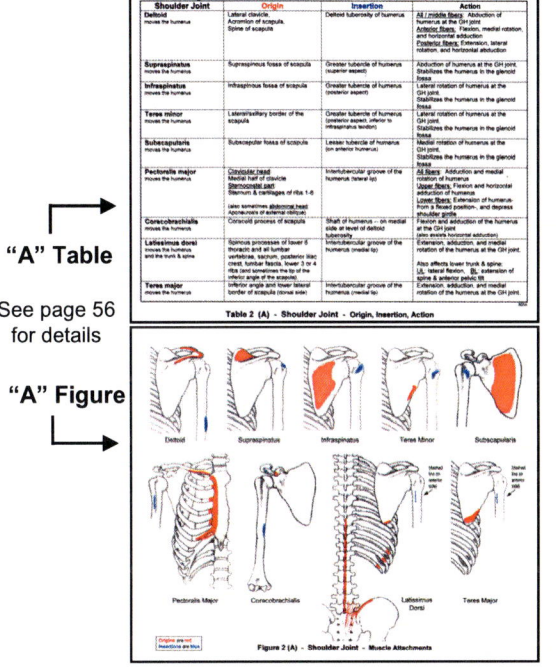

4. Verbal Data: Origins, Insertions, Actions
5. Visual depictions of the verbal data

"B" Table

See page 58 for details

"B" Figure

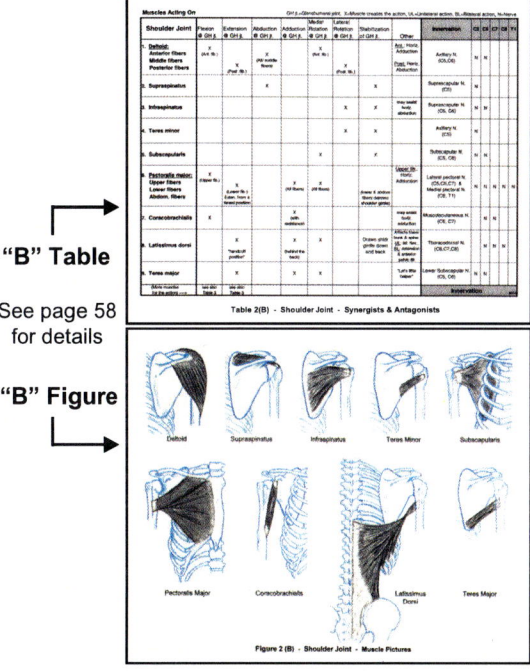

6. Relational Tables for Actions & Innervation
7. Drawings of the muscles on same scale

Notes Page

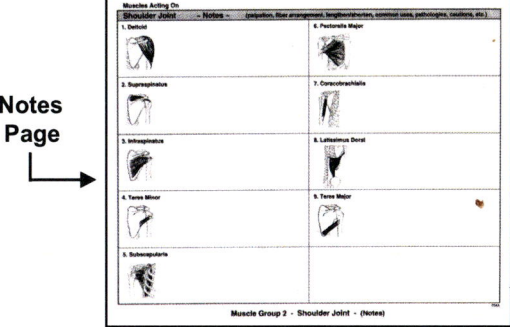

8. Page for writing notes on individual muscles

Mastering Muscles & Movement   © 2007   Chapter 3 – How to Use the Muscle Chapters

How To Use the General Information Pages

Each group of muscles in Chapters 4, 5, and 6 begins with three pages of general information. This overview of the group includes summaries of the joint(s) involved, the actions that are possible, and the bones, landmarks and other structures that are muscle attachments or are significantly involved in some other way. The General Information pages have the following components:

- List of muscles in the group
- Descriptions of the joint(s) moved by the muscles
- Movements available at those joints
- Action Drawings
- Bones and bony landmarks
- Overview drawings of the muscles and their attachments

Joints

The main joint or joints that are moved by the muscle group are given first, and other joints that are secondarily involved are listed second. A special symbol ◄► is used to indicate the meeting point (articulation) of the bones that make up the joint. Also included are the type of joint, the movements available at the joint, and other pertinent information. For example, the following information is given for the glenohumeral joint.

Glenohumeral Joint (GH)

Glenoid fossa of **scapula** ◄► Head of **humerus**

Ball and socket joint

Movements available:
Flexion, Extension
Abduction, Adduction
Medial Rotation, Lateral Rotation
Horizontal Abduction, Horizontal Adduction
Circumduction (a combination movement)

Notes:
Joint is highly mobile, and very unstable
Joint capsule is main support (few extrinsic ligaments)

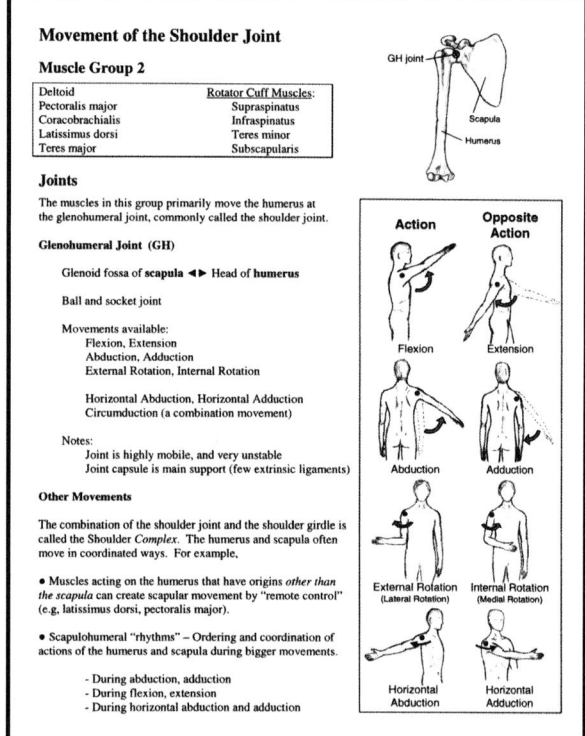

Joints & Actions (see p. 77 for full size page)

Action Drawings

For each muscle group, illustrations show the actions available for the main joints or structures moved when the muscles contract. These illustrations are organized in pairs to show how different actions oppose each other (opposite actions are done by antagonist muscles). Also, the precise point where the joint in question is moving around its axis is indicated with a symbol " ⊗ ", and the direction the body part is moving is indicated with an arrow. Here are some examples:

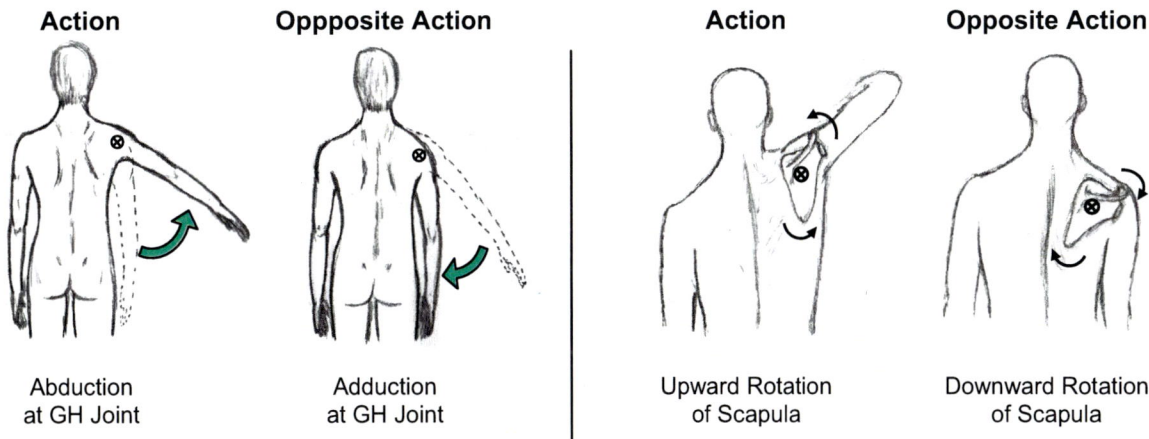

Examples of Action Pair Drawings

Bones and Bony Landmarks

This section (not shown here, see page 78 for example) lists the bones, bony landmarks, and other body structures that serve as muscle attachments or are significantly involved with the use of the muscles. It is a good idea to review the bony landmarks listed on this page before you begin studying the origins and insertions of the muscles (all bony landmark drawings are centralized in Chapter 2 of this book). Then, as you read and memorize each muscle, the names of landmarks will be familiar. This page also provides open space for writing notes.

Muscle Overview Drawings

These drawings provide a "big picture" for the muscle group to give an overall sense of the group before going on to study the individual muscles. Two types of drawings are provided:

1. All muscles shown together in place on the skeleton.

2. Skeleton pictures with all origins in red and insertions in blue.

Use these drawings to look for patterns to help you understand how the muscles in the group work together. For example, Muscle Group 2 has muscles that move the humerus at the glenohumeral joint. You can see that all the insertions (shown in blue) are gathered on the humerus, while the origins (shown in red) are on a large area on many other bones of the body.

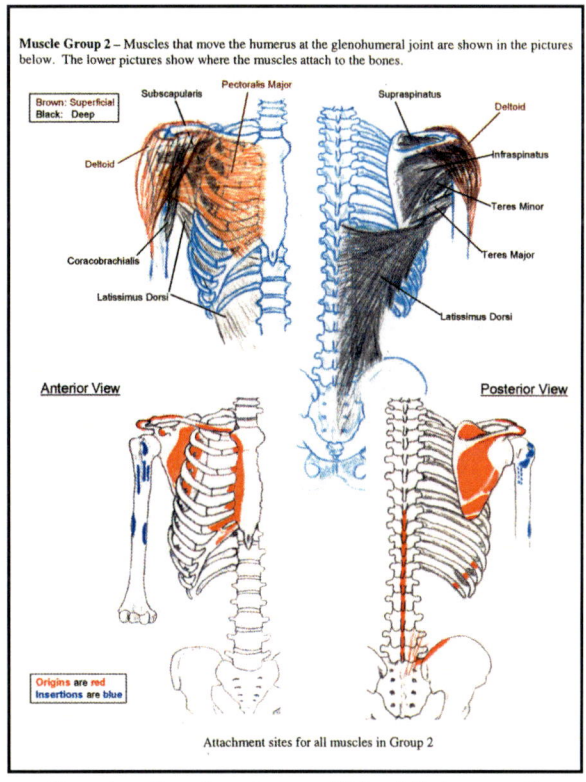

Overview Drawings (see p. 79 for full size page)

How To Use the Tables and Figures

Each group of muscles is presented in a consistent manner. Once you get used to this organization, you will find it easy to study and interrelate the information, as well as quickly look up individual muscles.

Each of the 13 groups of muscles is presented with two pairs of tables and figures: The "**A**" table and figure, and the "**B**" table and figure. All tables and figures for a given muscle group are enumerated with the group number, for example,

> Muscle Group   1: Table **1**(A), Figure **1**(A), Table **1**(B), Figure **1**(B)
> Muscle Group   2: Table **2**(A), Figure **2**(A), Table **2**(B), Figure **2**(B)
> .
> .
> Muscle Group 13: Table **13**(A), Figure **13**(A), Table **13**(B), Figure **13**(B)

The "A" Table and Figure

An "**A**" Table and "**A**" Figure for a muscle group are shown on pages facing each other, so you can easily refer back and forth as you study them. The example on the following page shows Table 2 (A) and Figure 2 (A), which are the muscles that move the humerus at the shoulder joint (glenohumeral joint).

Table 2 (A) – Origin, Insertion, Action

The "**A**" Table contains verbal descriptions of the origins, insertions, and actions for each muscle. As you study the muscles, occasionally look up and down each column to compare and contrast which muscles have similar attachments and actions, and which muscles differ. By continually looking for patterns as you learn the information, you will help anchor the words in your brain and make them easier to recall later.

As you read the Origins and Insertions for each muscle, look down to the facing page, Figure 2 (A), and observe the actual locations on the bones drawn in red and blue. This will help relate the words to an image of exactly what the words mean.

Note that portions of text in the tables are in smaller print and enclosed in parentheses. These parenthetical phrases add extra detail to the basic information in the box. The idea is to have a concise description of origin, insertion, and action that is sufficient for most readers, and then provide more details for those readers who require advanced information.

The actions named in the right hand "Action" column of this table are sorted out and re-listed across the top of the B table (see next section).

Figure 2 (A) – Muscle Attachments

The "**A**" Figure shows the areas where the muscles attach to the bones: Red=Origin, Blue=Insertion. Visualize lines of force (e.g., puppet strings, ropes,…) connecting the red area to the blue area and think about what happens when the blue point on the more moveable bone is pulled toward the red point on the more stable bone. Relate the words in the <u>Action</u> column on Table 2 (A) to the movement you visualize.

Note that you can lift the "A" Figure page to see pictures of individual muscles to go with each of the origin/insertion bone drawings. The "B" Figure – Muscle Pictures is always two pages after the "A" Figure, so the muscle pictures lie directly under the bone attachment drawings.

Example of an "A" Table and Figure

Each **row** gives all information for a single muscle. For example, row 2 gives Origin, Insertion & Action for the Supraspinatus muscle.

Each **column** gives a single feature for all muscles. For example, column 3 gives the Insertions for all the muscles.

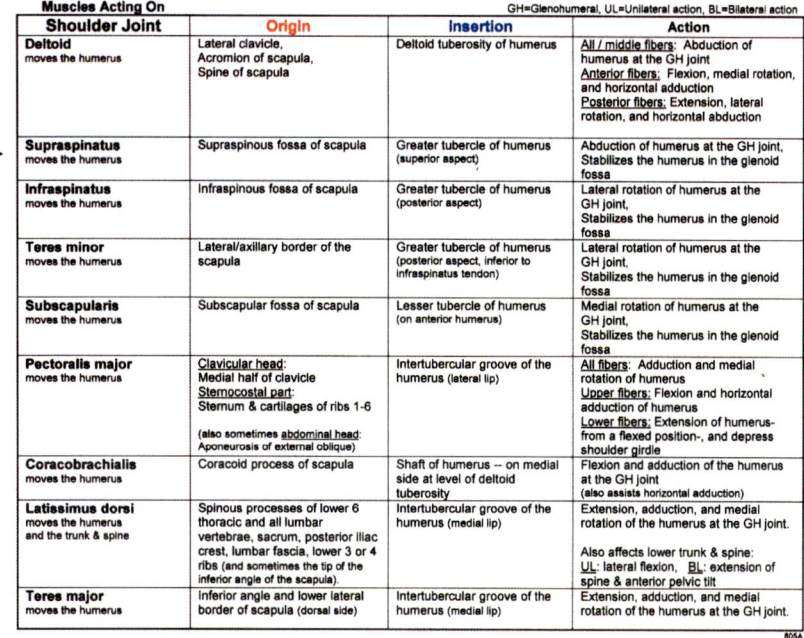

Table 2 (A) - Shoulder Joint - Origin, Insertion, Action

See page 80 for full size Table 2 (A)

Compare the words in the **A** Table above to the pictures in the **A** Figure below.

Origins: Shown in Red
Insertions: Shown in Blue
Actions: Visualize the blue being pulled toward the red

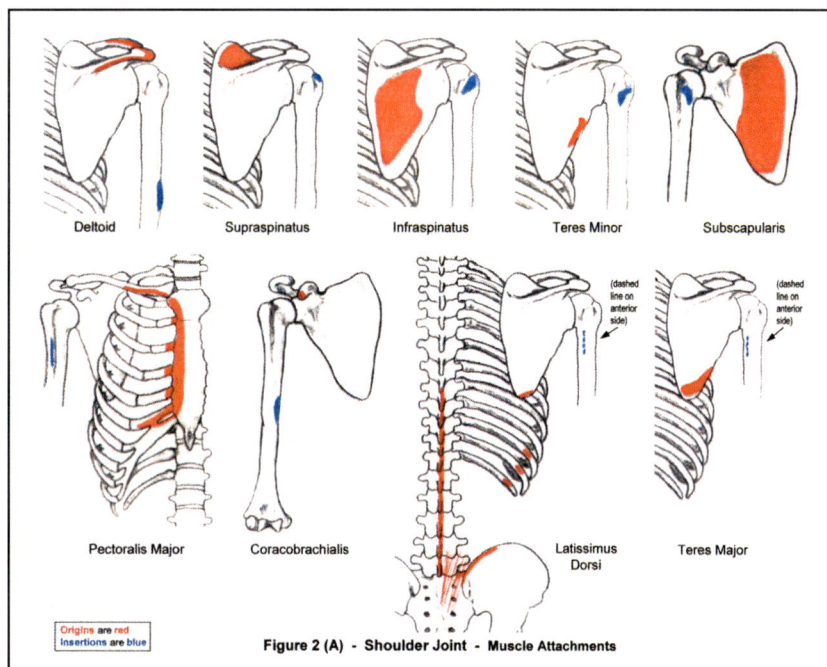

Figure 2 (A) - Shoulder Joint - Muscle Attachments

See page 81 for full size Figure 2 (A)

Each row of pictures is sized and oriented for easy comparison of muscles.

A dashed line indicates a muscle attachment on the opposite side of the bone from the side you are viewing.

Mastering Muscles & Movement © 2007

Chapter 3 - How to Use the Muscle Chapters

The "B" Table and Figure

The "**B**" Table and "**B**" Figure are on pages facing each other so you can easily refer back and forth as you study them. The example on the following page shows Table 2 (B) and Figure 2 (B), which are the muscles that move the humerus at the shoulder joint.

Table 2 (B) – Synergists & Antagonists

The **B** Table has the muscle names listed down the left-hand side, and across the top are all the actions that are available to the joint or structure being moved. In each "cell" of this grid are X's or other symbols that indicate how the muscle (the row) is involved with the action (the column). The arrangement of this table is particularly suitable for learning the relationships of muscles that work together to perform a given action (synergists), as well as which muscles oppose that action (antagonists).

The B table also has an area on the right hand side that gives the **Innervation** for each muscle. The names of the nerve(s) that supply each muscle are listed, and the spinal segments are indicated in table format.

Features of the B tables:

- Pairs of actions that are opposites are placed in adjacent columns. This allows looking down one column to see the synergists for an action, and then looking at the adjacent column to see the antagonists for the action.

- The following symbols are used in the cells:

X	:	The muscle creates the action
Assist	:	The muscle assists the action but is not a prime mover
May assist	:	The muscle may assist, depending on strength requirements or relative bone angles
"__ fibers"	:	Muscle *portion* creates the action (e.g., "anterior fibers")
UL	:	Unilateral contraction creates the action (applies for muscles of the axial skeleton)
BL	:	Bilateral contraction creates the action (applies for muscles of the axial skeleton)
(empty cell)	:	The muscle *does not* contribute to the action

- The "Other" column gives additional information that does not fit into one of the "action" categories.

- The bottom row, with title "More muscles for the action --->", may indicate that muscles in other tables also contribute to the action indicated in a column. If so, the words "see also Table #" are in the cell.

Figure 2 (B) – Muscle Pictures

The **B** Figure contains pictures of the muscles. This figure is presented on a page facing Table 2 (B) so you can look back and forth to relate the actions in the table with muscle positions, shapes, and fiber directions.

Note that the "B" Figure is on a page directly under the "A" Figure two pages prior, so you can simply lift the overlaid pages to directly relate the muscle shape and location with the red and blue muscle attachments shown in the "A" Figure.

Example of a "B" Table and Figure

Each **row** shows X's for the actions created by a single muscle. For example, row 2 indicates that the Suprapinatus muscle creates abduction and stabilizes the GH joint.

Each **column** shows X's for all muscles that create an action. For example, Lateral Rotation at the GH joint is caused by the Deltoid (posterior fibers), Infraspinatus, and Teres Minor muscles.

Muscles Acting On — GH jt.=Glenohumeral joint, X=Muscle creates the action, UL=Unilateral action, BL=Bilateral action, N=Nerve

Shoulder Joint	Flexion @ GH jt.	Extension @ GH jt.	Abduction @ GH jt.	Adduction @ GH jt.	Medial Rotation @ GH jt.	Lateral Rotation @ GH jt.	Stabilization of GH jt.	Other	Innervation	C5	C6	C7	C8	T1
1. **Deltoid:** Anterior fibers, Middle fibers, Posterior fibers	X (Ant. fib.)	X (Post. fib.)	X (All/ middle fibers)		X (Ant. fib.)	X (Post. fib.)		Ant.: Horiz. Adduction; Post: Horiz. Abduction	Axillary N. (C5,C6)	N	N			
2. **Suprapinatus**			X				X		Suprascapular N. (C5)	N				
3. **Infraspinatus**						X	X	may assist horiz. abduction	Suprascapular N. (C5, C6)	N	N			
4. **Teres minor**						X	X		Axillary N. (C5)	N				
5. **Subscapularis**					X		X		Subscapular N. (C5, C6)	N	N			
6. **Pectoralis major:** Upper fibers, Lower fibers, Abdom. fibers	X (Upper fib.)	X (Lower fib.) Exten. from a flexed position		X (All fibers)	X (All fibers)			(lower & abdom fibers depress shoulder girdle)	Upper fib.: Horiz. Adduction; Lateral pectoral N. (C5,C6,C7) & Medial pectoral N. (C8, T1)	N	N	N	N	N
7. **Coracobrachialis**	X			X (with resistance)				may assist horiz. adduction	Musculocutaneous N. (C6, C7)		N	N		
8. **Latissimus dorsi**		X "handcuff position"		X (behind the back)	X			Draws shldr girdle down and back; Affects lower trunk & spine: UL: lat. flex., BL: extension & anterior pelvic tilt	Thoracodorsal N. (C6,C7,C8)		N	N	N	
9. **Teres major**		X		X	X			"Lat's little helper"	Lower Subscapular N. (C5, C6)	N	N			
(More muscles for the action) --->	see also Table 3	see also Table 3							Innervation					

Table 2 (B) - Shoulder Joint - Synergists & Antagonists

Right-hand portion of Table shows nerves that supply each muscle (Innervations)

See page 82 for full size Table 2 (B)

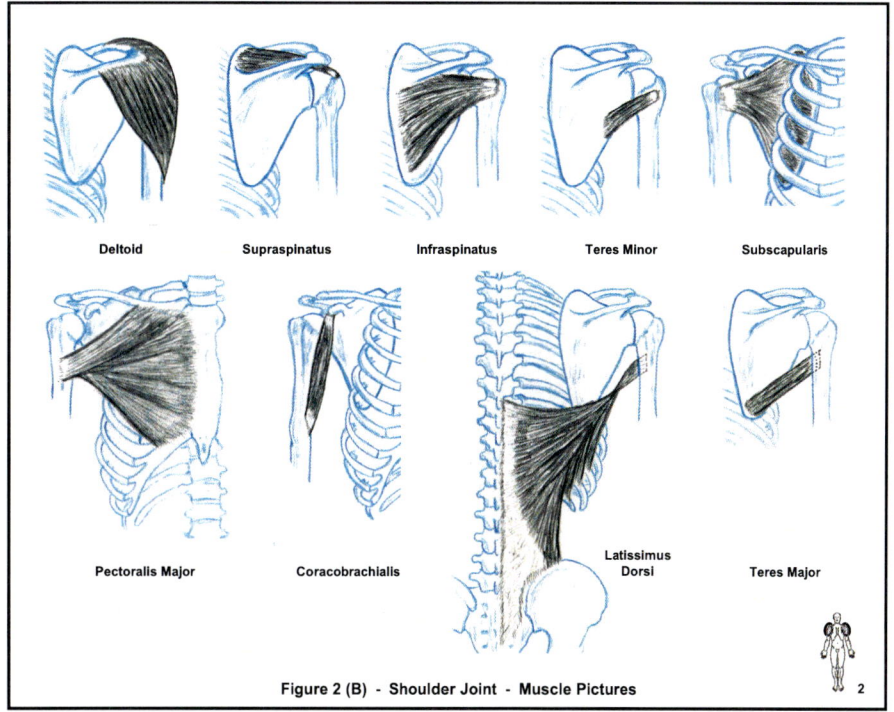

Figure 2 (B) - Shoulder Joint - Muscle Pictures

(Deltoid, Supraspinatus, Infraspinatus, Teres Minor, Subscapularis, Pectoralis Major, Coracobrachialis, Latissimus Dorsi, Teres Major)

See page 83 for full size Figure 2 (B)

Each row of pictures is sized and oriented for easy comparison of muscles.

Dashed lines indicates that part of a muscle is on the **opposite** side of the bone from the side you are viewing.

Mastering Muscles & Movement © 2007 — Chapter 3 - How to Use the Muscle Chapters

An Example: Learning the Deltoid Muscle

This section shows you how to use the tables and figures to study the Deltoid muscle and compare and contrast it to other muscles in its group. The components of this example are taken from the section "Muscle Group 2 – Movement of the Shoulder Joint" on pages 80-83, which is in Chapter 4 – Muscles That Move the Upper Extremity.

The figure below shows a blow up of the **Deltoid** muscle sections from Table 2(A), Figure 2(A), and Figure 2(B). This shows how to tie together the information.

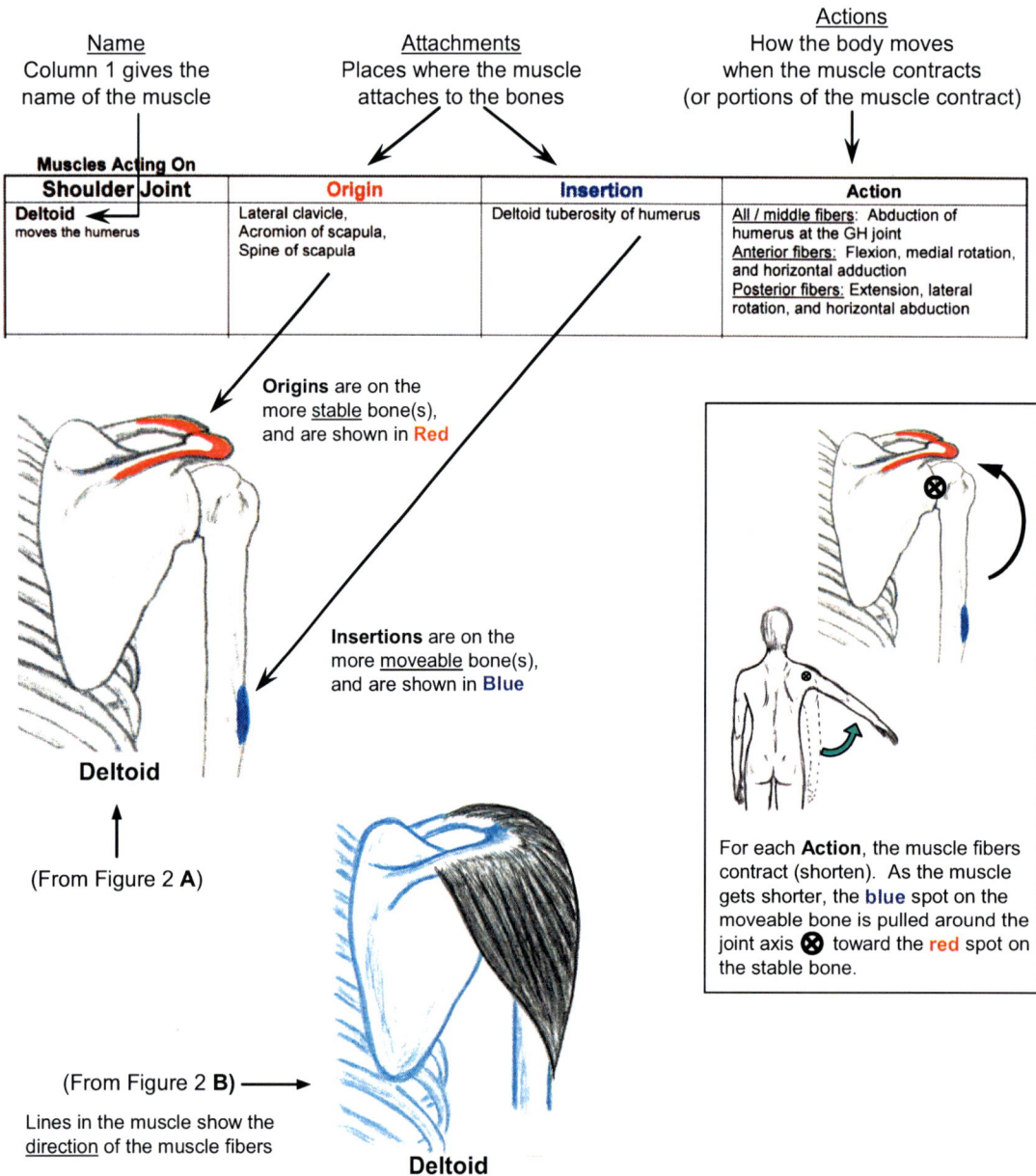

To strengthen your knowledge, compare the origins of the deltoid with its neighboring muscles. Do this verbally by comparing the words in the Origin column of the table, and visually by comparing the red markings on the bone drawings. Do the same for the insertions. Now, view the muscle drawings with clarity about where beneath the muscle fibers each muscle attaches to the bones.

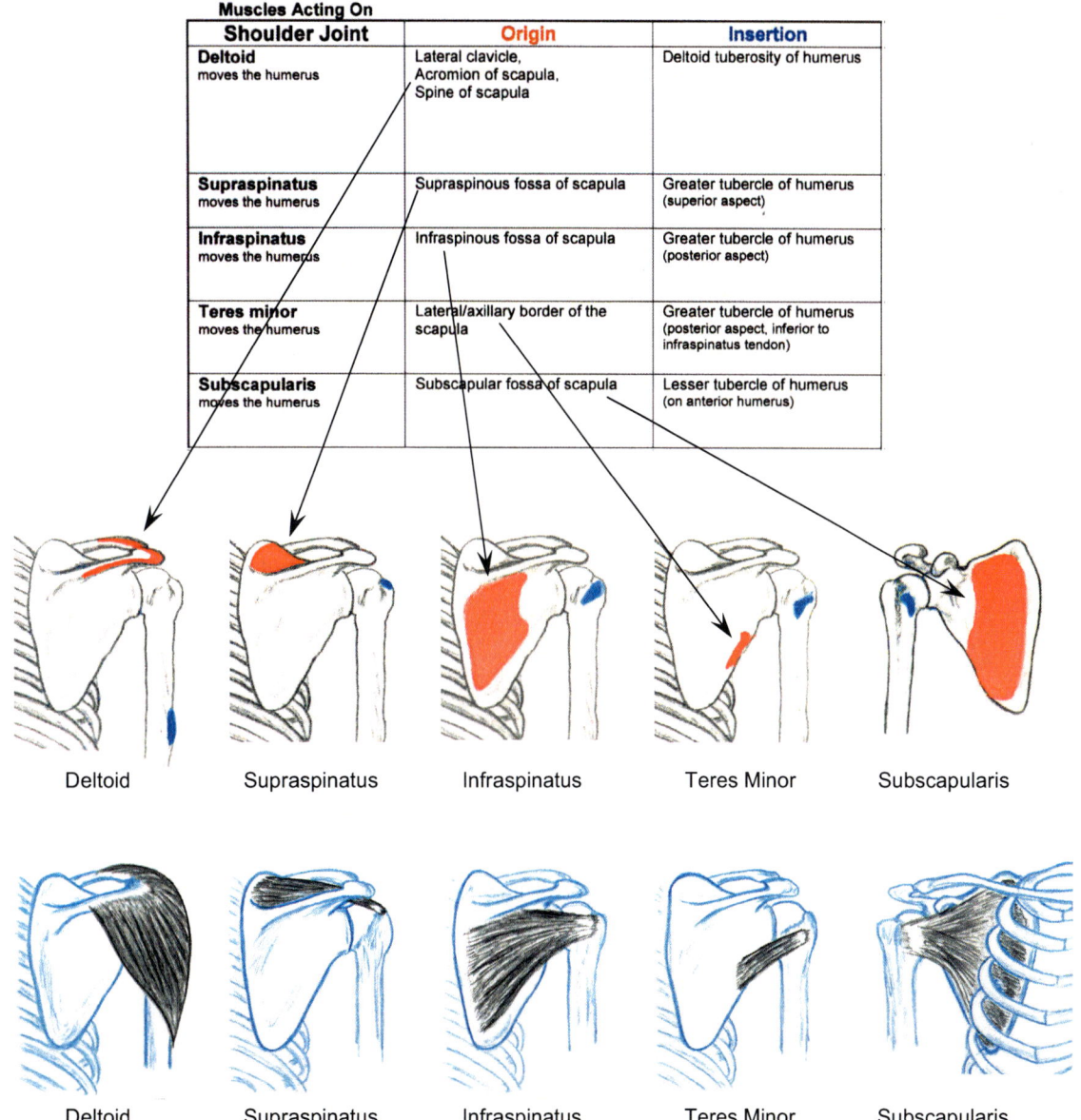

A Note on Actions:

The Actions given in the "A" Tables (for example, Table 2 (A)) are considered to be the primary or "usual" way that the muscle moves the bones, which is based on the muscle contracting under "normal" conditions. These actions are directly related to which bony attachments are called origins and which are called insertions. Remember, the definition of <u>origin</u> is the attachment on the more <u>stable</u> bone, and the definition of <u>insertion</u> is the attachment on the more <u>moveable</u> bone.

In many cases, there is a way that the so-called "more moveable" bone can be stabilized, which makes the so-called "more stable" bone then become the moveable one. In that case we can name a *different* action, called the **Reversed O/I Action**, that describes how the reverse-of-normal bone is moving.

About Mastering the Muscles

As you study each group of muscles, you will be building a "foundation" of basic knowledge that will enable you to communicate anatomical and movement information in a clear and efficient way. A few situations where this will be useful may be: Assessing and working with clients, coherently communicating with other health professionals (verbally and written), reading books and magazine/journal articles, and continuing your education in advanced classes relevant to your specialty.

In order to be fluent in the language of muscles and movement, you will need to learn a basic set of information about each muscle. Then, you will need to be able to communicate and use that information in a variety of ways.

What To Learn About Each Muscle

Use the following list as a guide to what you need to know about each muscle to master it. Items 1 through 6 are basic knowledge, and items 7 and 8 are advanced knowledge you will learn with practical applications.

1. Name and palpate the origin(s)

2. Name and palpate the insertion(s)

3. Trace the shape of the entire muscle on the body, and palpate the muscle.
    - Include knowing where the tendon is vs. the muscle belly

4. Know and touch the joint(s) the muscle acts on (there are sometimes more than one)

5. Indicate the fiber arrangement (shape and direction)
    - Show it on the body, describe it, draw a diagram showing it

6. Name and demonstrate the actions of the muscle

7. For the muscle's main action (or action<u>s</u>):
    - Passively shorten it
    - Passively lengthen it
    - Instruct client to actively shorten it (concentric contraction)
    - Instruct client to actively lengthen it (the antagonist is working)
    - Provide correct resistance to test the strength of the muscle
    - Name one or more synergists
        - (<u>must</u> include the name of the *action* that is being "synergized")
    - Name one or more antagonists
        - (<u>must</u> include the name of the *action* that is being "antagonized")

8. Know something about the muscle as it applies to the daily life of a person
    - Activities and exercises that use this muscle (as agonist, antagonist, and stabilizer)
    - Movements in which the muscle contracts concentrically, and contracts eccentrically
    - Problems or pathologies that may apply to this muscle

How to Use the Information You've Learned

Here are some ways that you will use your knowledge of a muscle. You may have to recall or communicate the information from any of three main directions: **verbal**, **visual**, or **kinesthetic**. The triangle below illustrates this concept. In any given situation, you may need to access your knowledge from one of the corners of the triangle. You then need to be able to connect to the types of information represented by the other corners as you pursue the requirements of the situation at hand. In addition, you may need to think **relationally**, i.e., for a muscle or action, be able to think of related muscles or actions.

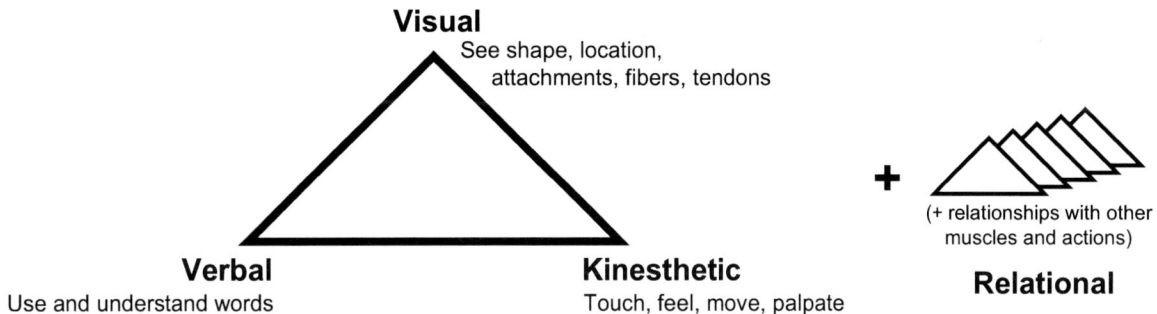

Here are some examples:

Legs of Triangle	If This Happens…	Can You:
Visual to --- verbal	You look at a picture of a muscle.	Name the muscle, and verbally describe its characteristics.
Verbal to -- visual	You hear/read the name of the muscle.	Visualize or draw where it is on the body and show what it looks like.
Visual to -- verbal - to - visual	Your client points to a place on their body.	Recall the names of the muscles there, and visualize what they look like at different depths.
Kinesthetic to -- verbal -- to -- relational	Your client moves a body part and says it hurts when they move.	Based on the joint and the movement: - Identify the muscles that are shortening (agonists) - Identify the muscles that are lengthening (antagonists)
	Your client exhibits a limitation in a certain movement.	Figure out what is stopping it.
	Your client is weak and needs to strengthen a muscle.	Provide the correct resistance to activate the muscle.

Palpation Techniques

Here are some ideas to consider when you are palpating the body to identify, assess, or treat muscles.

(Locate, isolate, engage, palpate, observe cautions)

- Be aware of the tactile: Touch, pressure, texture.

- Find landmarks to locate the full length of the muscle.

- Try to feel different depths, being aware of changes in fiber direction and knowing that layers of fascia separate muscles that lie at different depths.

- Use movement (passive, active) to help locate muscles.

- Consider main movements vs. secondary (or assist) movements.

- Provide correct contact point and direction of resistance to engage or activate a muscle.

- Remember gravity is always there to provide resistance (position the body so the muscle has to work against gravity).

- To palpate on yourself, look for a way to "self-engage" the muscle – press against a table, wall, chair, or a place on your own body.

- A muscle is easier to feel upon initial muscle activation rather than full-out contraction – have your partner initiate the action and then release it several times to help isolate the muscle.

Summary and Generalizations

Here are some generalizations to use while you are learning the muscles in Chapters 4, 5 and 6. Keeping these in mind will help you stay oriented, see useful patterns, and reduce the amount of rote memorization you have to do.

Summary:

The movement that occurs when a given muscle contracts can be summarized as the sum of six factors:

1. Direction and arrangement of its fibers + ...
2. Locations of its attachment sites + ...
3. The mechanical capability of the joint(s) being moved + ...
4. Stuff in the way (muscle tissue, fascia, bones, ligaments, skin, fat, organs, etc.) + ...
5. What *other* muscles are doing at the same time (opposing, stabilizing, etc.) + ...
6. Which bones are most moveable at the moment (what is weight bearing, current direction of gravity, what is held in place by outside forces like wall, table, another person, etc.)

Generalizations:

Muscles are strongest at their resting or neutral position.

Muscles on the anterior body usually cause flexion (except below knee).

Muscles on the posterior body usually cause extension (except below knee).

Muscles that have an oblique (diagonal) angle usually create or control rotations.

Muscles that attach to the lateral side of limbs abduct.

Muscles that attach to the medial side of limbs adduct.

For muscles of the axial skeleton, an antagonist for a *unilateral* action is the same muscle on the other side of the body (see Chapter 5, page 112).

Chapter 4
Muscles That Move the Upper Extremity

Introduction	**68**
Movement of the Scapula/Clavicle ......(Muscle Group 1)	**69**
Movement of the Shoulder Joint ......(Muscle Group 2)	**77**
Movement of the Elbow and Forearm ......(Muscle Group 3)	**85**
Movement of the Wrist, Hand, and Fingers ......(Muscle Group 4)	**93**
Movement of the Thumb ......(Muscle Group 5)	**101**
Intrinsic Muscles of the Hand	**108**

Group 1 – Scapula / Clavicle

p. 69

Trapezius
Levator scapula
Rhomboid major & minor
Serratus anterior
Pectoralis minor
Subclavius

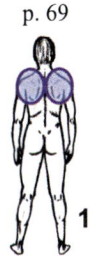

Group 2 – Shoulder Joint

p. 77

Deltoid
Supraspinatus
Infraspinatus
Teres minor
Subscapularis
Pectoralis major
Coracobrachialis
Latissimus dorsi
Teres major

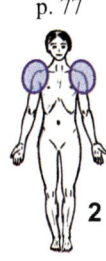

Group 3 – Elbow, Forearm

p. 85

Biceps brachii
Brachialis
Brachioradialis
Pronator teres
Pronator quadratus
Triceps brachii
Anconeus
Supinator

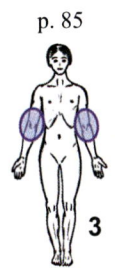

Group 4 – Wrist, Hand, Fingers

p. 93

Flexor carpi radialis
Palmaris longus
Flexor carpi ulnaris
Flexor digitorum superficialis
Flexor digitorum profundus
Extensor carpi radialis longus
Extensor carpi radialis brevis
Extensor carpi ulnaris
Extensor digitorum
Extensor indicis

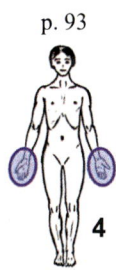

Group 5 – Thumb

p. 101

Flexor pollicis longus
Flexor pollicis brevis
Opponens pollicis
Adductor pollicis
Abductor pollicis longus
Abductor pollicis brevis
Extensor pollicis longus
Extensor pollicis brevis

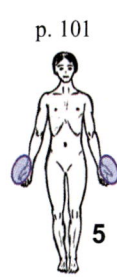

Introduction

The **upper extremity** is the upper-body portion of the appendicular skeleton (p. 34). It is made up of the scapula and clavicle (shoulder girdle), upper arm, forearm, and hand. The sternoclavicular joint, where the medial end of the clavicle articulates with the sternum, is the *only* synovial joint connecting each upper extremity to the trunk.

This chapter describes the muscles that move the scapula on the torso, and move the various joints within the upper extremity. The muscles are separated into five functional groups, with some overlap of function between groups for muscles that cross multiple joints:

Group 1 – Movement of the shoulder girdle, which is the scapula and clavicle moving together on the torso
Group 2 – Movement of the humerus at the shoulder joint
Group 3 – Movement of the forearm, bending at the elbow and rotating on its lengthwise axis
Group 4 – Movement of the wrist, hand and fingers
Group 5 – Movement of the thumb

At the end of the chapter, additional illustrations present the intrinsic muscles of the hand.

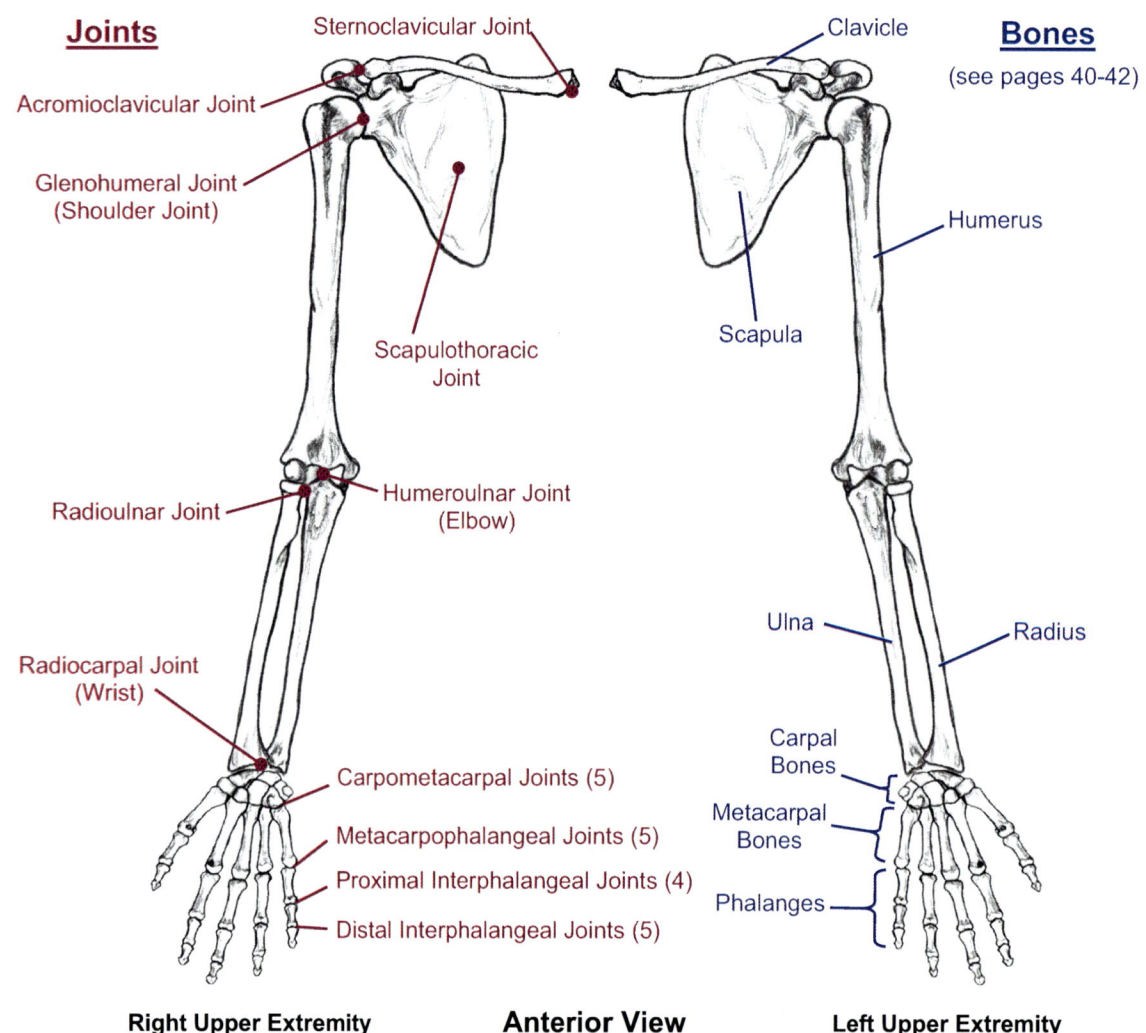

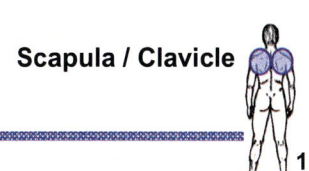

Movement of the Scapula/Clavicle

Muscle Group 1:

Trapezius	Serratus anterior
Levator scapula	Pectoralis minor
Rhomboid major and minor	Subclavius

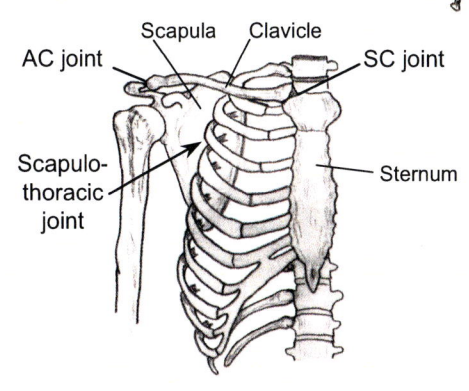

Joints

The scapula and clavicle combined are called the **shoulder girdle**. As the shoulder girdle moves, the scapula glides on the back of the ribcage. Three joints are involved with movement of the shoulder girdle.

Sternoclavicular Joint (SC)

Manubrium of **sternum** ◄► medial end of **clavicle**
Modified ball and socket joint
(Complex joint - may also be classified in other ways)

The SC joint is the only bony connection of the upper-body appendicular skeleton with the axial skeleton.

Acromioclavicular Joint (AC)

Acromion of **scapula** ◄► lateral end of **clavicle**
Gliding joint

Scapulothoracic Joint (ST)

Anterior surface of **scapula** ◄► Posterior thoracic wall
False joint (the scapula suspended by muscles)

Movements available

The scapula and clavicle move together, so all three joints (SC, AC, ST) are said to have the same movements, which are the actions of the scapula.

Scapular actions:
  Elevation
  Depression
  Abduction (Protraction)
  Adduction (Retraction)
  Upward Rotation – glenoid fossa goes up   (also called Lateral Rotation – inferior angle moves laterally)
  Downward Rotation – glenoid fossa goes down (also called Medial Rotation – inferior angle moves medially)

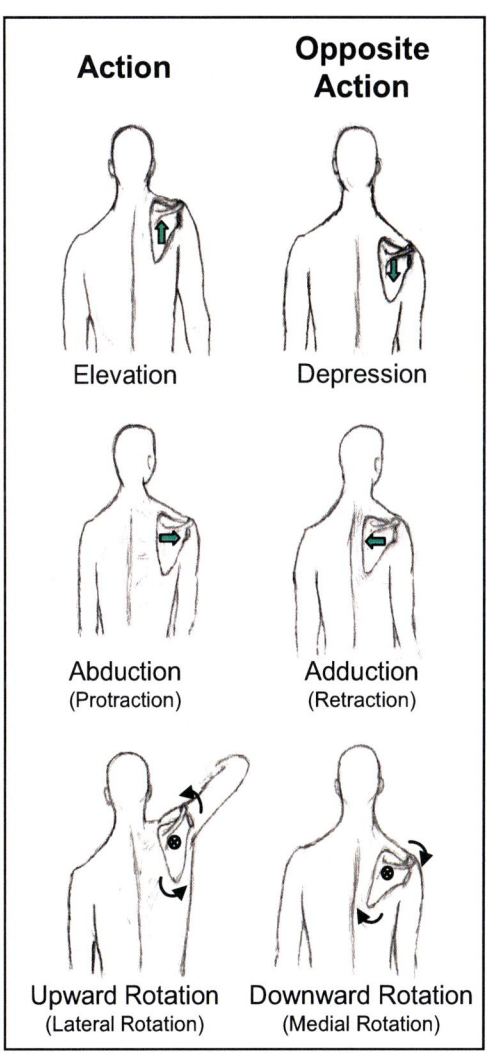

Note: As the scapula abducts, its lateral edge also slides *forward*, following the curve of the ribs. This is why abduction is sometimes called protraction.

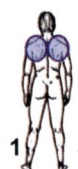 Scapula / Clavicle

Bones and bony landmarks

Muscles that move the Scapula/Clavicle have attachments on the upper trunk, spine, and head, and of course the scapula and clavicle. Review the bony landmarks and other structures listed below, referring to the drawings in Chapter 2, pages 40 and 43-45.

Scapula (p. 40)
    Acromion
    Coracoid process
    Superior angle
    Inferior angle
    Spine of scapula
    Root of spine
    Medial (vertebral) border
    Lateral (axillary) border
    Subscapular fossa

Notes:
- There are bony landmarks on both the posterior and anterior sides of the scapula

- The anterior side of the scapula is also known as the costal side (root "costo"=rib)

Clavicle (p. 40)
    Medial end is called the sternal end
    Lateral end is called the acromial end

Sternum (p. 46)
    Manubrium

Occiput (p. 43)
    External occipital protuberance
    Superior nuchal line

Spine (p. 44)
    Cervical Vertebrae
        Spinous Processes (SP's)
        Transverse Processes (TVP's)

    Thoracic Vertebrae
        Spinous Processes (SP's)

Ribs (p. 46)
    Anterior and anteriolateral surfaces
    Costal cartilage

Tendinous Structures:

Nuchal ligament
(also called the
Ligamentum Nuchae)

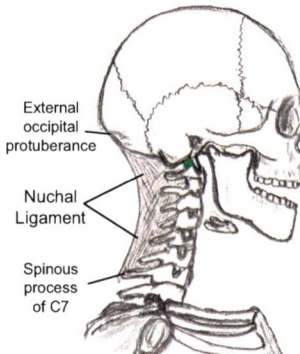

Notes

Scapula / Clavicle

Muscle Group 1 – Muscles that move the shoulder girdle (scapula/clavicle) are illustrated as a group on this page. The next four pages have tables and figures that describe each muscle individually, and provide many ways of comparing and contrasting the muscles to each other.

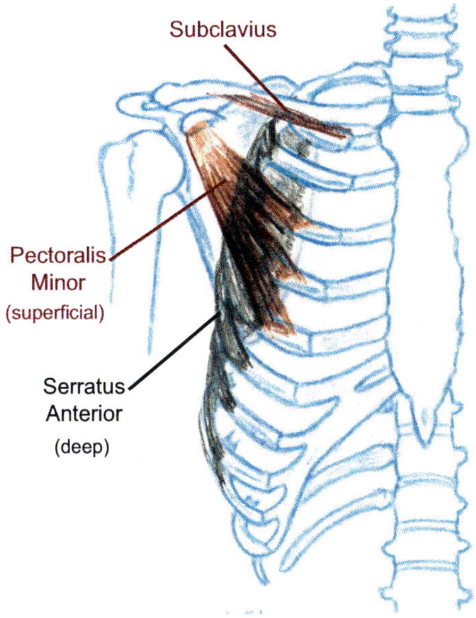

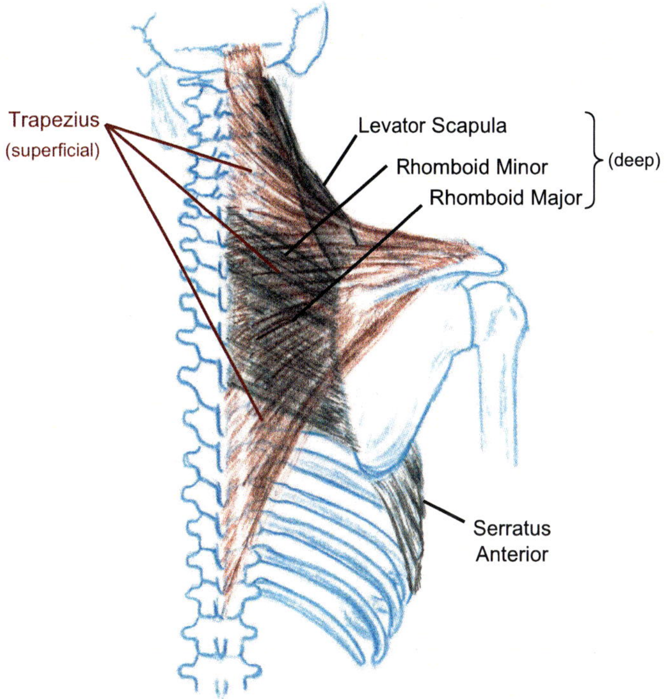

Anterior View

Posterior View

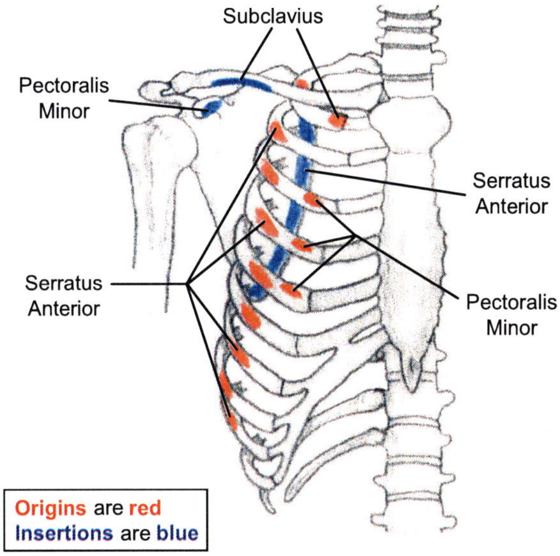

Origins are red
Insertions are blue

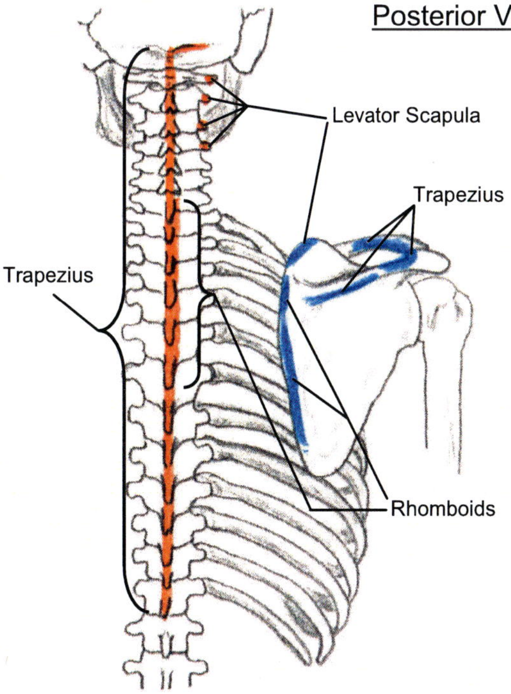

Attachment sites for all muscles in Group 1

Mastering Muscles & Movement © 2007   Chapter 4 – Muscles That Move the Upper Extremity

Scapula / Clavicle

TVP=Transverse process of vertebra, SP=Spinous process of vertebra

Muscles Acting On Scapula / Clavicle	Origin	Insertion	Action
Trapezius — moves the scapula (*also moves the neck*) Has 3 parts: Upper Fibers, Middle Fibers, Lower Fibers	Upper: Occiput, nuchal ligament, and SP of C7 Middle: SP's of T1-T4 Lower: SP's of T5-T12 *Overall description*: Occiput, nuchal ligament, and SP's of C7-T12 (Occipital attachment is the external occipital protuberance and medial 1/3 of superior nuchal line)	Upper: Lateral ⅓ of clavicle Middle: Acromion of scapula and spine of scapula (lateral superior portion) Lower: Spine of scapula (medial inferior portion) *Overall description*: Lateral clavicle, acromion and spine of scapula	Upper fibers: Elevation & upward rotation of the scapula Middle fibers: Adduction (retraction) Lower fibers: Depression & upward rotation All fibers: Stabilize and adduct the scapula (*When the scapula is fixed*: The *upper trapezius* moves the head & neck -- See Muscle Group 7)
Levator Scapula — moves the scapula (*also moves the neck*)	TVP's of C1-C4	Superior angle of the scapula (and upper medial border of scapula)	Elevation and downward rotation of the scapula. (*When scapula is fixed*: Moves the neck -- See Group 7)
Rhomboids — move the scapula: Rhomboid Minor, Rhomboid Major	Minor: SP's of C7, T1 Major: SP's of T2-T5 *Overall description*: SP's of C7-T5 (deep to mid. trapezius)	Medial (vertebral) border of the scapula	Adduction (retraction) and elevation of the scapula. (Also assists downward rotation of the scapula)
Serratus Anterior — moves the scapula	Ribs #1 - 8 or 9 (anterolateral surfaces)	Medial (vertebral) border of the scapula (on the <u>anterior</u> side)	Abduction (protraction) and upward rotation of the scapula. Stabilizes the scapula.
Pectoralis Minor — moves the scapula	Anterior ribs #3-5 (deep to pec major)	Coracoid process of scapula	Anterior tilt (i.e., draws scapula forward, downward, and inward). Assists depression of scapula. *When scapula is fixed*: Assists in forced inhalation.
Subclavius — stabilizes the clavicle	Rib #1 (medial portion at its junction with costal cartilage)	Inferior aspect of clavicle	Depresses the clavicle. Stabilizes the sternoclavicular joint.

(larger illustrations on page 75)

Table 1 (A) - Scapula / Clavicle - Origin, Insertion, Action

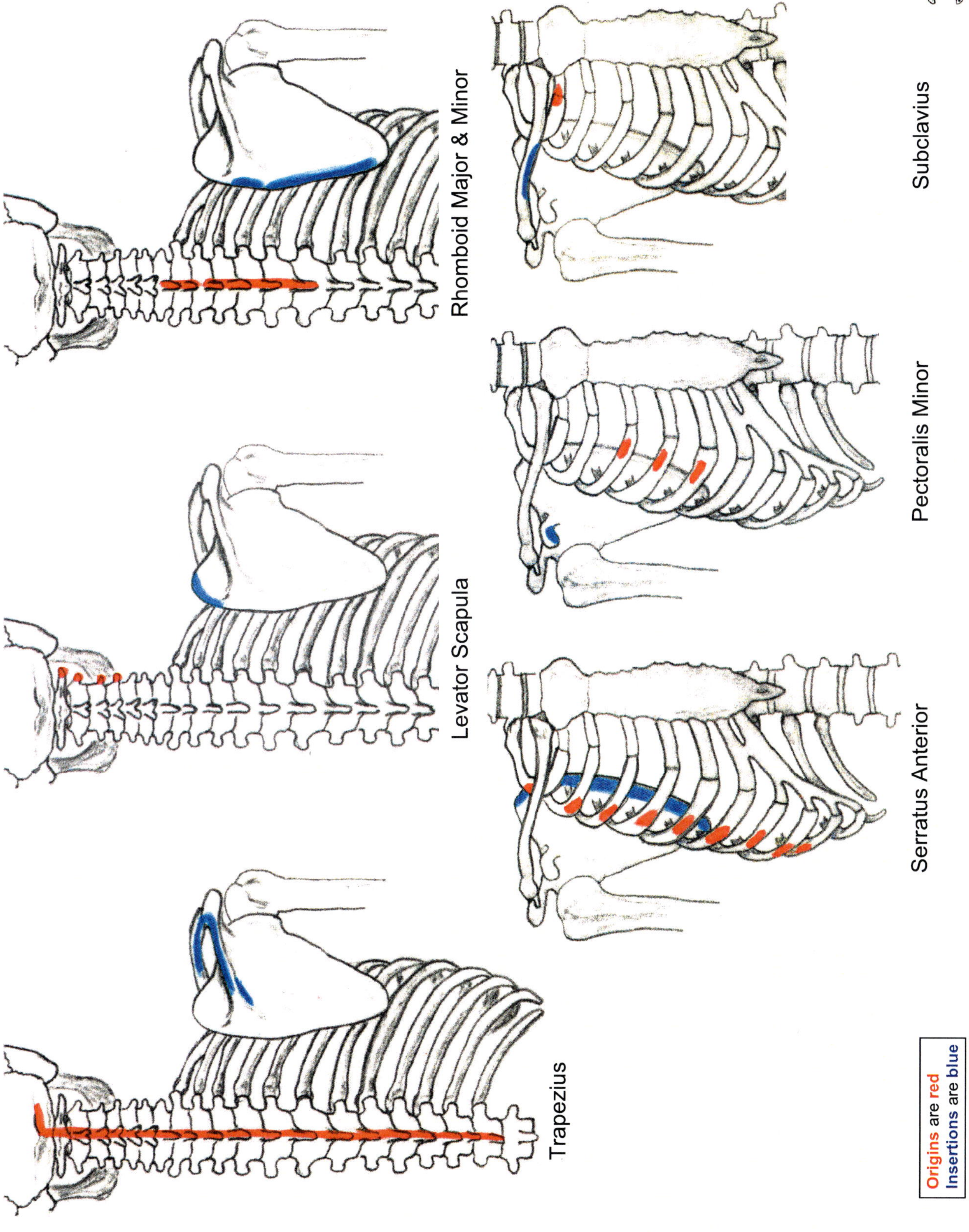

Scapula / Clavicle

Rhomboid Major & Minor

Subclavius

Levator Scapula

Pectoralis Minor

Serratus Anterior

Trapezius

Figure 1 (A) - Scapula / Clavicle - Muscle Attachments

Origins are red
Insertions are blue

Lift page to see muscle pictures

Mastering Muscles & Movement © 2007

Chapter 4 - Muscles That Move the Upper Extremity

Scapula / Clavicle

All actions are the shoulder girdle (scapula+clavicle) moving as a unit, X=Muscle creates the action, N=Nerve

Muscles Acting On Scapula / Clavicle

Scapula / Clavicle	Elevation	Depression	Protraction/ Abduction	Retraction/ Adduction	Lateral/ Upward Rotation	Medial/ Downward Rotation	Stabili-zation of scapula	Other	Innervation
1. Trapezius:									
Upper fibers	X (upper fibers)							Upper fibers: See also Table 7 for reversed O/I actions	Spinal Accessory N. (Cranial N. XI), and C3, C4
Middle fibers		X (lower fibers)		X (middle fibers)			X (All fibers)		
Lower fibers					X (lower fibers)				
2. Levator scapula	X					X		See also Table 7 for reversed O/I actions	Dorsal scapular N. (C5), and C3, C4
3. Rhomboids: Rhomboid major Rhomboid minor	X			X		assist			Dorsal scapular N. (C5)
4. Serratus anterior			X		X		X	"Punching" motion	Long thoracic N. (C5-C7)
5. Pectoralis minor		assist (with anterior tilt)						Creates "Anterior tilt" = Draws scapula forward, down, and in	Medial pectoral N. (C8, T1)
6. Subclavius		assist (depresses clavicle)						Stabilizes sterno-clavicular jt.	Subclavian N. (C5, C6)
(More muscles for the action) --->									

Innervation

	Cr. XI	C3	C4	C5	C6	C7	C8	T1
Trapezius	N	N	N					
Levator scapula		N	N	N				
Rhomboids				N				
Serratus anterior				N	N	N		
Pectoralis minor							N	N
Subclavius				N	N			

Table 1 (B) - Scapula / Clavicle - Synergists & Antagonists

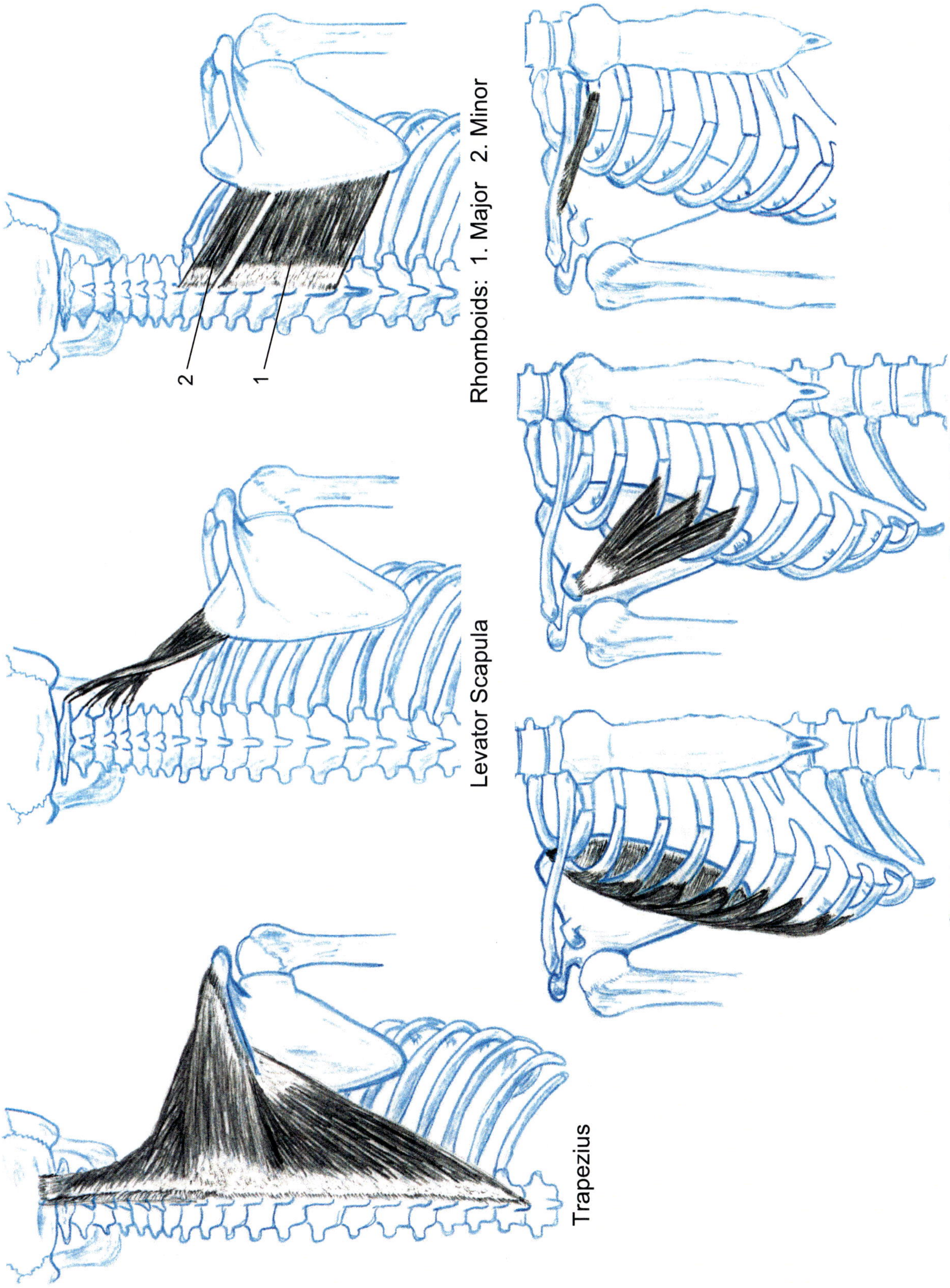

Scapula / Clavicle

Rhomboids: 1. Major 2. Minor

Subclavius

Levator Scapula

Pectoralis Minor

Serratus Anterior

Trapezius

Figure 1 (B) - Scapula / Clavicle - Muscle Pictures

Muscles Acting On Scapula / Clavicle ~ Notes ~
(palpation, fiber arrangement, lengthen/shorten, common uses, pathologies, cautions, etc.)

1. Trapezius
(Upper fibers, Middle fibers, Lower fibers)

2. Levator Scapula

3. Rhomboid Major & Minor

4. Serratus Anterior

5. Pectoralis Minor

6. Subclavius

Muscle Group 1 - Scapula / Clavicle - (Notes)

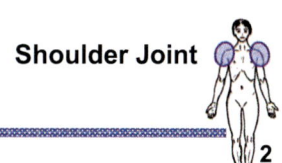

Movement of the Shoulder Joint

Muscle Group 2

Deltoid	Rotator Cuff Muscles:
Pectoralis major	Supraspinatus
Coracobrachialis	Infraspinatus
Latissimus dorsi	Teres minor
Teres major	Subscapularis

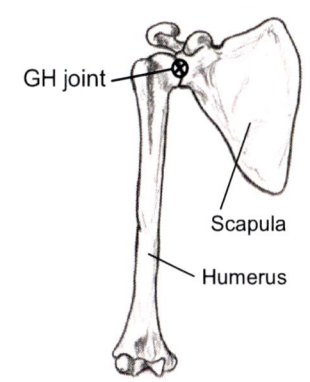

Joints

The muscles in this group primarily move the humerus at the glenohumeral joint, commonly called the shoulder joint.

Glenohumeral Joint (GH) – The Shoulder Joint

    Glenoid fossa of **scapula** ◄► Head of **humerus**

    Ball and socket joint

    Movements available:
        Flexion, Extension
        Abduction, Adduction
        Lateral Rotation, Medial Rotation

        Horizontal Abduction, Horizontal Adduction
        Circumduction (a combination movement)

    Notes:
        The GH joint is highly mobile, and very unstable.
        Joint capsule is main support (few extrinsic ligaments)

Other Movements

The combination of the shoulder joint and the shoulder girdle is called the Shoulder *Complex*. The humerus and scapula often coordinate their movements. For example,

● Muscles acting on the humerus which have origins that are *not on the scapula* can create scapular movement by "remote control" (e.g, latissimus dorsi, pectoralis major).

● Scapulohumeral "rhythms" – For big arm motions, humerus and scapula movements are sequenced and coordinated.

    - During abduction, adduction
    - During flexion, extension
    - During horizontal abduction and adduction

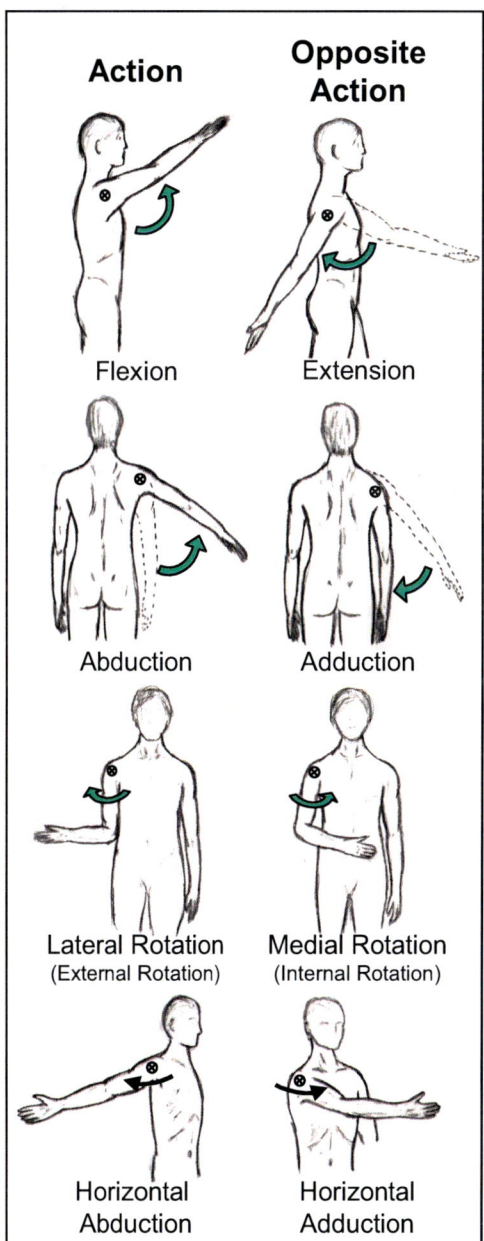

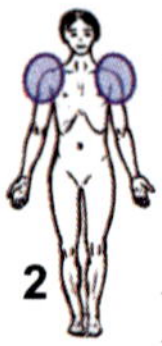

Shoulder Joint

Bones and bony landmarks

Muscles that move the humerus at the shoulder joint have attachments over a large area of the skeleton. Review the bony landmarks and other structures listed below, referring to the drawings in Chapter 2, p. 40-47.

Humerus (p. 41)

    Head
    Greater tubercle
    Lesser tubercle
    Intertubercular groove (Bicipital groove)
        - Medial edge and lateral edge of the groove are separate landmarks
    Deltoid tuberosity
    Shaft

Scapula (p. 40)

    Glenoid fossa (also called glenoid cavity)
    Supraspinous fossa
    Infraspinous fossa
    Subscapular fossa
    Lateral (axillary) border
    Spine (of scapula)
    Acromion
    Coracoid process
    Inferior angle

Clavicle (p. 40)

Sternum (p. 46)
    Manubrium
    Body
    Xiphoid process

Ribs (p. 46)
    Anterior: Costal cartilages
    Posterior: Lower 3-4 ribs

Vertebrae (p. 44)
    Thoracic spinous processes (SP's)
    Lumbar spinous processes (SP's)

Sacrum (p. 44)

Pelvis (p. 47) – Posterior iliac crest

<u>Tendinous Structures:</u>

    Lumbar Fascia
        Also called Thoracolumbar Fascia
        Also called Common Tendon
        Also called Thoracolumbar Aponeurosis (TLA)

    Rotator Cuff
        Also called Musculotendinous Cuff

Notes

Shoulder Joint

Muscle Group 2 – Muscles that move the humerus at the glenohumeral joint are illustrated as a group on this page. The four pages that follow have tables and figures that describe each muscle individually, and provide many ways of comparing and contrasting the muscles to each other.

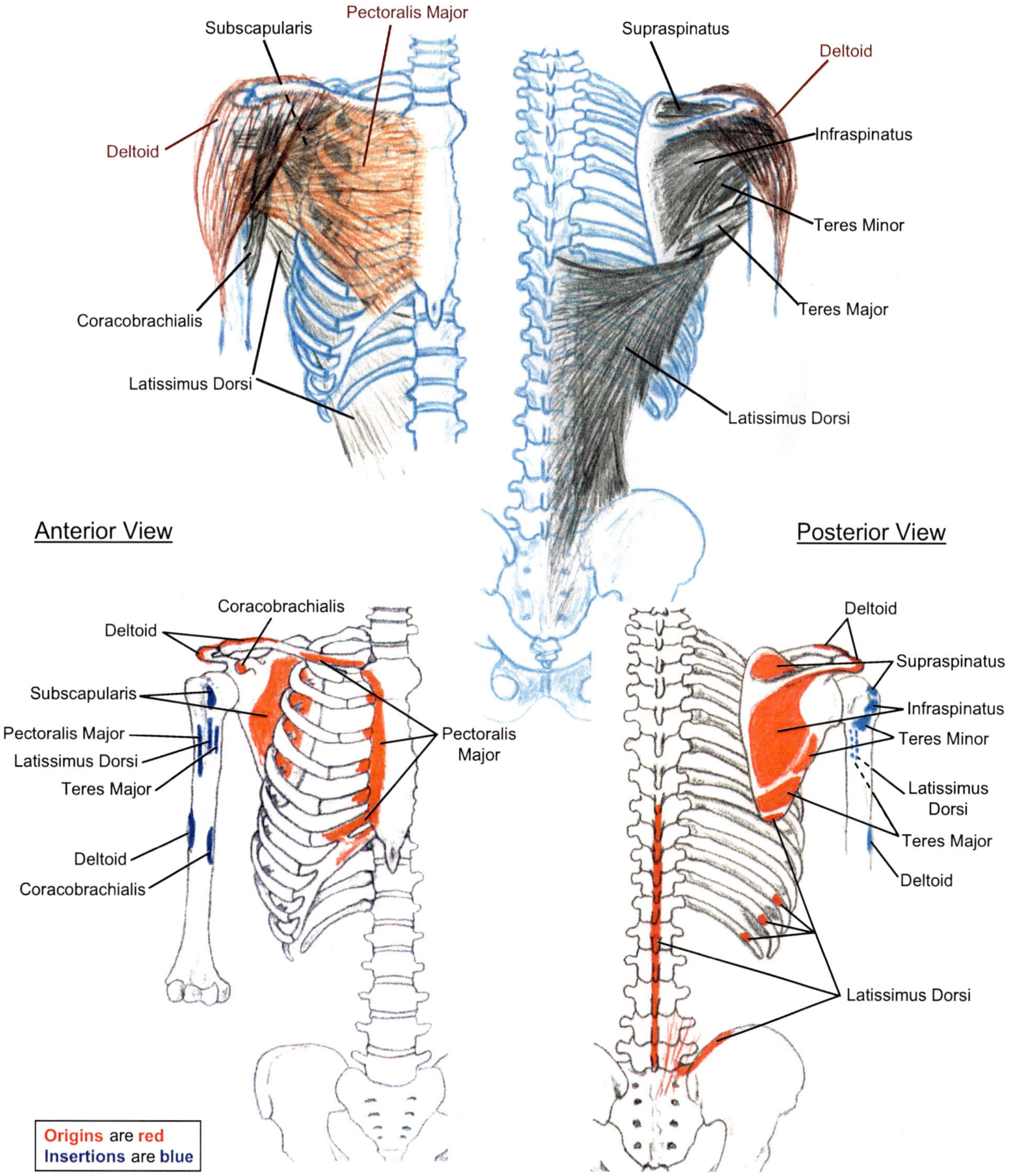

Attachment sites for all muscles in Group 2

Origins are red
Insertions are blue

Mastering Muscles & Movement © 2007 — Chapter 4 – Muscles That Move the Upper Extremity

Shoulder Joint

Muscles Acting On Shoulder Joint — GH=Glenohumeral joint (shoulder joint), UL=Unilateral action, BL=Bilateral action

Shoulder Joint	Origin	Insertion	Action
Deltoid — moves the humerus	Lateral clavicle, Acromion of scapula, Spine of scapula	Deltoid tuberosity of humerus	All / middle fibers: Abduction of humerus at the GH joint Anterior fibers: Flexion, medial rotation, and horizontal adduction Posterior fibers: Extension, lateral rotation, and horizontal abduction
Supraspinatus — moves the humerus	Supraspinous fossa of scapula	Greater tubercle of humerus (superior aspect)	Abduction of humerus at the GH joint, Stabilizes the humerus in the glenoid fossa
Infraspinatus — moves the humerus	Infraspinous fossa of scapula	Greater tubercle of humerus (posterior aspect)	Lateral rotation of humerus at the GH joint, Stabilizes the humerus in the glenoid fossa
Teres minor — moves the humerus	Lateral/axillary border of the scapula	Greater tubercle of humerus (posterior aspect, inferior to infraspinatus tendon)	Lateral rotation of humerus at the GH joint, Stabilizes the humerus in the glenoid fossa
Subscapularis — moves the humerus	Subscapular fossa of scapula	Lesser tubercle of humerus (on anterior humerus)	Medial rotation of humerus at the GH joint, Stabilizes the humerus in the glenoid fossa
Pectoralis major — moves the humerus	Clavicular head: Medial half of clavicle Sternocostal part: Sternum & cartilages of ribs 1-6 (also sometimes abdominal head: Aponeurosis of external oblique)	Intertubercular groove of the humerus (lateral lip)	All fibers: Adduction and medial rotation of humerus Upper fibers: Flexion and horizontal adduction of humerus Lower fibers: Extension of humerus - from a flexed position-, and depression of shoulder girdle
Coracobrachialis — moves the humerus	Coracoid process of scapula	Shaft of humerus -- on the medial side half way down	Flexion and adduction of the humerus at the GH joint (also assists horizontal adduction)
Latissimus dorsi — moves the humerus and the trunk & spine	Spinous processes of lower 6 thoracic and all lumbar vertebrae, sacrum, posterior iliac crest, lumbar fascia, lower 3 or 4 ribs (and sometimes the tip of the inferior angle of the scapula).	Intertubercular groove of the humerus (medial lip)	Extension, adduction, and medial rotation of the humerus at the GH joint. Also affects lower trunk & spine: UL: lateral flexion, BL: extension of spine & anterior pelvic tilt
Teres major — moves the humerus	Inferior angle and lower lateral border of scapula (dorsal side)	Intertubercular groove of the humerus (medial lip)	Extension, adduction, and medial rotation of the humerus at the GH joint.

Rotator Cuff Muscles (Supraspinatus, Infraspinatus, Teres minor, Subscapularis)

(larger illustrations on page 83)

Table 2 (A) - Shoulder Joint - Origin, Insertion, Action

Shoulder Joint

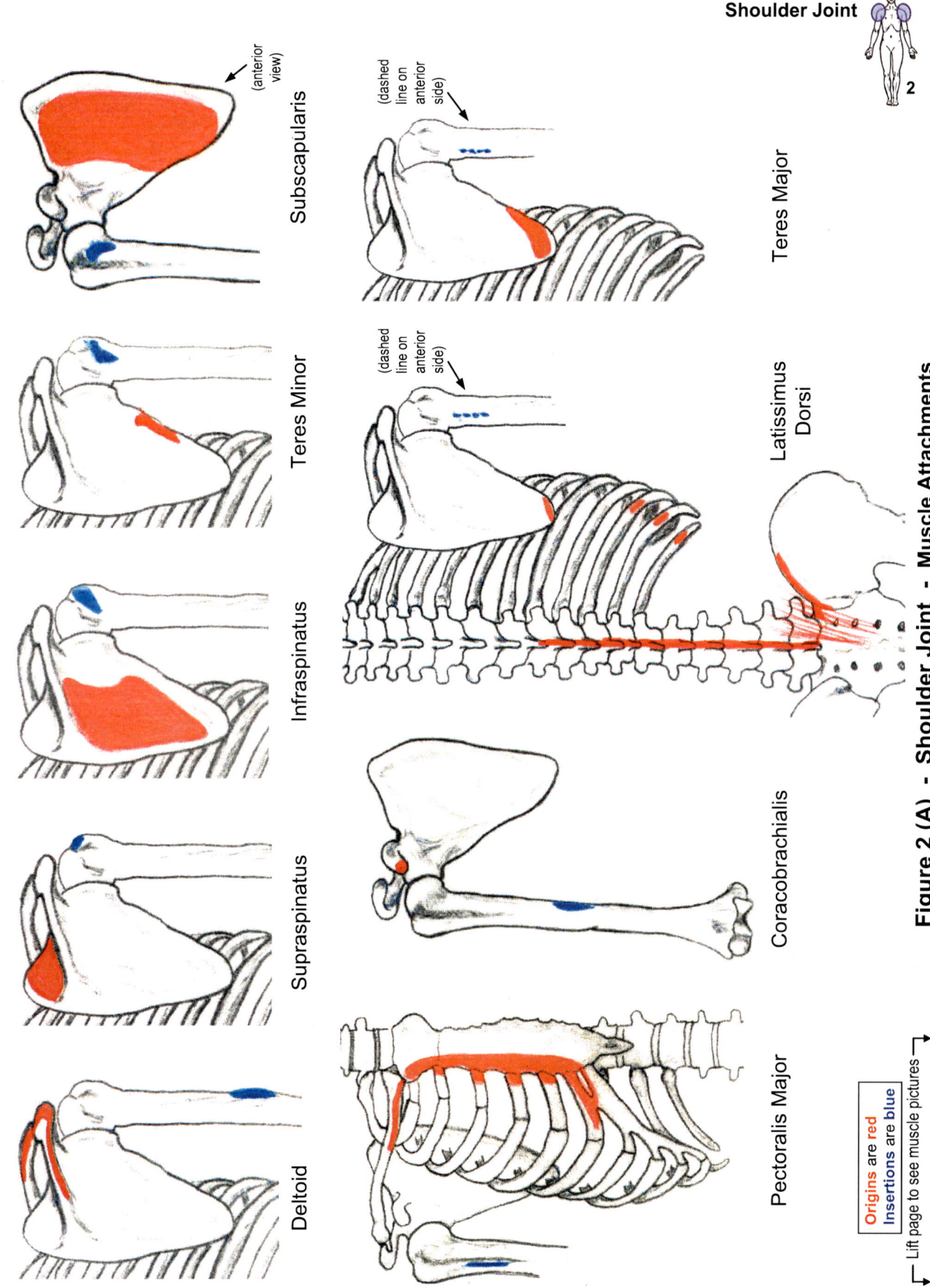

Figure 2 (A) - Shoulder Joint - Muscle Attachments

Chapter 4 - Muscles That Move the Upper Extremity

Shoulder Joint

GH jt.=Glenohumeral joint (shoulder joint), X=Muscle creates the action, UL=Unilateral action, BL=Bilateral action, N=Nerve

Muscles Acting On Shoulder Joint

Shoulder Joint	Flexion @ GH jt.	Extension @ GH jt.	Abduction @ GH jt.	Adduction @ GH jt.	Medial Rotation @ GH jt.	Lateral Rotation @ GH jt.	Stabilization of GH jt.	Other	Innervation	C5	C6	C7	C8	T1
1. Deltoid: Anterior fibers / Middle fibers / Posterior fibers	X (Ant. fib.)	X (Post. fib.)	X (All/ middle fibers)		X (Ant. fib.)	X (Post. fib.)		Ant.: Horiz. Adduction / Post.: Horiz. Abduction	Axillary N. (C5,C6)	N	N			
2. Supraspinatus			X				X		Suprascapular N. (C5)	N				
3. Infraspinatus						X	X	may assist horiz. abduction	Suprascapular N. (C5, C6)	N	N			
4. Teres minor						X	X		Axillary N. (C5)	N				
5. Subscapularis					X		X		Subscapular N. (C5, C6)	N	N			
6. Pectoralis major: Upper fibers / Lower fibers / Abdom. fibers	X (Upper fib.)	X (Lower fib.) Exten. from a flexed position		X (All fibers)	X (All fibers)		(lower & abdom fibers depress shoulder girdle)	Upper fib.: Horiz. Adduction	Lateral pectoral N. (C5,C6,C7) & Medial pectoral N. (C8, T1)	N	N	N	N	N
7. Coracobrachialis	X			X (with resistance)				may assist horiz. adduction	Musculocutaneous N. (C6, C7)	N	N			
8. Latissimus dorsi		X "handcuff position"		X (behind the back)	X		Draws shldr girdle down and back	Affects lower trunk & spine: UL: lat. flex., BL: extension & anterior pelvic tilt	Thoracodorsal N. (C6,C7,C8)	N	N	N	N	
9. Teres major		X		X	X			"Lat's little helper"	Lower Subscapular N. (C5, C6)	N	N			
(More muscles for the action) --->	see also Table 3	see also Table 3							Innervation					

Table 2 (B) - Shoulder Joint - Synergists & Antagonists

Shoulder Joint

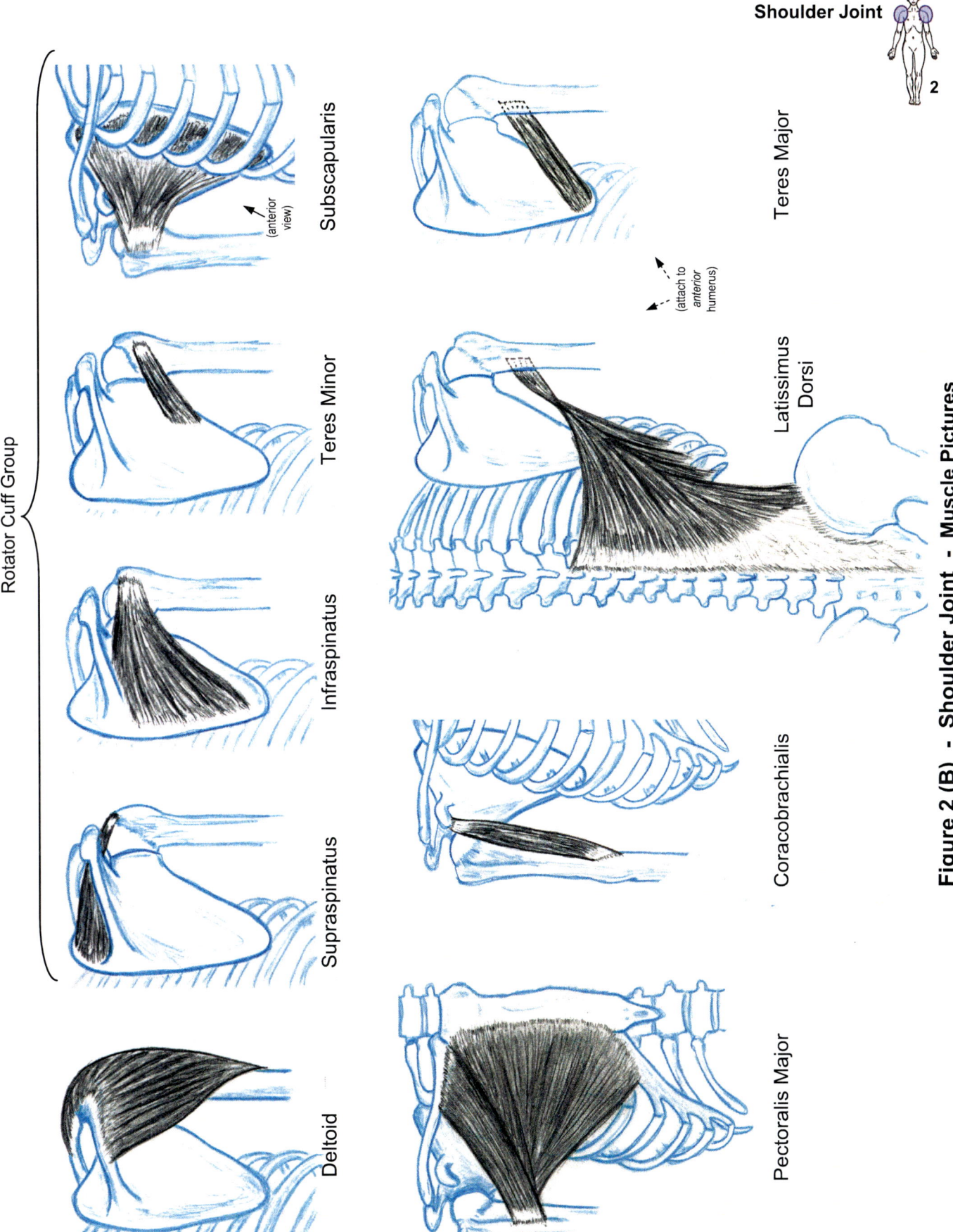

Figure 2 (B) - Shoulder Joint - Muscle Pictures

Shoulder Joint

Muscles Acting On Shoulder Joint ~ Notes ~ (palpation, fiber arrangement, lengthen/shorten, common uses, pathologies, cautions, etc.)

1. Deltoid	6. Pectoralis Major
2. Supraspinatus	7. Coracobrachialis
3. Infraspinatus	8. Latissimus Dorsi
4. Teres Minor	9. Teres Major
5. Subscapularis	

Muscle Group 2 - Shoulder Joint - (Notes)

84 Chapter 4 - Muscles That Move the Upper Extremity

Elbow, Forearm

Movement of the Elbow and Forearm

Muscle Group 3:

Biceps brachii	Pronator quadratus
Brachialis	Triceps brachii
Brachioradialis	Anconeus
Pronator teres	Supinator

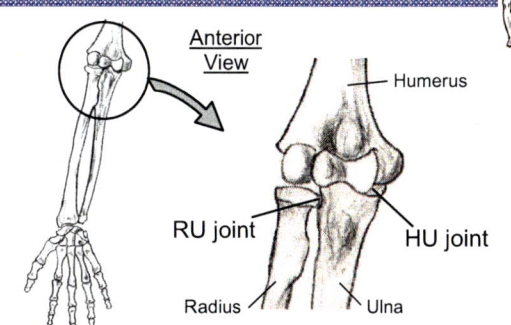

Joints

The muscles in this group primarily move the forearm at the humeroulnar joint and the radioulnar joints. In addition, some muscles that move the elbow also move the shoulder joint described in the previous section.

Humeroulnar Joint (HU) – The Elbow Joint

Trochlea of **humerus** ◄► Trochlear notch of **ulna**
Hinge Joint
Movements available: Flexion, Extension

Radioulnar Joint (RU) (proximal)

Head of **radius** ◄► Radial notch of **ulna**
Pivot joint
Movements available: Supination, Pronation

Supination and pronation of the forearm occur with lateral and medial rotation at this pivot joint (moving coupled with rotation at the *distal* radioulnar joint).

Glenohumeral Joint (GH)

(covered in previous section – Group 2: Shoulder Joint)

Other Joints

The following joints are included here for completeness, but will not be used when naming actions of the elbow/forearm.

Distal Radioulnar Joint
    Distal radius ◄► Head of ulna (distal end of ulna)
    Pivot joint: The radius rotates around the ulna

Radiohumeral Joint (sometimes called the humeroradial joint)
    Head of radius ◄► Capitulum of humerus
    During sup. & pronation: Movement = rotations
    With elbow joint: Movement = flexion, extension

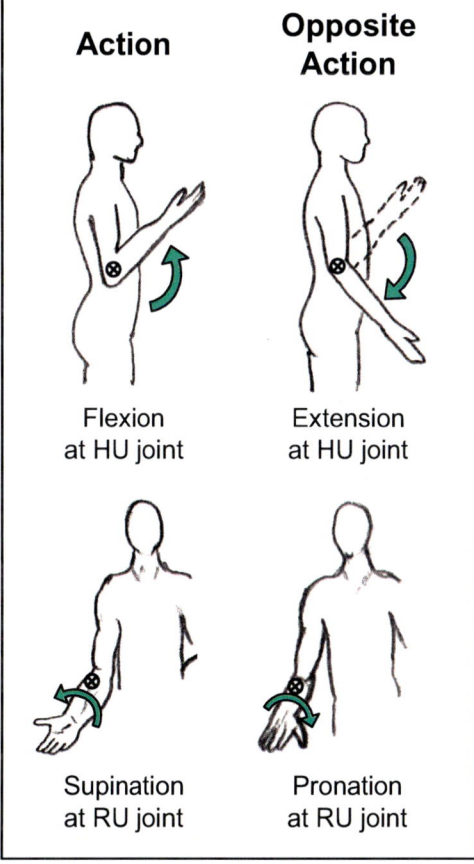

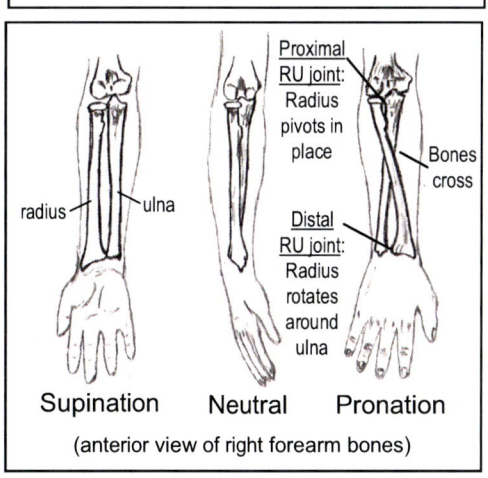

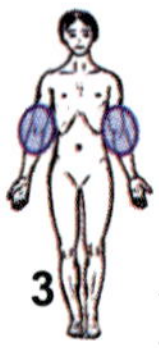

Elbow, Forearm

Bones and bony landmarks

Muscles that move the elbow and forearm have attachments on the scapula and the long bones of the upper extremity. Review the bony landmarks and other structures listed below, referring to the drawings in Chapter 2, pages 40-41.

Humerus (p. 41)

    Intertubercular groove (Bicipital groove)
    Epicondyles (lateral and medial)
    Supracondylar ridge (lateral)
    Trochlea
    Olecranon fossa
    Shaft (locations on it)

Scapula (p. 40)

    Supraglenoid tubercle
    Infraglenoid tubercle
    Coracoid process

Radius (p. 41)

    Head
    Radial tuberosity
    Styloid process
    Shaft (locations on it like mid lateral, etc.)

Ulna (p. 41)

    Olecranon process
    Coronoid process
    Ulnar tuberosity
    Trochlear notch
    Radial notch
    Shaft (locations on it like proximal posterior, etc.)

<u>Tendinous Structures</u>
    Bicipital aponeurosis

Notes

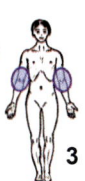

Elbow, Forearm

Muscle Group 3 – Muscles that move the elbow and forearm (and sometimes the shoulder joint) are illustrated as a group on this page. The next four pages have tables and figures that describe each muscle individually, and provide many ways of comparing and contrasting the muscles to each other.

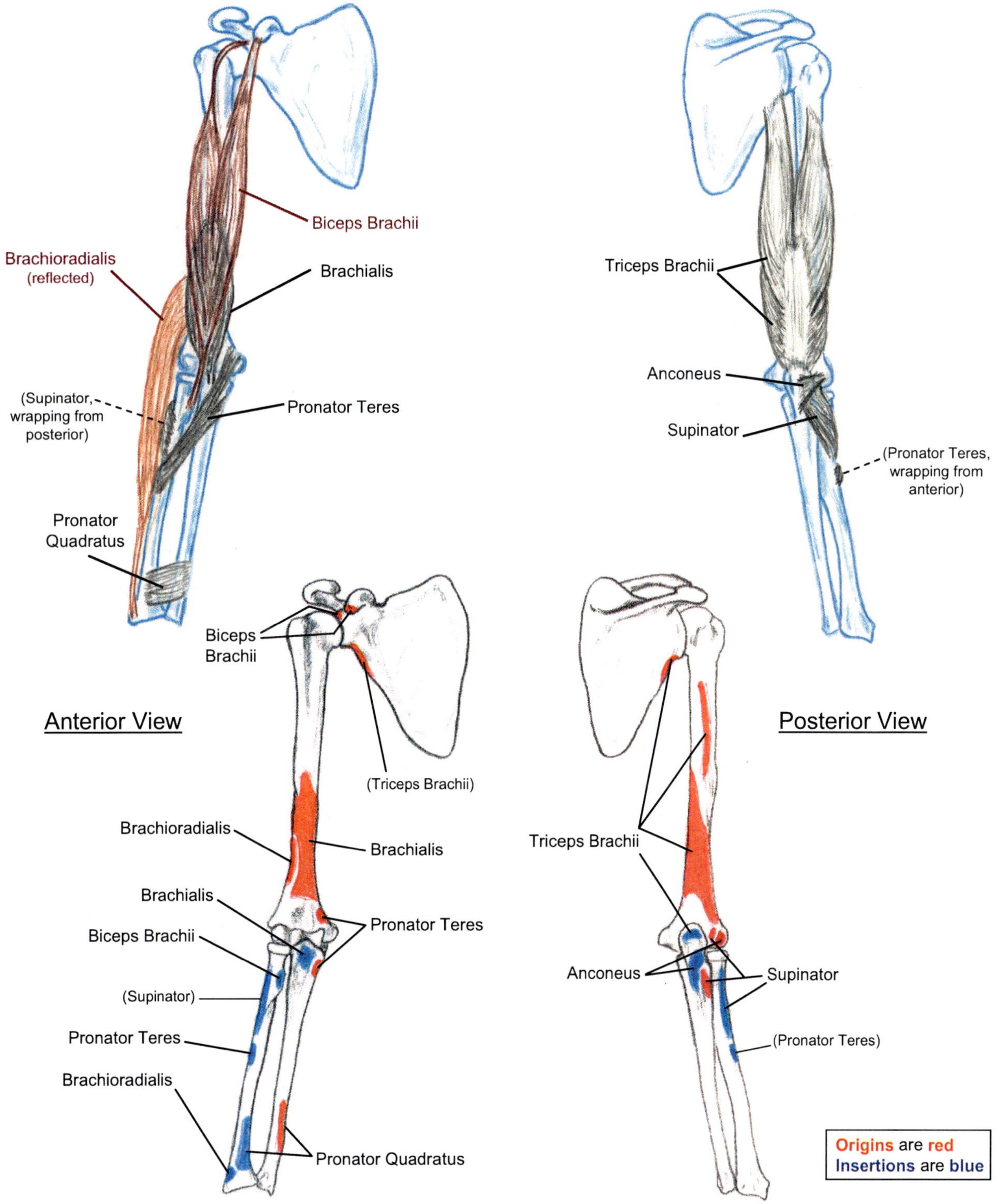

Attachment sites for all muscles in Group 3

Origins are red
Insertions are blue

Mastering Muscles & Movement © 2007   Chapter 4 – Muscles That Move the Upper Extremity   **87**

Elbow, Forearm

Joints: HU=Humeroulnar (elbow, hinge), RU=Radioulnar (pivot), GH=Glenohumeral (shoulder)

Muscles Acting On Elbow, Forearm	Origin	Insertion	Action
Biceps brachii — moves the elbow and forearm and shoulder joint	Long head: Supraglenoid tubercle of scapula (via the intertubercular groove) Short head: Coracoid process of scapula	Radial tuberosity of radius (ignore the deceiving non-attachment toward ulna via bicipital aponeurosis)	Flexion at the elbow (HU joint), Supination at the radioulnar joint, Flexion at the GH joint (Short head: Assists horizontal adduction)
Brachialis — moves the elbow	Anterior humerus (distal half of anterior surface)	Ulnar tuberosity (and coronoid process of ulna)	Flexion at the elbow (HU joint)
Brachioradialis — moves the elbow	Lateral supracondylar ridge of humerus	Styloid process of radius	Flexion at the elbow (HU joint), especially when forearm is in a neutral/handshake position (Also assists: pronation from a supinated position to neutral, and supination from a pronated position to neutral)
Pronator teres — moves the forearm and elbow	Medial epicondyle of humerus, Coronoid process of ulna	Mid lateral shaft of radius	Pronation at the radioulnar joint, Assists flexion at the elbow
Pronator quadratus — moves the forearm	Distal anterior ulna	Distal anterior radius	Pronation at the radioulnar joint
Triceps brachii — moves the elbow and shoulder joint	Long head: Infraglenoid tubercle of scapula Lateral head: Proximal posterior humerus Medial head: Distal half of posterior humerus	Olecranon process of ulna	All heads: Extension at the elbow (HU jt.), Long head: Extension at the GH joint, Assists adduction at the GH joint
Anconeus — moves the elbow	Lateral epicondyle of humerus (posterior aspect)	Olecranon process and proximal posterior ulna	Assists extension at the elbow (HU joint) (Also helps stabilize the elbow during pronation and supination at RU joint)
Supinator — moves the forearm	Lateral epicondyle of humerus, and proximal posterior ulna	Proximal lateral shaft of radius (& wraps around to cover part of anterior and posterior surfaces)	Supination at the radioulnar joint

(larger illustrations on page 91)

Table 3 (A) - Elbow, Forearm - Origin, Insertion, Action

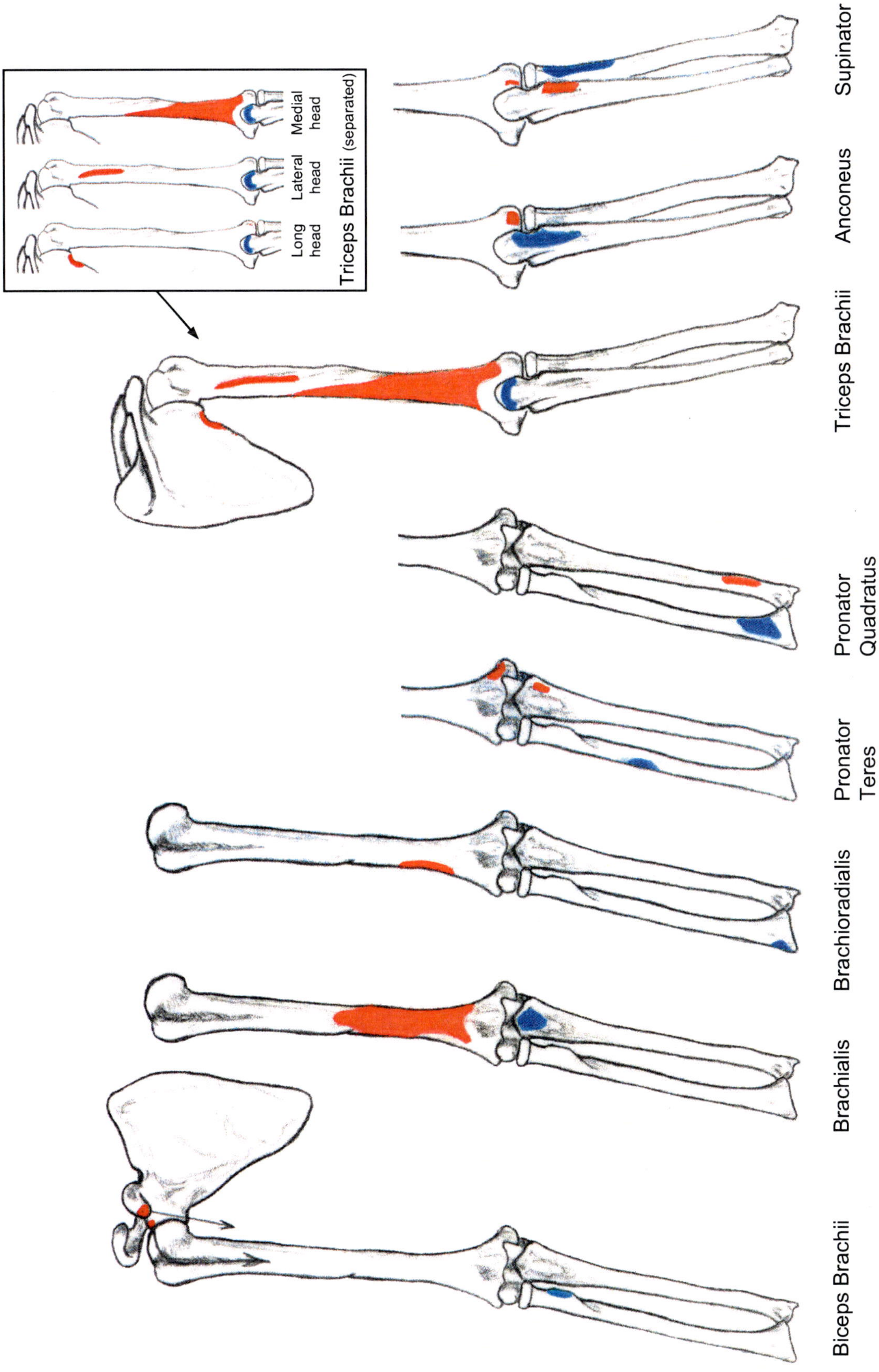

Elbow, Forearm

3

Figure 3 (A) - Elbow, Forearm - Muscle Attachments

Supinator
Anconeus
Triceps Brachii
 - Medial head
 - Lateral head
 - Long head
Triceps Brachii (separated)
Pronator Quadratus
Pronator Teres
Brachioradialis
Brachialis
Biceps Brachii

Origins are red
Insertions are blue
Lift page to see muscle pictures

Mastering Muscles & Movement © 2007   Chapter 4 - Muscles That Move the Upper Extremity   89

Elbow, Forearm

Joints: HU jt.=Humeroulnar joint (elbow, hinge), RU jt.=Radioulnar joint (pivot), GH jt.=Glenohumeral joint (shoulder, ball&socket)

Muscles Acting On

Elbow, Forearm	Flexion @ elbow	Extension @ elbow	Pronation @ RU jt.	Supination @ RU jt.	Flexion @ GH jt.	Extension @ GH jt.	Other	Innervation	C5	C6	C7	C8	T1
1. Biceps brachii:													
Long head	X			X	X		May assist abduction @ GH joint when externally rotated	Musculocutaneous N. (C5, C6)	N	N			
Short head	X				X		Short head: Assists horizontal adduction	Musculocutaneous N. (C5, C6)	N	N			
2. Brachialis	X (in neutral/ handshake position)						(the "true" flexor)	Musculocutaneous N. (C5, C6)	N	N			
3. Brachioradialis	assist		assist (moving from supinated position to neutral)	assist (moving from pronated position to neutral)				Radial N. (C5, C6)	N	N			
4. Pronator teres	assist		X					Median N. (C6, C7)		N	N		
5. Pronator quadratus			X					Median N. (C8, T1)				N	N
6. Triceps brachii:													
Long head		X				X (long head)	Long head assists adduction @ GH jt.	Radial N. (C7, C8)			N	N	
Lateral head		X						Radial N. (C7, C8)			N	N	
Medial head		X						Radial N. (C7, C8)			N	N	
7. Anconeus		assist					Helps stabilize elbow during RU jt. rotations	Radial N. (C7, C8, T1)			N	N	N
8. Supinator				X				Radial N. (C6)		N			
(More muscles for the action) --->	see also Table 4				see also Table 2	see also Table 2		Innervation					

Table 3 (B) - Elbow, Forearm - Synergists & Antagonists

Elbow, Forearm

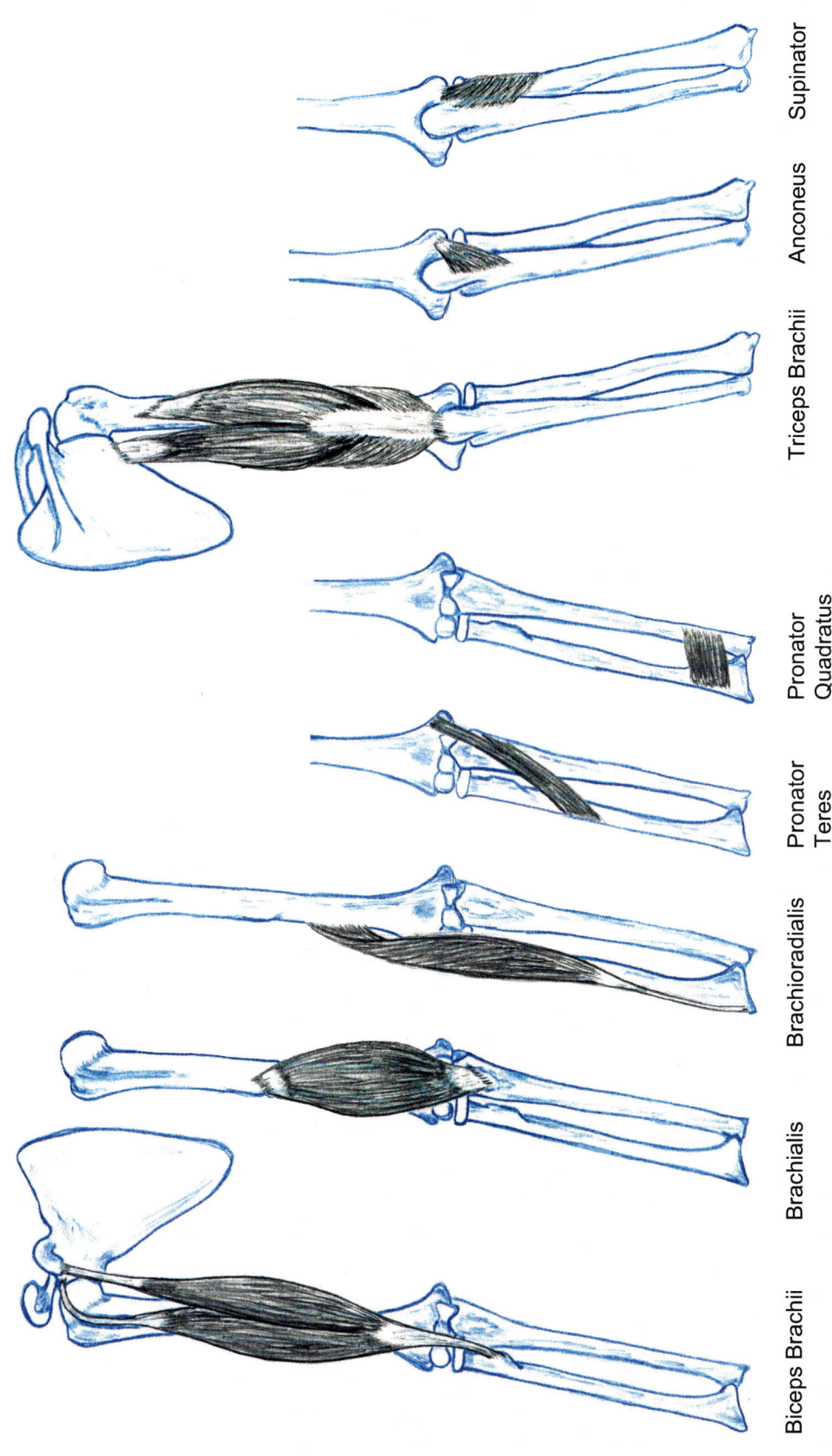

Figure 3 (B) - Elbow, Forearm - Muscle Pictures

Elbow, Forearm

3

Muscles Acting On Elbow, Forearm ~ Notes ~ (palpation, fiber arrangement, lengthen/shorten, common uses, pathologies, cautions, etc.)

1. Biceps Brachii	5. Pronator Quadratus
2. Brachialis	6. Triceps Brachii
3. Brachioradialis	7. Anconeus
4. Pronator Teres	8. Supinator

Muscle Group 3 - Elbow, Forearm - (Notes)

Movement of the Wrist, Hand and Fingers

Muscle Group 4:

Flexors:	Extensors:
Flexor carpi radialis	Extensor carpi radialis longus
Palmaris longus	Extensor carpi radialis brevis
Flexor carpi ulnaris	Extensor carpi ulnaris
Flexor digitorum superficialis	Extensor digitorum
Flexor digitorum profundus	Extensor indicis

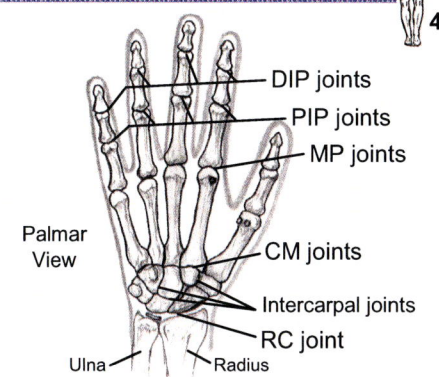

Joints

The muscles in this group move the wrist, or combinations of the wrist, hand, and fingers if they span multiple joints. Also, many of these muscles cross the elbow joint, but most have only minimal action there because they attach very close to the joint. (Note that joints specific to the *thumb* (digit #1) are *not* listed here -- the thumb is covered in the next muscle group - Group 5).

Radiocarpal Joint (RC) – The Wrist Joint
   Distal end of **radius** ◄► Proximal row of **carpals**
   Ellipsoid joint
   Movements: Flexion, Extension, Abduction, Adduction
   (Abduction= Radial deviation, Adduction= Ulnar deviation)
   (Carpal bones involved: Scaphoid, lunate, triquetrum (not pisiform)

Intercarpal Joints (all articulations between carpal bones)
   Any **carpal** surface ◄► Any **carpal** surface
   Gliding joints (movement = gliding)

Carpometacarpal Joints (CM) #2 - #5
   Distal **carpals** ◄► Bases of **metacarpals** #2-5
   Gliding joints (movement = gliding)
   (Carpal bones involved: Trapezoid, capitate, hamate (not trapezium)

Metacarpophalangeal Joints (MP) #2 - #5
   Heads of **metacarpal** bones ◄► Bases of proximal **phalanges**
   Condyloid joints
   Movements: Flexion, Extension, Abduction, Adduction
   (Abduction=spreading the fingers, Adduction=closing the fingers)

Interphalangeal Joints #2 - #5
   Joints between the **phalanges** of the fingers (PIP & DIP, see below)
   Hinge joints
   Movements: Flexion, Extension

   Proximal Interphalangeal Joints (PIP)
       Head of proximal phalanx ◄► Base of middle phalanx

   Distal Interphalangeal Joints (DIP)
       Head of middle phalanx ◄► Base of distal phalanx

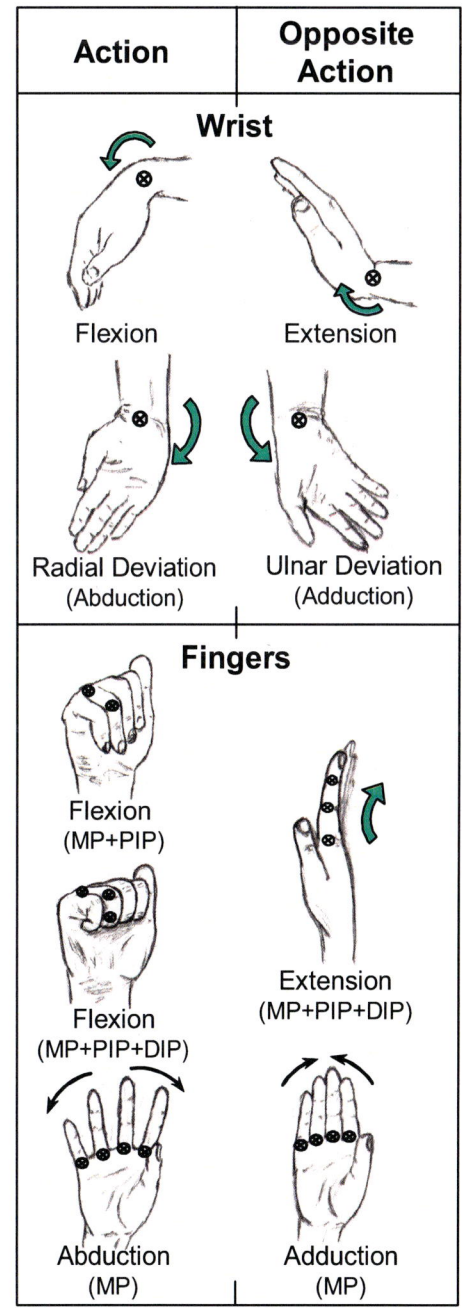

Wrist, Hand, Fingers

Bones and bony landmarks

Muscles that move the wrist, hand, and fingers have attachments near the elbow and on the bones of the forearm, hand, and fingers. Review the bony landmarks and other structures listed below, referring to the drawings in Chapter 2, pages 41-42.

Humerus (p. 41)
    Lateral and medial epicondyles
    Lateral supracondylar ridge

Radius (p. 41)
    Shaft (locations on it)

Ulna (p. 41)
    Coronoid process
    Shaft (locations on it)

Carpal Bones (p. 42)

	Proximal Row	Distal Row
Radial side:	Scaphoid	Trapezium
| :	Lunate	Trapezoid
| :	Triquetrum	Capitate
Ulnar side:	Pisiform	Hamate

Metacarpal Bones (p. 42)

    #1 on radial side to #5 on ulnar side
    Base of…, Head of…

Phalanges of Hand (p. 42)

    #1   = Pollux   = Thumb (2 phalanges)
    #2-#5 = Digits   = Fingers (3 phalanges)

    Proximal, Middle, and Distal phalanges
    Base of …
    Head of …

    (#1=Radial side <------> #5=Ulnar side)

Tendinous Structures:

Interosseus membrane-
    -between radius and ulna
Common flexor tendon
Common extensor tendon   (p. 95)
Palmar aponeurosis
Flexor retinaculum   (p. 108)
Extensor retinaculum
Tendon sheaths
Carpal tunnel

Concepts for Muscles That Cross Multiple Joints

- **Passive insufficiency:** When stretching a muscle, all the joints crossed by the muscle cannot be opened to their full range at the same time (the muscle can't get long enough). One joint must be "backed off" to allow the other joints to open further.

- **Active insufficiency:** When contracting a muscle, it cannot get short enough to close all the joints to their full range at the same time. One joint must be "opened back up" to allow the other joints close further.

Intrinsic Muscles of the Hand (see page 108)

There are several muscles that reside within the structure of the hand itself. These *intrinsic* muscles of the hand are not included in the Group 4 tables, but detailed illustrations are given on page 108.

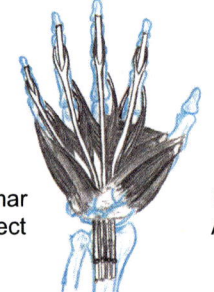
Palmar Aspect

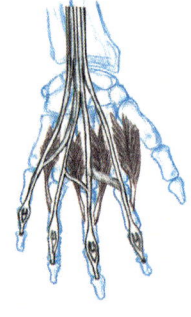
Dorsal Aspect

Wrist, Hand, Fingers

Muscle Group 4 – Muscles that move the wrist, hand and fingers (and sometimes the elbow) are illustrated as a group on this page. The following four pages have tables and figures that describe each muscle individually, and provide many ways of comparing and contrasting the muscles to each other.

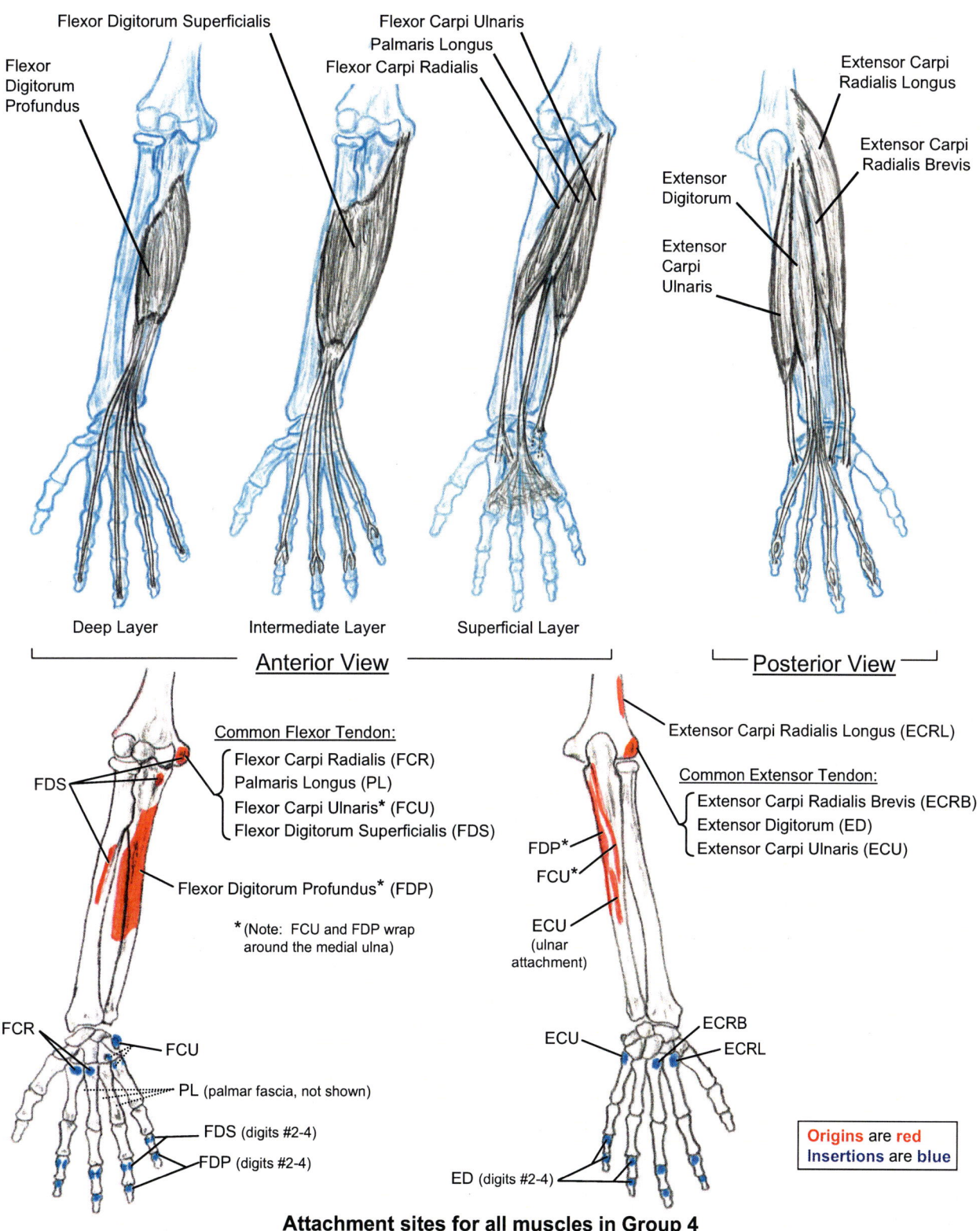

Attachment sites for all muscles in Group 4

Wrist, Hand, Fingers

Muscles Acting On — Joints: RC=Radiocarpal (wrist), MP=Metacarpophalangeal (knuckles), PIP & DIP=Proximal & Distal Interphalangeal (fingers)

Wrist, Hand, Fingers	Origin	Insertion	Action
Flexor carpi radialis — moves the wrist	Medial epicondyle of the humerus	Bases of metacarpals #2 & #3 (palmar side)	Flexion at the wrist (RC joint), Radial Deviation (abduction) at the wrist (Also may assist flexion at the elbow, and pronation of the forearm)
Palmaris longus — moves the palm and wrist	Medial epicondyle of the humerus	Palmar aponeurosis/ fascia	Flexion at the wrist, Assists cupping the hand by tensing the palmar fascia (Also may assist flexion at the elbow)
Flexor carpi ulnaris — moves the wrist	Medial epicondyle of humerus, Proximal *posterior* ulna	Pisiform (also by ligaments to hamate & 5th metacarpal)	Flexion at the wrist, Ulnar Deviation (adduction) at the wrist (Also may assist flexion at the elbow)
Flexor digitorum superficialis — moves the fingers and wrist	Medial epicondyle of humerus, Coronoid process of ulna, Anterior shaft of radius	Sides of middle phalanges of digits #2 - 5 (palmar side, the tendon splits)	Flexion of fingers at the PIP and MP joints, Flexion at the wrist (Also may assist flexion at the elbow)
Flexor digitorum profundus — moves the fingers and wrist	Proximal half of the anterior and medial ulna (wraps around) (and interosseus membrane)	Bases of *distal* phalanges of digits #2 - 5 (palmar side)	Flexion of fingers at the DIP, PIP, and MP joints (closing hand into a full fist), Assists flexion at the wrist
Extensor carpi radialis longus — moves the wrist	Lateral supracondylar ridge of the humerus	Base of metacarpal #2 (dorsal side)	Extension at the wrist, Radial Deviation (abduction) at wrist, Assists flexion at the elbow (when forearm is in neutral/handshake position)
Extensor carpi radialis brevis — moves the wrist	Lateral epicondyle of the humerus	Base of metacarpal #3 (dorsal side)	Extension at the wrist, Radial Deviation (abduction) at the wrist
Extensor carpi ulnaris — moves the wrist	Lateral epicondyle of humerus, Posterior middle shaft of ulna	Base of metacarpal #5 (ulnar side)	Extension at the wrist, Ulnar Deviation (adduction) at the wrist
Extensor digitorum *(including extensor digiti minimi)* — moves the fingers and wrist	Lateral epicondyle of the humerus	Bases of middle & distal phalanges #2-5 (dorsal side)	Extension of the fingers at the DIP, PIP, MP joints, Extension at the wrist (Also assists abduction of the fingers)
Extensor indicis — moves the index finger and wrist	Distal posterior ulna and the interosseus membrane	Merges with the extensor digitorum tendon near the base of the index finger	Extension of the index finger, Assists extension at the wrist

Table 4 (A) - Wrist, Hand, Fingers - Origin, Insertion, Action

(larger illustrations on page 99)

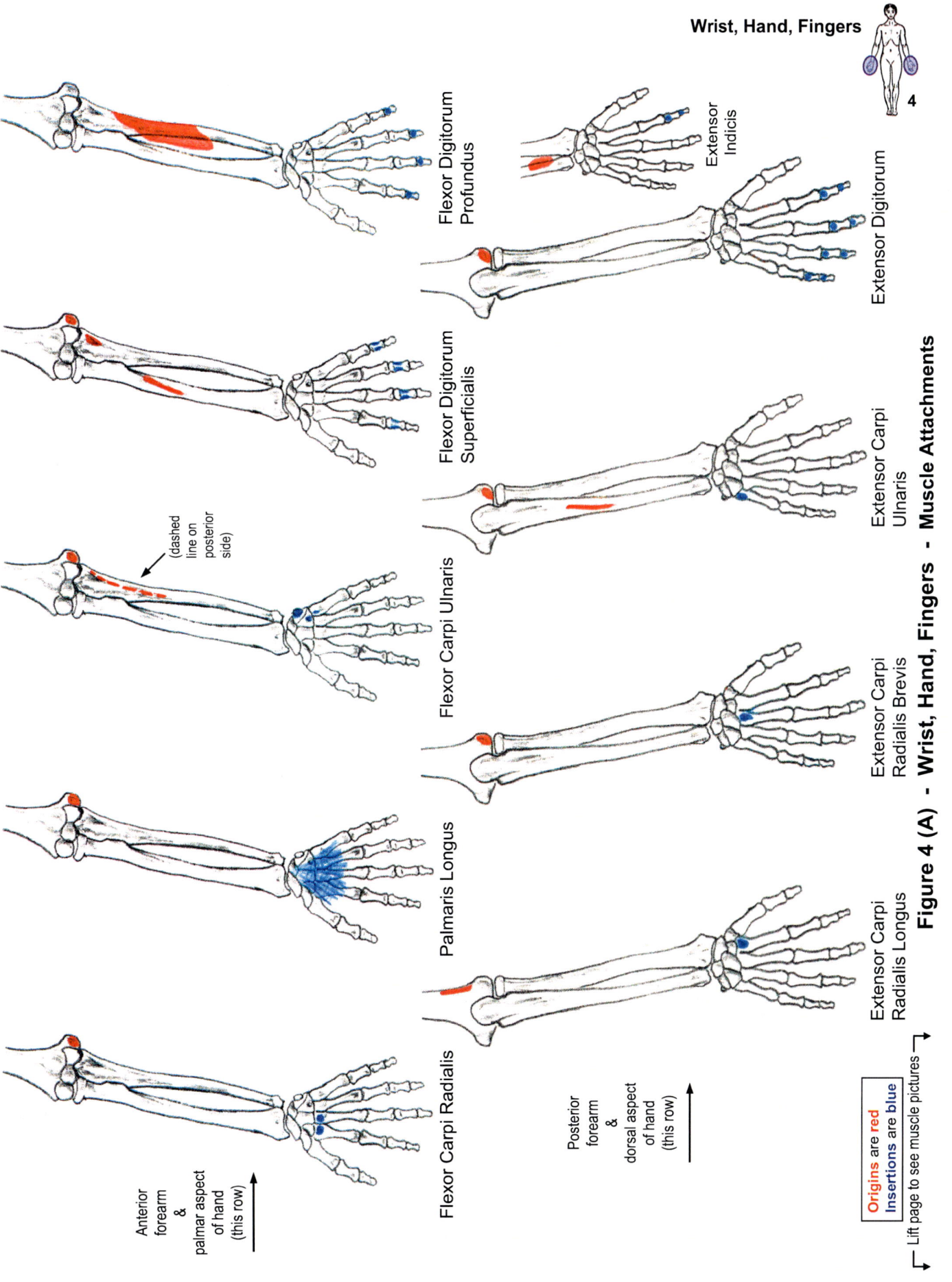

Figure 4 (A) - Wrist, Hand, Fingers - Muscle Attachments

Wrist, Hand, Fingers

Joints: RC=Radiocarpal (wrist), MP=Metacarpophalangeal (knuckles), PIP & DIP=Proximal & Distal Interphalangeal (fingers)

Table 4 (B) - Wrist, Hand, Fingers - Synergists & Antagonists

Muscles Acting On: Wrist, Hand, Fingers	Flexion @ wrist	Extension @ wrist	Abduction/ Radial Deviation	Adduction/ Ulnar Deviation	Flexion of fingers	Extension of fingers	Flexion @ elbow	Other	Innervation	C6	C7	C8	T1
1. Flexor carpi radialis	X		X				may assist	may assist pronation	Median N. (C6, C7)	N	N		
2. Palmaris longus	X						may assist	Assists cupping the hand	Median N. (C6, C7)	N	N		
3. Flexor carpi ulnaris	X			X			may assist		Ulnar N. (C8, T1)			N	N
4. Flexor digitorum superficialis	X				X PIP joints (+ MP)		may assist		Median N. (C7, C8, T1)		N	N	N
5. Flexor digitorum profundus	assist				X DIP joints (+ PIP, MP)			Closes hand into full fist	Median N. (C8, T1) to digits 2 and 3, Ulnar N. (C8, T1) to digits 4 and 5			N	N
6. Extensor carpi radialis longus		X	X				assist (in neutral/ handshake position)		Radial N. (C6, C7)	N	N		
7. Extensor carpi radialis brevis		X	X						Radial N. (C6, C7)	N	N		
8. Extensor carpi ulnaris		X		X					Radial N. (C6, C7, C8)	N	N	N	
9. Extensor digitorum (including extensor digiti minimi)		X				X DIP, PIP (+ MP)		works with lumbricals and interossei	Radial N. (C6, C7, C8)	N	N	N	
10. Extensor indicis		assist				X (index finger)			Radial N. (C7, C8)		N	N	
(More muscles for the action) --->	see also Table 5	see also Table 5					see also Table 3		Innervation				

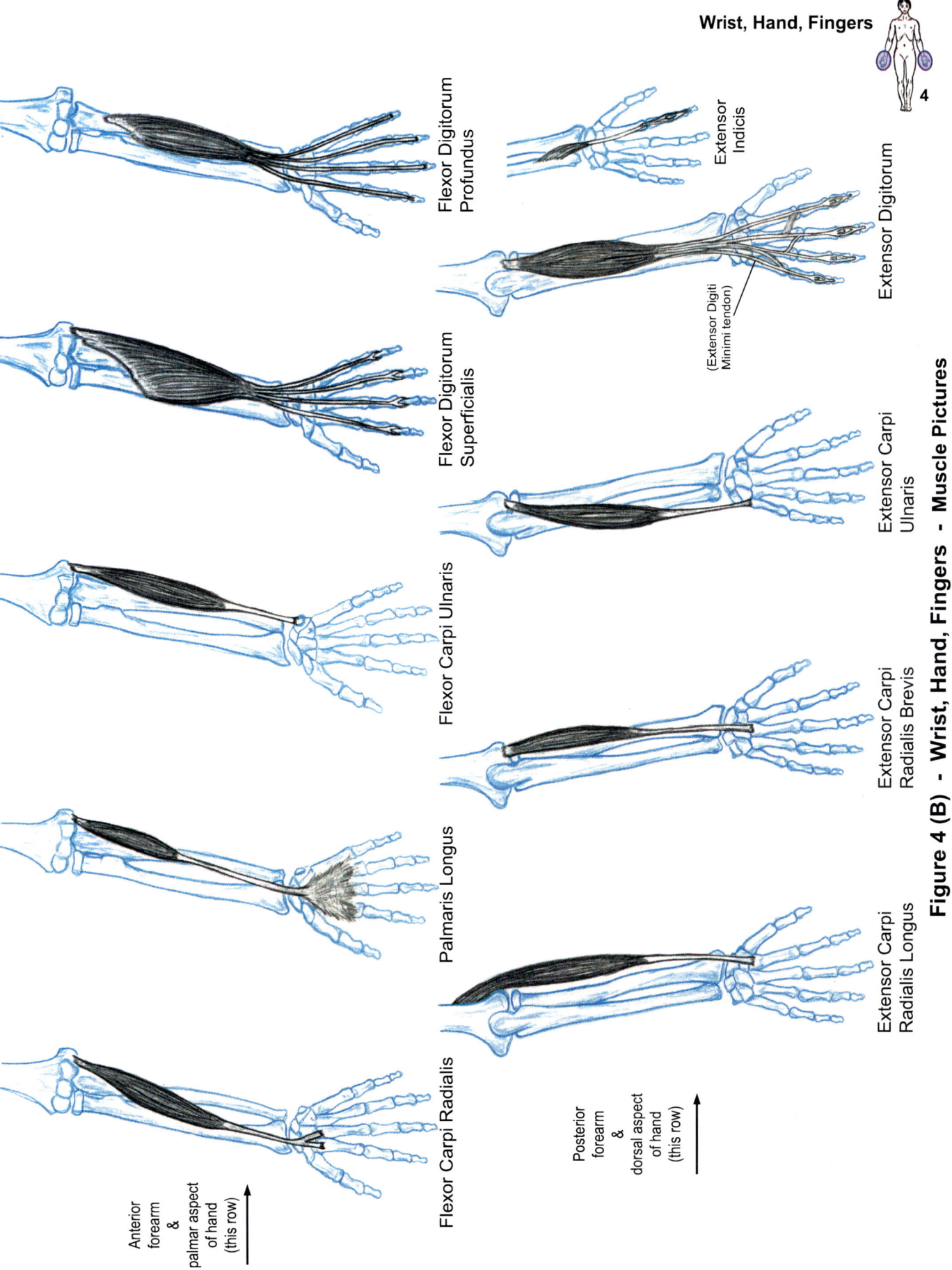

Figure 4 (B) - Wrist, Hand, Fingers - Muscle Pictures

Muscles Acting On
Wrist, Hand, Fingers ~ Notes ~ (palpation, fiber arrangement, lengthen/shorten, common uses, pathologies, cautions, etc.)

1. Flexor Carpi Radialis	6. Extensor Carpi Radialis Longus
2. Palmaris Longus	7. Extensor Carpi Radialis Brevis
3. Flexor Carpi Ulnaris	8. Extensor Carpi Ulnaris
4. Flexor Digitorum Superficialis	9. Extensor Digitorum
5. Flexor Digitorum Profundus	10. Extensor Indicis

Muscle Group 4 - Wrist, Hand, Fingers - (Notes)

100   Chapter 4 - Muscles That Move the Upper Extremity   Mastering Muscles & Movement © 2007

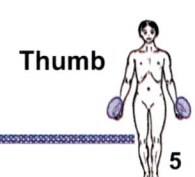

Thumb 5

Movement of the Thumb

Muscle Group 5:

Flexor pollicis longus	Abductor pollicis brevis
Flexor pollicis brevis	Abductor pollicis longus
Opponens pollicis	Extensor pollicis longus
Adductor pollicis	Extensor pollicis brevis

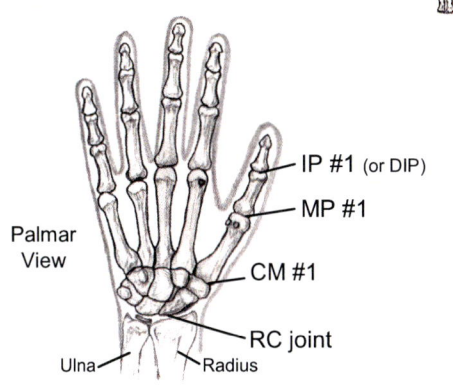

Joints

The muscles in this group move the thumb and the wrist. The anatomical word for thumb is *pollux*, hence "pollicis" in all the muscle names. The distinctive movements of the thumb are possible because of the special-shaped saddle joint at the carpometacarpal joint, which is at the base of the 1st metacarpal bone (near the wrist – <u>not</u> at base of thumb).

Thumb movements are different – visualize the thumb as a finger rotated 90 degrees.

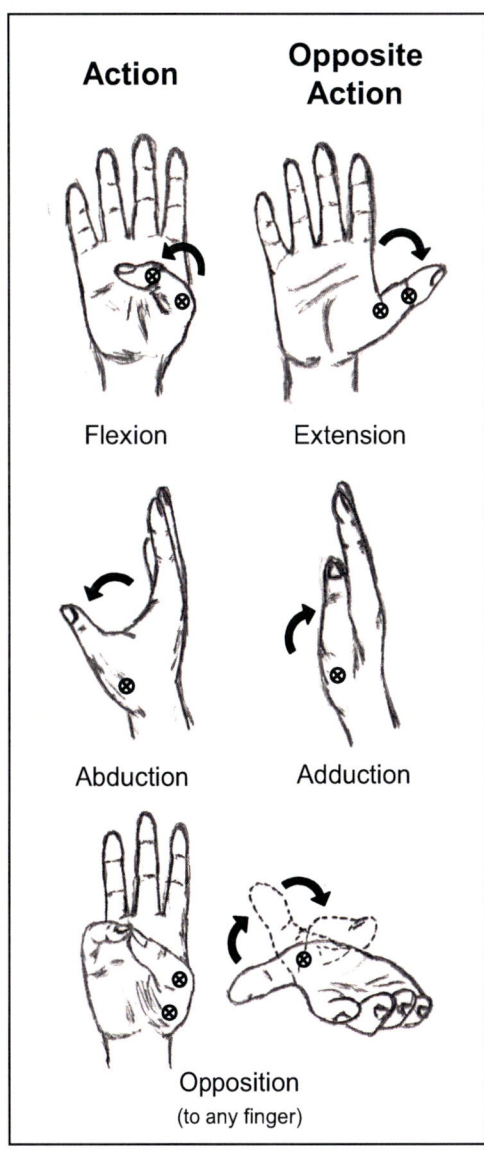

Carpometacarpal Joint (CM) #1

Distal **carpal** (trapezium) ◄► Base of 1st **metacarpal**
Saddle joint
Movements available:
    Flexion, Extension
    Abduction, Adduction
    Opposition, (Reposition)

Metacarpophalangeal Joint (MP) #1

Head of **metacarpal** bone ◄► Base of proximal **phalange**
Condyloid joint
Movements available: Flexion, Extension
                                Abduction, Adduction (very limited)

Interphalangeal (IP) #1

Head of proximal **phalange** ◄► Base of distal **phalange**
Hinge joint
Movements available: Flexion, Extension

The thumb has only two phalanges (fingers have three).
The IP joint is usually called the DIP rather than the PIP.

Radiocarpal Joint (RC) – Wrist
    (covered in previous section: Group 4 – Wrist, Hand, Fingers)

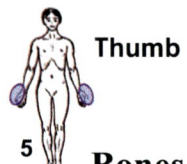 **Thumb**

5 Bones and bony landmarks

Muscles that move the thumb have attachments on the radius and ulna and their interosseus membrane, and on the thumb and radial side of the hand and wrist. Review the bony landmarks and other structures listed below, referring to the drawings in Chapter 2, pages 41-42.

Radius (p. 41)

Ulna (p. 41)

Carpals (p. 42):
   Carpals on radial side of wrist: Trapezium, scaphoid

Metacarpal #1 (p. 42)

Distal and proximal phalanges, digit #1 (thumb) (p. 42)

Tendinous Structures

   Interosseus membrane between Radius and Ulna
   Flexor Retinaculum
   Anatomical "snuffbox" – created by the tendons of EPL and EPB muscles

Terminology
   Palmar aspect of the hand
   Dorsum of the hand
   Thenar eminence
   Hypothenar eminence

Intrinsic Muscles of the Hand (see page 108)

The Group 5 tables include intrinsic muscles of the hand that move the *thumb*. There are other intrinsic muscles that move the fingers and shape the palm. Detailed drawings showing *all* intrinsic muscles of the hand are presented on page 108. Here are some features and groups that apply to the intrinsic muscles:

- Thenar muscles – are named "_____ pollicis brevis"

- Hypothenar muscles – are named "_____ digiti minimi"

- Muscles of the mid-hand – Dorsal Interossei, Palmar Interossei, Lumbricals

- Muscles in the web between the index finger and thumb –
      Adductor Pollicis and 1st Dorsal Interosseus

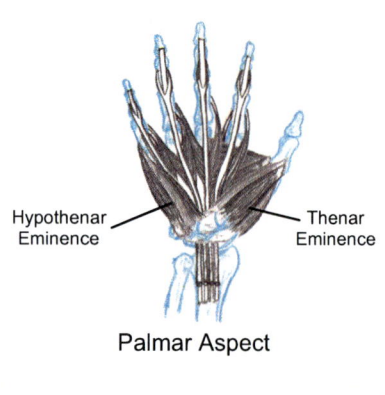

Thumb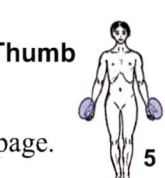

Muscle Group 5 – Muscles that move the thumb (and wrist) are illustrated as a group on this page. The next four pages have tables and figures that describe each muscle individually, and provide many ways of comparing and contrasting the muscles to each other.

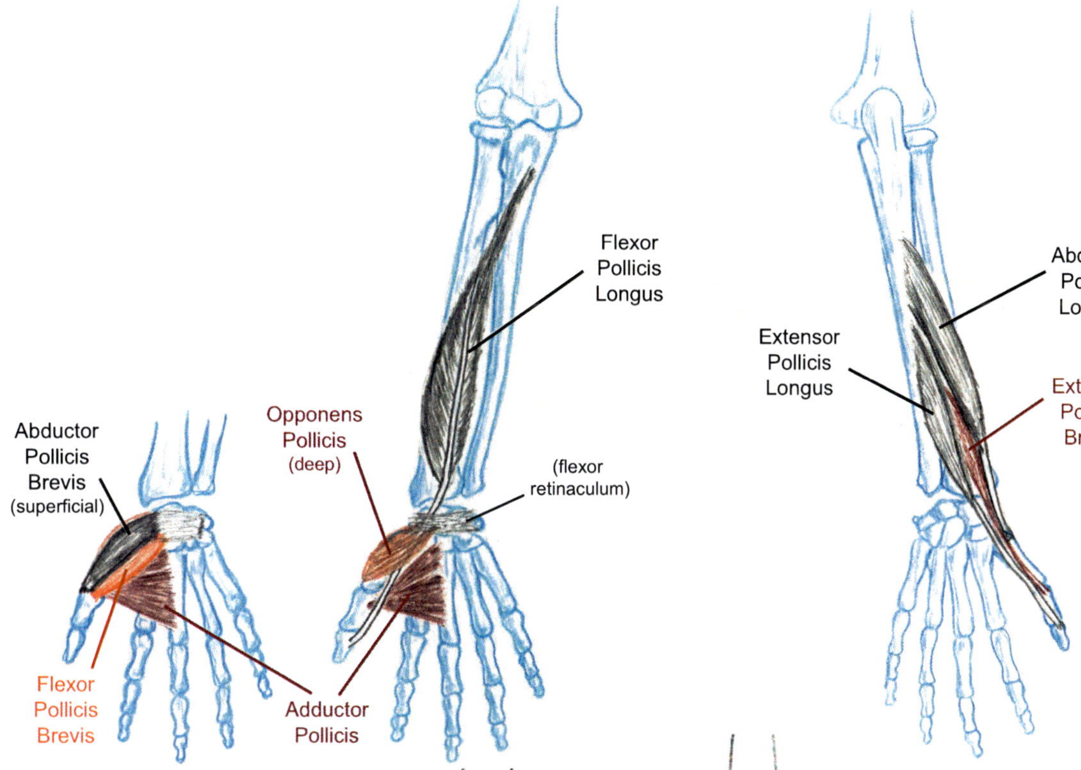

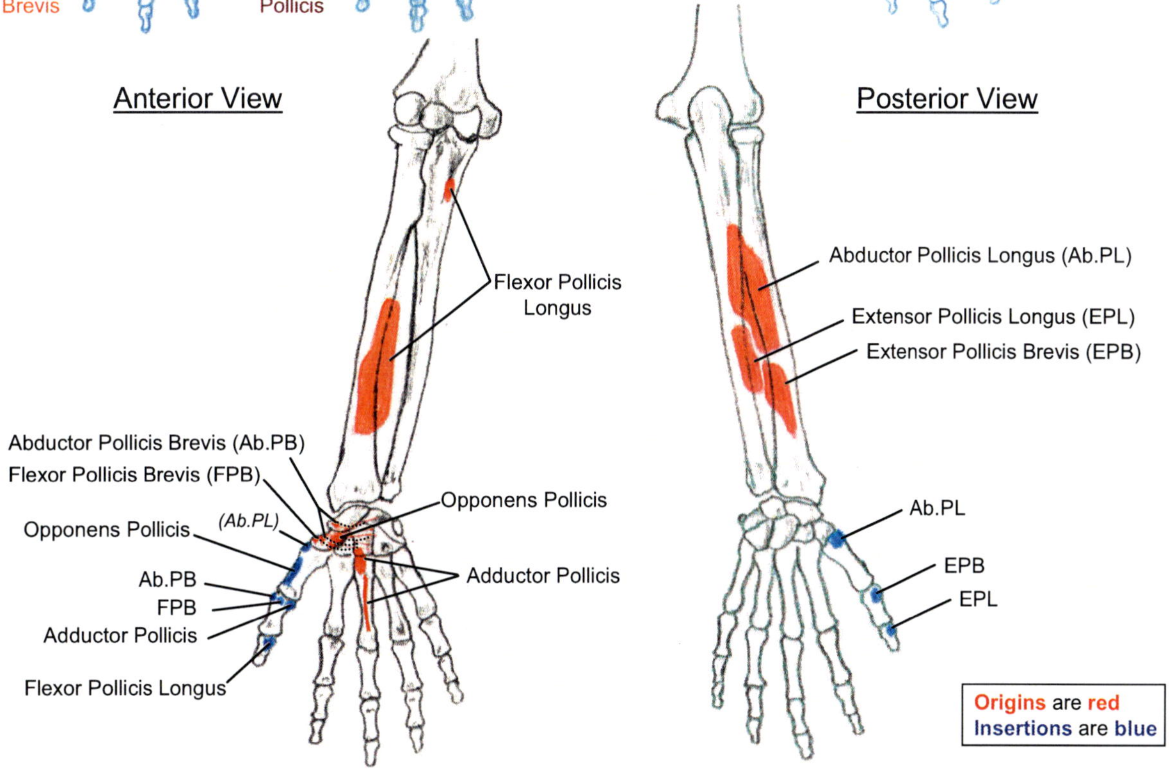

Anterior View

Posterior View

Attachment sites for all muscles in Group 5

Origins are red
Insertions are blue

Thumb

RC=Radiocarpal (wrist), CM=Carpometacarpal, MP=Metacarpophalangeal, IP=Interphalangeal

Muscles Acting On Thumb

Thumb	Origin	Insertion	Action
Flexor pollicis longus — moves the thumb	Anterior shaft of radius, and interosseus membrane (also by a slip to coronoid process of ulna)	Base of distal phalanx of thumb (palmar side)	Flexion of the thumb (distal phalanx at the IP joint) (Also assists flexion of the thumb at the MP and CM joints, and may assist flexion at the wrist)
Flexor pollicis brevis — moves the thumb	Flexor retinaculum and carpal bones (Superficial head to flexor retinaculum & trapezium, Deep head to trapezoid and capitate)	Base of proximal phalanx of thumb (palmar side)	Flexion of the thumb (at the MP and CM joints)
Opponens pollicis — moves the thumb	Trapezium and flexor retinaculum	Metacarpal #1 (whole length of radial side)	Opposition of thumb to the fingers (Opposition = abduction + flexion + rotation at the 1st CM joint (saddle joint). The finger pads face the thumb pad.)
Adductor pollicis — moves the thumb	Shaft of 3rd metacarpal bone, and capitate bone (palmar side) (+ bases of metacarpals 2 & 3)	Base of proximal phalanx of thumb (ulnar side)	Adduction of the thumb (also assists flexion of the thumb at the MP joint)
Abductor pollicis brevis — moves the thumb	Flexor retinaculum, trapezium and scaphoid	Base of proximal phalanx of thumb (radial side)	Abduction of the thumb
Abductor pollicis longus — moves the thumb and the wrist	Posterior ulna and radius, and interosseus membrane	Base of metacarpal #1 (radial side)	Abduction and extension of the thumb (at the CM joint), and Radial deviation (abduction) at the wrist
Extensor pollicis longus — moves the thumb	Middle posterior ulna, and interosseus membrane	Base of distal phalanx of thumb (dorsal side)	Extension of the thumb (at the IP, MP & CM joints) (Also assists extension and radial deviation at the wrist)
Extensor pollicis brevis — moves the thumb	Distal posterior radius, and interosseus membrane	Base of proximal phalanx of thumb (dorsal side)	Extension of the thumb (at the MP & CM joints) (Also assists radial deviation at the wrist)

Intrinsic Hand Muscles

(larger illustrations on page 107)

Table 5 (A) – Thumb – Origin, Insertion, Action

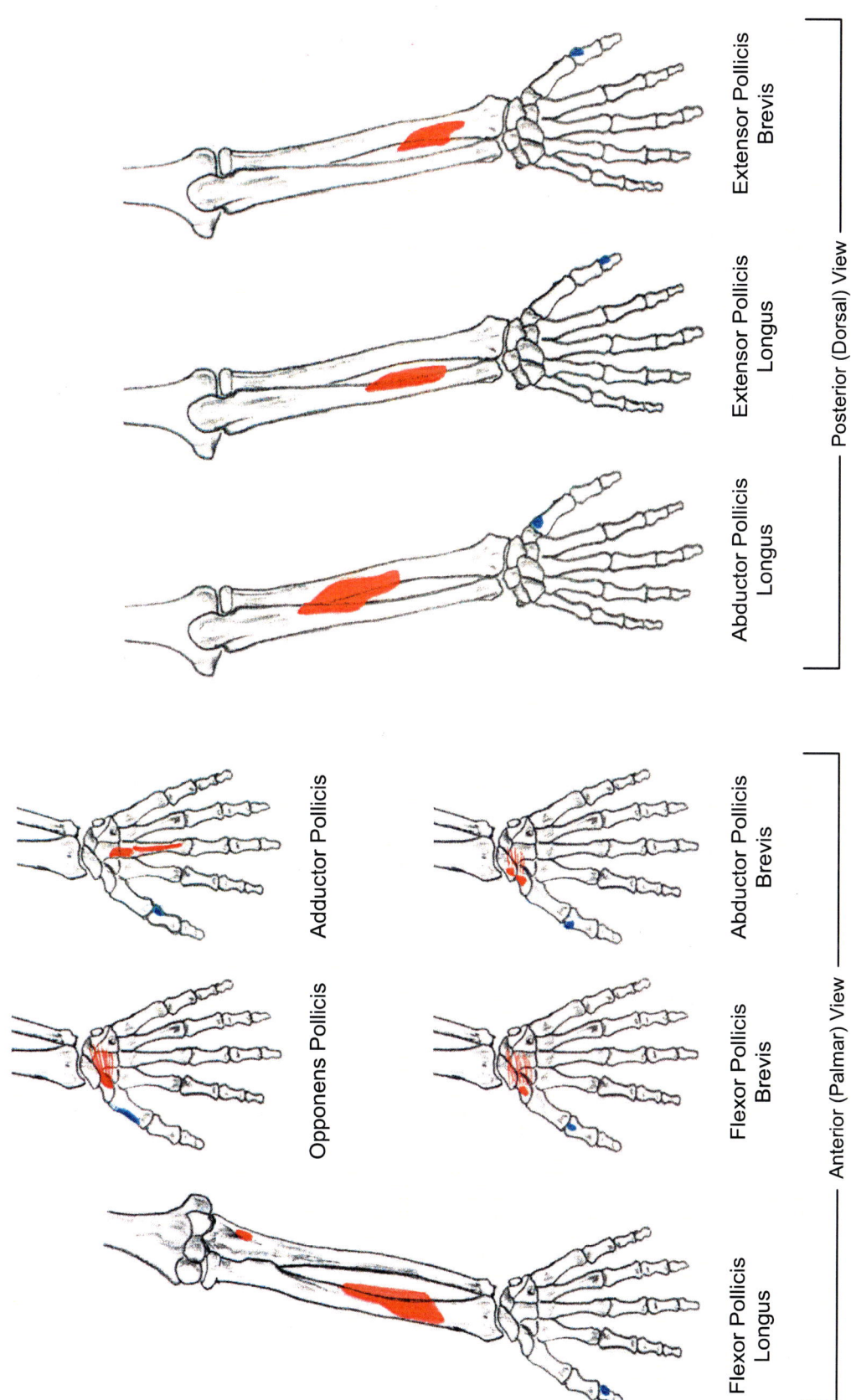

Figure 5 (A) - Thumb - Muscle Attachments

Thumb 5

Joints: RC=Radiocarpal (wrist, ellipsoid), CM=Carpometacarpal #1 (thumb, saddle), MP=Metacarpophalangeal #1

Muscles Acting On Thumb

Thumb	Flexion of thumb	Extension of thumb	Adduction of thumb	Abduction of thumb	Opposition of thumb	Radial Deviation @ wrist	Flexion @ wrist	Extension @ wrist	Innervation	C6	C7	C8	T1
1. Flexor pollicis longus	X						may assist		Median N. (C8, T1)			N	N
2. Flexor pollicis brevis	X								Sup. Head: Median N. Deep Head: Ulnar N. (C8, T1)			N	N
3. Opponens pollicis					X (abduct+flex + rotate at CM & MP jts)				Median N. (C8, T1)			N	N
4. Adductor pollicis			X (at CM jt.)						Ulnar N. (C8, T1)			N	N
5. Abductor pollicis brevis				X					Median N. (C8, T1)			N	N
6. Abductor pollicis longus		X (at CM jt.)		X (at CM jt.)		X			Radial N. (C7, C8)		N	N	
7. Extensor pollicis longus		X				assist		assist	Radial N. (C6, C7, C8)	N	N	N	
8. Extensor pollicis brevis		X (at MP & CM jts)				assist			Radial N. (C7, C8)		N	N	
(More muscles for the action) ---->						see also Table 4	see also Table 4	see also Table 4	Innervation				

— Intrinsic Hand Muscles (3–5) —

Table 5 (B) – Thumb – Synergists & Antagonists

106 Chapter 4 – Muscles That Move the Upper Extremity

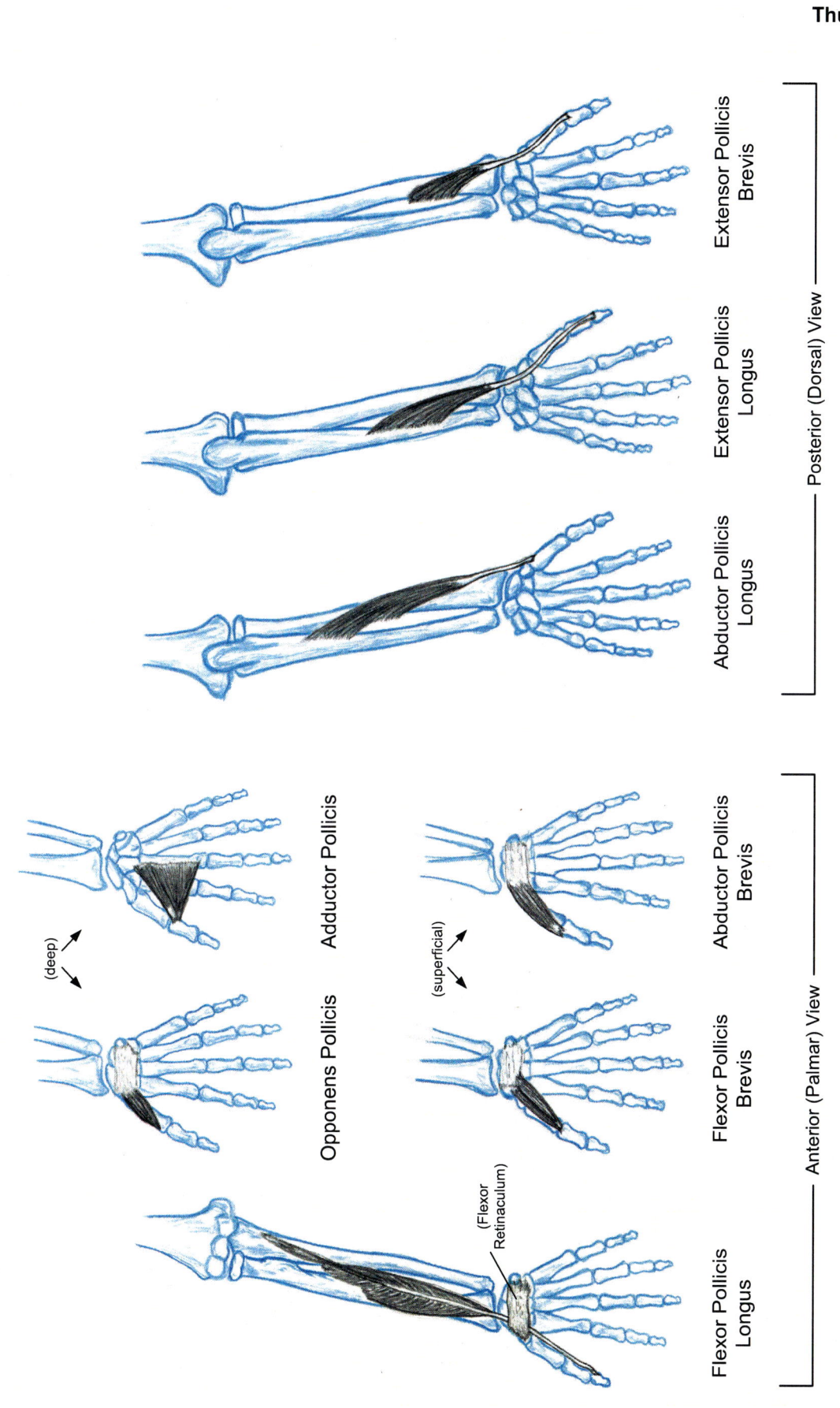

Figure 5 (B) - Thumb - Muscle Pictures

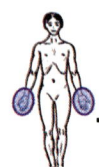

Intrinsic Muscles of the Hand

Right Hand – Layers of the Palmar Aspect

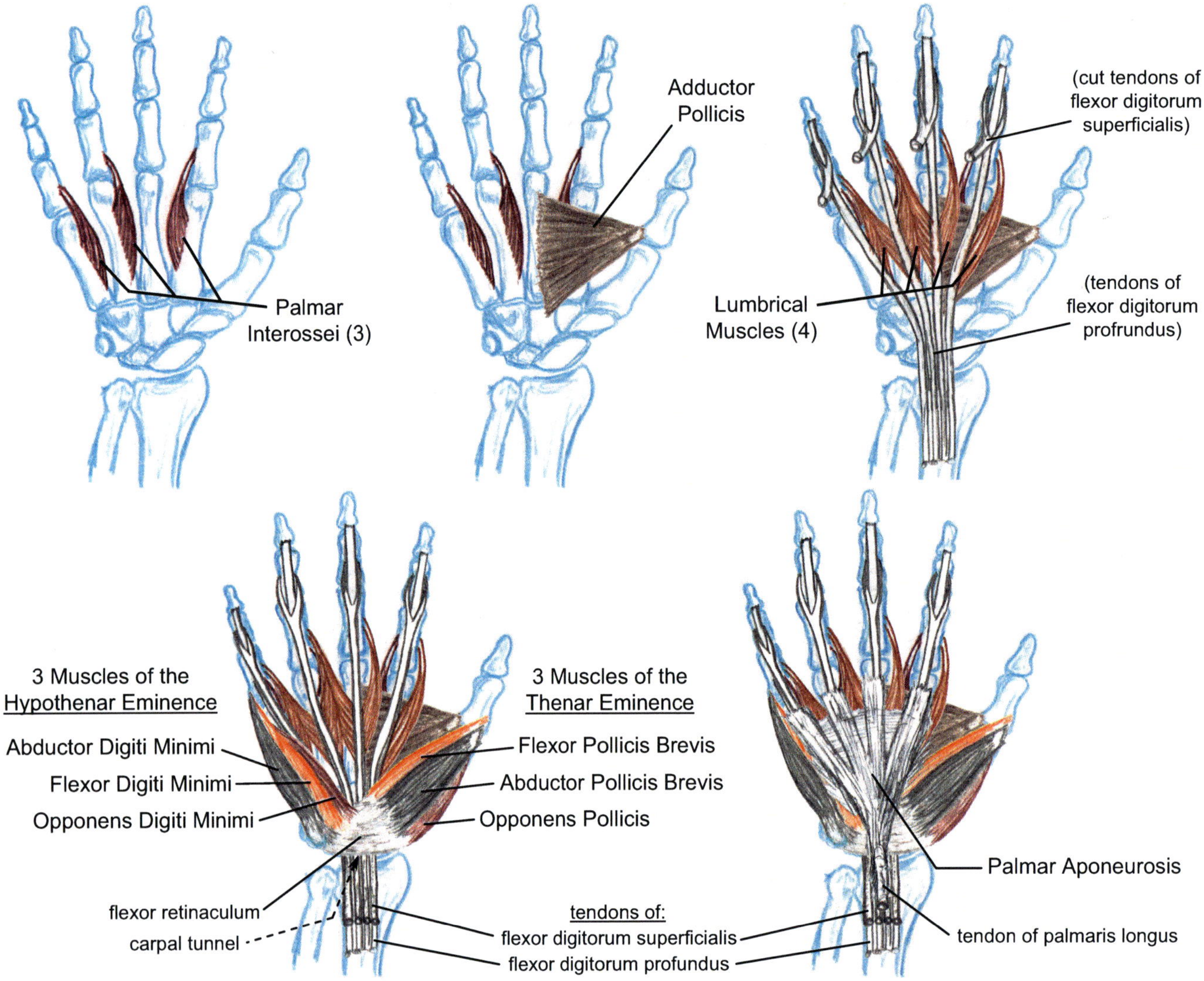

Palmar Interossei (3)

Adductor Pollicis

Lumbrical Muscles (4)

(cut tendons of flexor digitorum superficialis)

(tendons of flexor digitorum profrundus)

3 Muscles of the **Hypothenar Eminence**
- Abductor Digiti Minimi
- Flexor Digiti Minimi
- Opponens Digiti Minimi

flexor retinaculum
carpal tunnel

3 Muscles of the **Thenar Eminence**
- Flexor Pollicis Brevis
- Abductor Pollicis Brevis
- Opponens Pollicis

tendons of:
flexor digitorum superficialis
flexor digitorum profundus

Palmar Aponeurosis

tendon of palmaris longus

Dorsal Aspect of the Right Hand

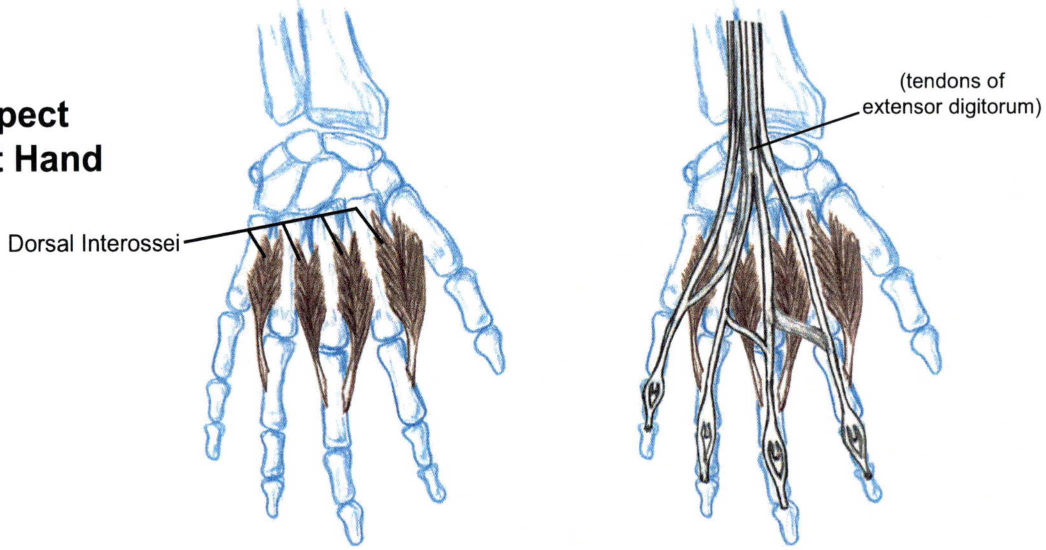

Dorsal Interossei

(tendons of extensor digitorum)

108   Chapter 4 – Muscles That Move the Upper Extremity   Mastering Muscles & Movement   © 2007

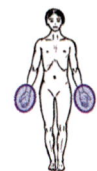

Intrinsic Muscles of the Hand – Palmar Aspect

Joints: CM=Carpometacarpal, MP=Metacarpophalangeal (knuckles), IP=Interphalangeal (fingers)

Muscle	Origin	Insertion	Action	Innervation
Muscles of the Palm (move the fingers or thumb)				
Palmar Interossei (3) Deep Layer	Shafts of metacarpal bones #2, #4 and #5 (each muscle arises from the middle finger side of the metacarpal bone)	Bases of the proximal phalanges of fingers #2, #4 and #5 (and the dorsal digital expansions of fingers #2, #4 and #5)	Adduction of fingers #2, #4 and #5, Assist flexion of fingers #2, #4 and #5 at the MP joints, Assist extension of fingers #2, #4 and #5 at the IP joints	Ulnar N. (C8, T1)
Adductor Pollicis Intermediate Layer	Shaft of 3rd metacarpal bone, & capitate bone (palmar side) (+ bases of 2nd and 3rd metacarpal bones)	Base of proximal phalanx of thumb (ulnar side)	Adduction of the thumb (also assists flexion of the thumb at the MP joint)	Ulnar N. (C8, T1)
Lumbrical Muscles (4) Superficial Layer	The four tendons of the flexor digitorum profundus muscle	The four tendons of the extensor digitorum mm. (attach via radial side of the dorsal digital exansions)	Flexion of fingers #2-5 at the MP joints, and Extension of fingers #2-5 at the IP joints	Digits #2 and #3: Median N. Digits #4 and #5: Ulnar N.
Muscles of the Hypothenar Eminence (move the little finger, digit #5)				
Opponens Digiti Minimi Deep Layer	Hook of hamate and flexor retinaculum	Shaft of 5th metacarpal bone (ulnar side)	Opposition of the little finger (move its finger pad around to face the thumb)	Ulnar N. (C8, T1)
Flexor Digiti Minimi Intermediate Layer	Hook of hamate and flexor retinaculum	Proximal phalanx of little finger (base of phalanx on the palmar side)	Flexion of the little finger	Ulnar N. (C8, T1)
Abductor Digiti Minimi Superficial & Ulnar Layer	Pisiform and tendon of the flexor carpi ulnaris	Proximal phalanx of little finger (base of phalanx on the ulnar side)	Abduction of the little finger	Ulnar N. (C8, T1)
Muscles of the Thenar Eminence (move the thumb, digit #1)				
Opponens Pollicis Deep Layer	Trapezium and flexor retinaculum	Metacarpal #1 (whole length of radial side)	Opposition of thumb to the fingers (Opposition = abduction + flexion + rotation at the 1st CM joint (saddle joint). The finger pads face the thumb pad.)	Median N. (C8, T1)
Flexor Pollicis Brevis Intermediate Layer	Flexor retinaculum and carpal bones (Superficial head to flexor retinaculum & trapezium, Deep head to trapezoid and capitate)	Base of proximal phalanx of thumb (radial side)	Flexion of the thumb (at the MP and CM joints)	Sup. Head: Median N. Deep Head: Ulnar N. (C8, T1)
Abductor Pollicis Brevis Superficial Layer	Flexor retinaculum, trapezium and scaphoid	Base of proximal phalanx of thumb (radial side)	Abduction of the thumb	Median N. (C8, T1)

Intrinsic Muscles of the Hand – Dorsal Aspect

Muscle	Origin	Insertion	Action	Innervation
Dorsal Layer #1				
Dorsal Interossei (4)	Shafts of metacarpal bones #1-5 (each muscle arises from the sides of two adjacent metacarpal bones)	Bases of the proximal phalanges of fingers #2-4 (and the dorsal digital expansions of fingers #2-4)	Abduction of fingers #2-4, Assist flexion of fingers #2-4 at the MP joints, Assist extension of fingers #2-4 at the IP joints	Ulnar N. (C8, T1)

Chapter 5
Muscles That Move the Axial Skeleton

Introduction..110
Movement of the Face and Jaw ....(Muscle Group 6) 113
Movement of the Neck and Head ....(Muscle Group 7) 121
Movement of the Spine ....(Muscle Group 8) 129
Movement of the Thorax, Abdomen, Breathing ...(Muscle Group 9) 137
(pages 145-150 reserved for expansion of this chapter) 145

Group 6 – Face, Jaw

Masseter
Temporalis
Lateral pterygoid
Medial pterygoid
Occipitofrontalis
Platysma
Suprahyoids Group
Infrahyoids Group

p. 113

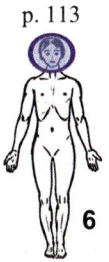

Group 7 – Neck, Head

Sternocleidomastoid
Scalenes group*
Longus capitis & longus colli
Suboccipital group
Splenius capitis
Splenius cervicis
Semispinalis capitis
Levator scapula*
Trapezius, upper fibers*
    *(reversed O/I actions)

p. 121

Group 8 – Spine

Spinalis
Longissimus
Iliocostalis
Semispinalis
Multifidus
Rotatores
Quadratus lumborum

p. 129

Group 9 – Thorax, Abdomen, Breathing

Rectus abdominis
External oblique
Internal oblique
Transverse abdominis
Diaphragm
External intercostals
Internal intercostals
Serratus posterior superior
Serratus posterior inferior
Levator costae

p. 137

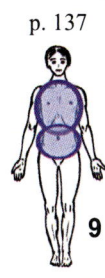

Introduction

The **axial skeleton** comprises the center or "core" of the body, and includes the spine, skull, mandible, hyoid, sternum, and ribs. The axial skeleton articulates with the appendicular skeleton at the sternoclavicular joints for the upper extremities and at the sacroiliac joints for the lower extremities (see pages 34 and 37).

The major functional divisions of the axial skeleton are the head, neck, back, thorax, and abdomen. The thorax is made up of the thoracic spine, ribs, and sternum, and contains the heart and lungs. The abdomen is the area from the bottom of the ribs down to the pelvic bowl, and it contains digestive, blood processing, reproductive, and elimination organs. The respiratory diaphragm muscle creates a boundary between the thorax and abdomen.

The spine (also called the vertebral column) is made up of 24 vertebrae, the sacrum, and the coccyx. Viewed from the side, it has four normal front-to-back curves that allow it to absorb shocks and move more freely - two kyphotic (primary) curves and two lordotic (secondary) curves. A healthy spine has no lateral curves.

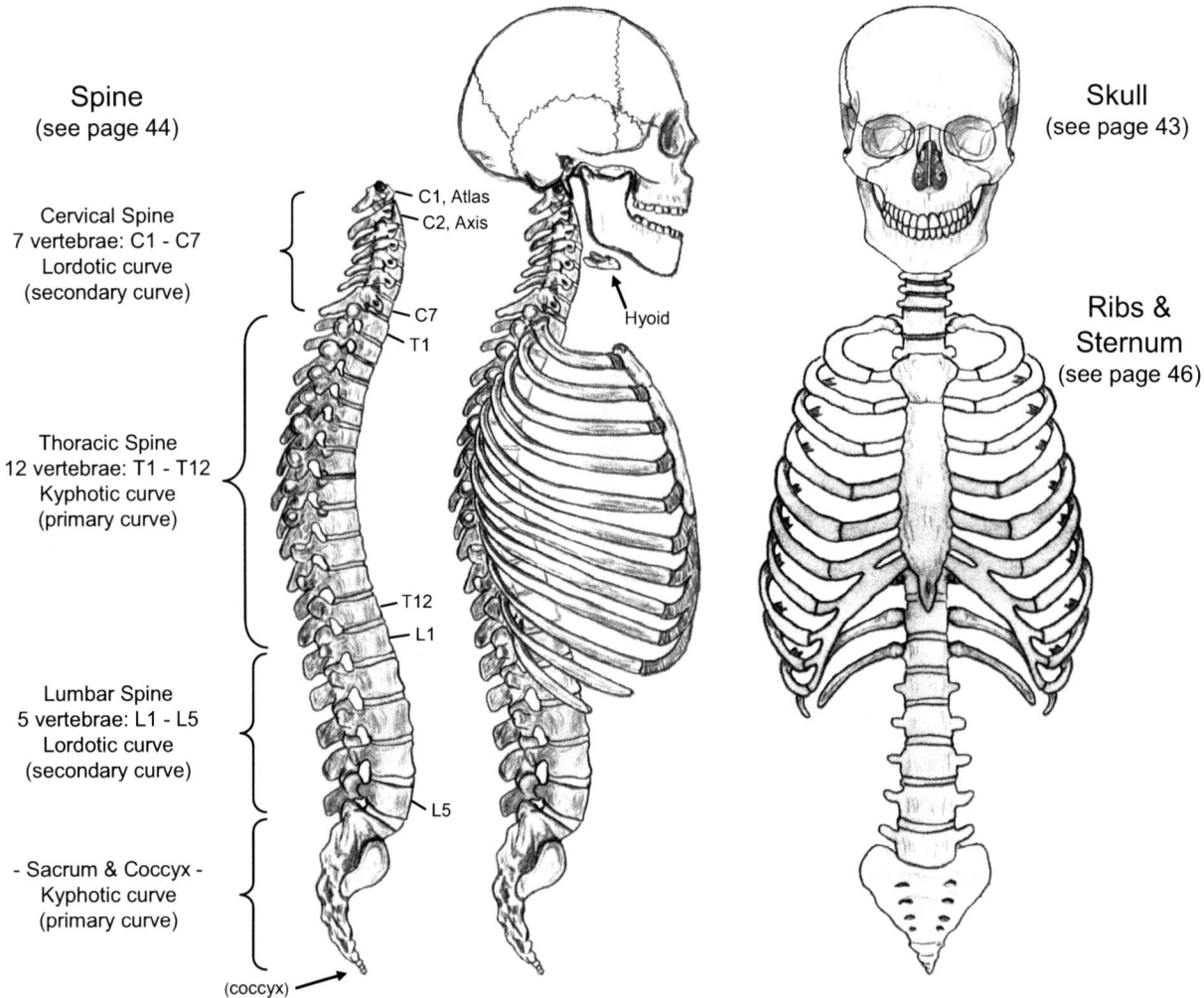

The Axial Skeleton: Spine, skull, mandible, hyoid, ribs, and sternum

At birth, the entire spine has a kyphotic curvature (hence the name *primary* curve). The cervical and lumbar lordotic curves develop *secondary* to the child raising the head to crawl and standing erect to walk.

Bony Landmarks of a Typical Vertebra

This drawing shows a basic set of features that are common to all vertebrae. The three types of vertebrae (cervical, thoracic, and lumbar) also have special features that distinguish them from each other (see Chapter 2, page 45).

The common landmarks are:
- Body
- Transverse process   (TVP)
- Spinous process   (SP)
- Pedicle
- Facets (a pair above, a pair below)
    The *articular process* is the bony protrusion, the *facet* is the smooth cartilage surface.
- Lamina
- Vertebral foramen

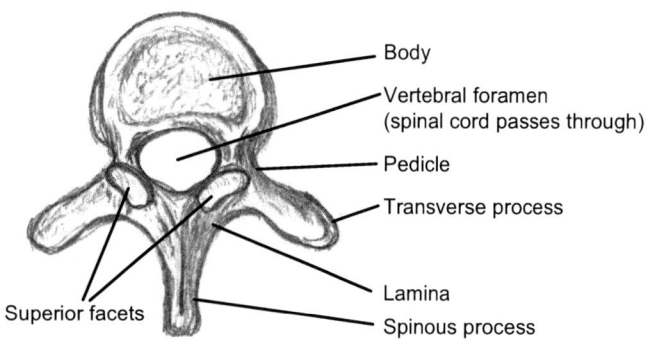

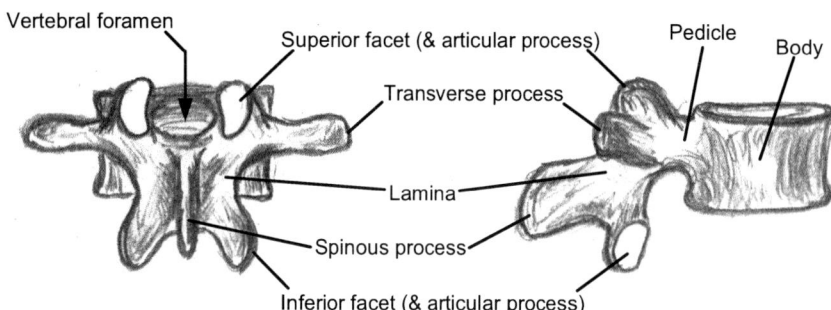

How Vertebrae "Stack Up" to Make Up the Spine

An **intervertebral joint** between two vertebrae is really three joints – one at the body and two at the facets. The inferior facets of each vertebra match up with the superior facets of vertebra below. The facets are synovial joints and their type is gliding. The body of each vertebra sits on the body of the vertebra below it and has an intervertebral disc between. This joint with its disc is an amphiarthrotic/cartilaginous joint.

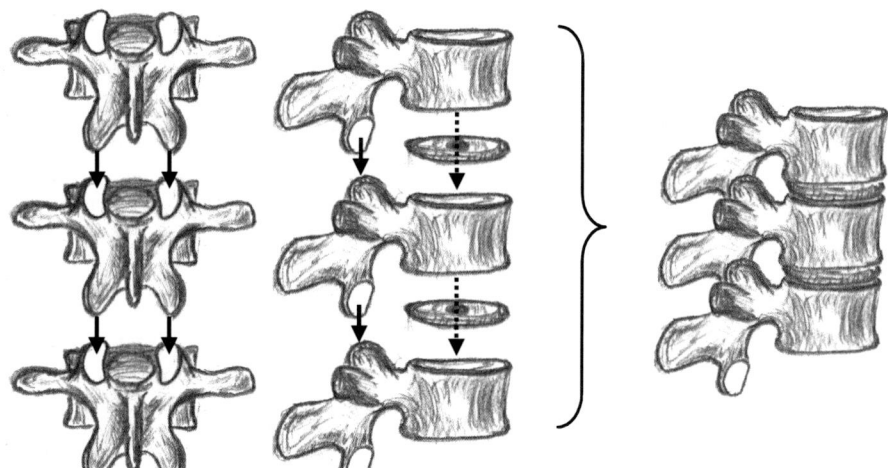

The vertebral bodies in the lower spine are large and stable because they must support more weight than the vertebrae near the top of the spine. Conversely, the vertebral foramen (the holes that create the channel for the spinal cord) are large near the top of the spine because the spinal cord is thicker.

New Kinesiology Concepts

Some additional kinesiology concepts are required when studying the muscles and actions of the *axial skeleton* vs. the actions of the limbs. This is because pairs of muscles on either side of the spine act on the spine from opposite directions.

Naming of Actions for Lateral and Rotational Movements

For limbs, described in Chapter 4–Upper Extremity and Chapter 6–Lower Extremity, movements that are lateral and rotational are named to describe whether the bones move *away from* the midline or *toward* the midline. Side-to-side limb movements are called abduction and adduction, and rotational movements are called lateral rotation and medial rotation.

For the axial skeleton the bones are *centered on* the midline, so the terminology used for the limbs does not make sense. Side-to-side movements are defined as right and left lateral flexion (instead of abduction and adduction as used for limbs), and rotational movements are defined as right and left rotation (instead of medial and lateral rotation as used for limbs).

Unilateral vs. Bilateral Use of Muscles

There are two of each muscle – one on each side of the spine. Unilateral action is when a muscle on only one side of the spine contracts. Bilateral action occurs when the same muscle on both sides of the spine contract at the same time.

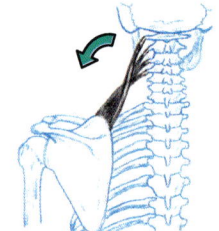
Muscle acting unilaterally on left side pulls neck to left

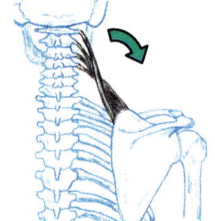
Muscle acting unilaterally on right side pulls neck to right

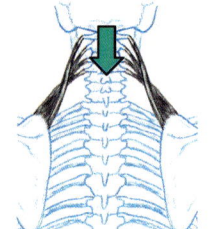
Muscle acting bilaterally pulls neck to the rear

Notice that for a unilateral contraction, an antagonist to a given action is the same muscle on the *other* side of the spine.

Rotational Actions to "Same Side" vs. to "Opposite Side"

The direction of a rotation movement is defined by the direction the anterior surface of the body moves (right vs. left when describing the axial skeleton - the head, neck, trunk). If a muscle that is on the *left* side of the spine contracts, and the body rotates to the *left*, then it's an action "to the same side". Conversely, if a muscle on the *right* side causes rotation to the *left* side, then it is creating a movement "to the opposite side".

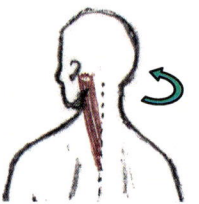
Rotation of the head/neck to the same side

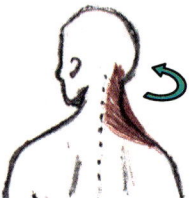
Rotation of the head/neck to the opposite side

Note that this concept only applies to rotations; lateral flexions are always "to the same side".

Face, Jaw

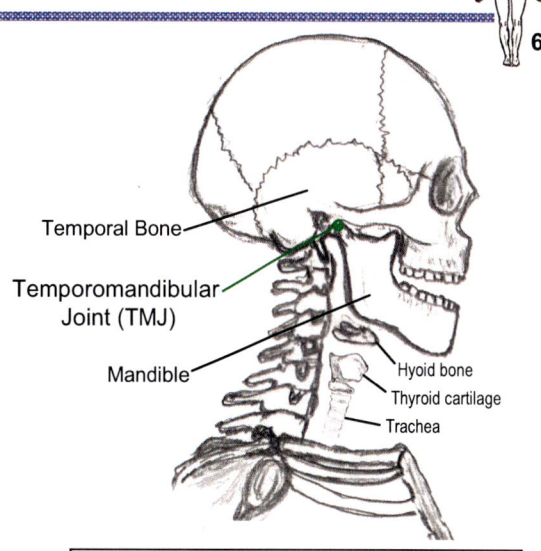

Movement of the Face and Jaw

Muscle Group 6:

Masseter	Occipitofrontalis
Temporalis	Platysma
Lateral pterygoid	Suprahyoids group
Medial pterygoid	Infrahyoids group

Joint

The first four of the muscles in this group move the mandible at the temporomandibular joint. The remaining muscles are involved with moving the face, jaw, and anterior neck/throat.

Temporomandibular Joint (TMJ)

Condyle of **mandible** ◄► mandibular fossa of **temporal** bone

Complex Type: Condyloid/gliding/hinge
Movements of the jaw available at the TMJ:
 Elevation, Depression
 Protraction, Retraction (Protrusion, Retrusion)
 Lateral Deviation

Note: There is a moveable articular disc (cartilage) between the bones at the TMJ.

Other Structures and Movements

The following are not individual actions of bones at joints, but are body functions that involve coordinated contractions of the muscles of the face, jaw, and anterior neck.

Mastication (chewing)
 Involves movement of TMJ (see above) along with movement of the tongue, cheeks, and lips.

Facial Expression
 Many muscles move the tissue of the face.

Deglutition (swallowing)
 Swallowing involves a complex series of movements of the tongue, throat, and anterior neck muscles.

Speech
 Muscles of the jaw, mouth, tongue, face, and soft palate are used when forming spoken words.

Sight
 Muscles close and open the eyelids, and small muscles move the eyeballs in their sockets.

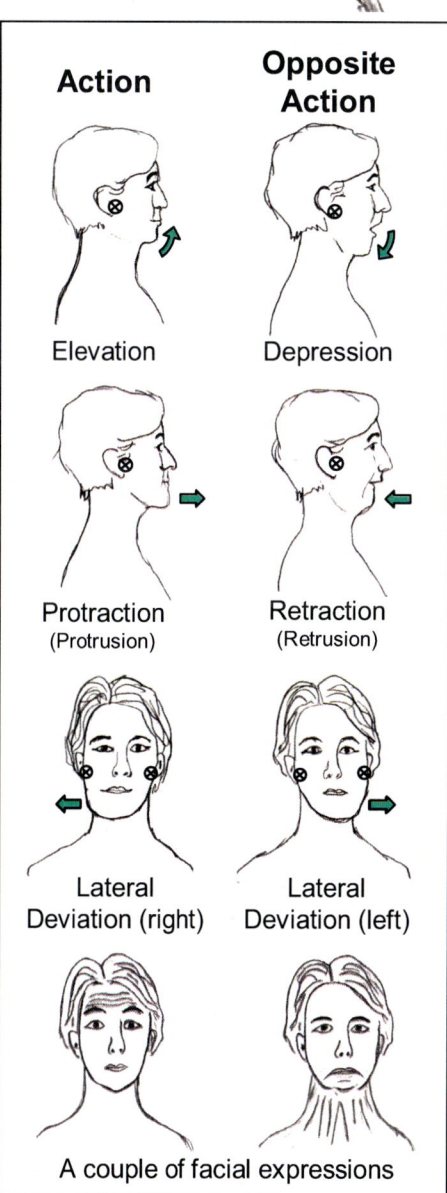

A couple of facial expressions

Mastering Muscles & Movement © 2007   Chapter 5 – Muscles That Move the Axial Skeleton

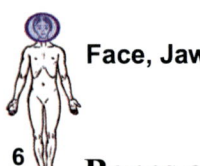

Face, Jaw

6

Bones and bony landmarks

Muscles that move the face and jaw have attachments on the skull, mandible, hyoid bone, fascia, sternum, and others. Review the bony landmarks and other structures listed below, referring to the drawings in Chapter 2, page 43-46.

Cranial Bones: (p. 43)

Occiput
    External occipital protuberance
    Superior nuchal line

Temporal (2)
    Styloid process
    Mastoid process
    Zygomatic arch
    Mandibular fossa

Parietal (2)
Frontal

Sphenoid
    Lateral pterygoid plate

Temporal Fossa:
    Flat area on side of skull covering portions of the temporal, parietal, frontal, and sphenoid bones

Facial Bones: (p. 43)

Maxilla (2)
Zygomatic (2)
Nasal (2)

Mandible
    Condyle of the mandible
    Coronoid process
    Ramus
    Angle
    Body

Hyoid Bone (p. 43)

Sternum (p. 46)
    Manubrium

Tendinous/Other Structures
    Galea aponeurotica
    Thyroid cartilage

Muscles of Facial Expression

Humans have a multitude of muscles that move the tissue of the face – more than any other mammal. Most of these muscles of *facial expression* are shown in this illustration. Note that two of these muscles, occipitofrontalis and platysma, are also included in the Group 6 tables.

In addition to expressing emotions, these muscles provide important functions such as closing the eyes and mouth, holding in the cheeks while chewing food, and shaping sounds while speaking.

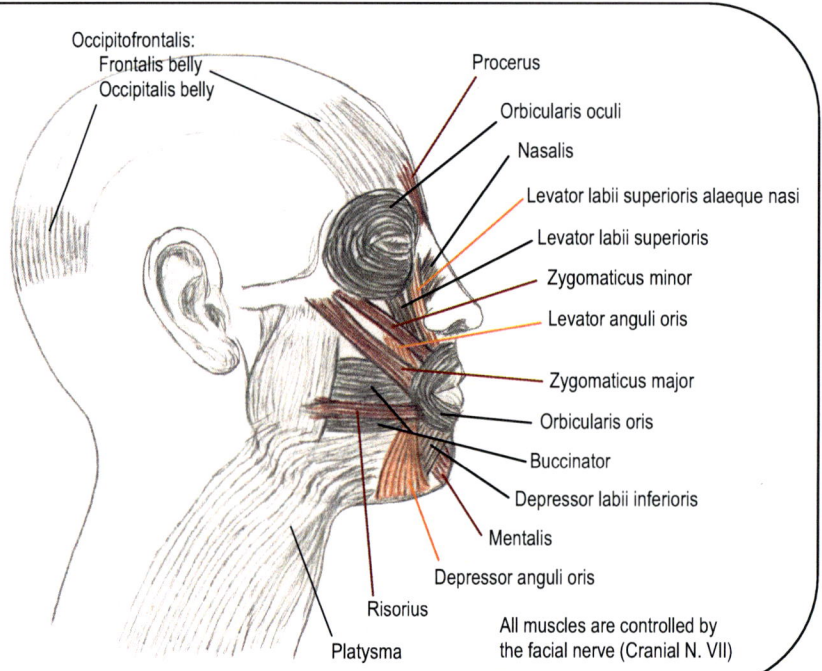

Face, Jaw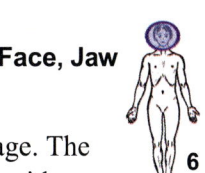

Muscle Group 6 – Muscles that move the face and jaw are illustrated as a group on this page. The four pages that follow have tables and figures that describe each muscle individually, and provide many ways of comparing and contrasting the muscles to each other.

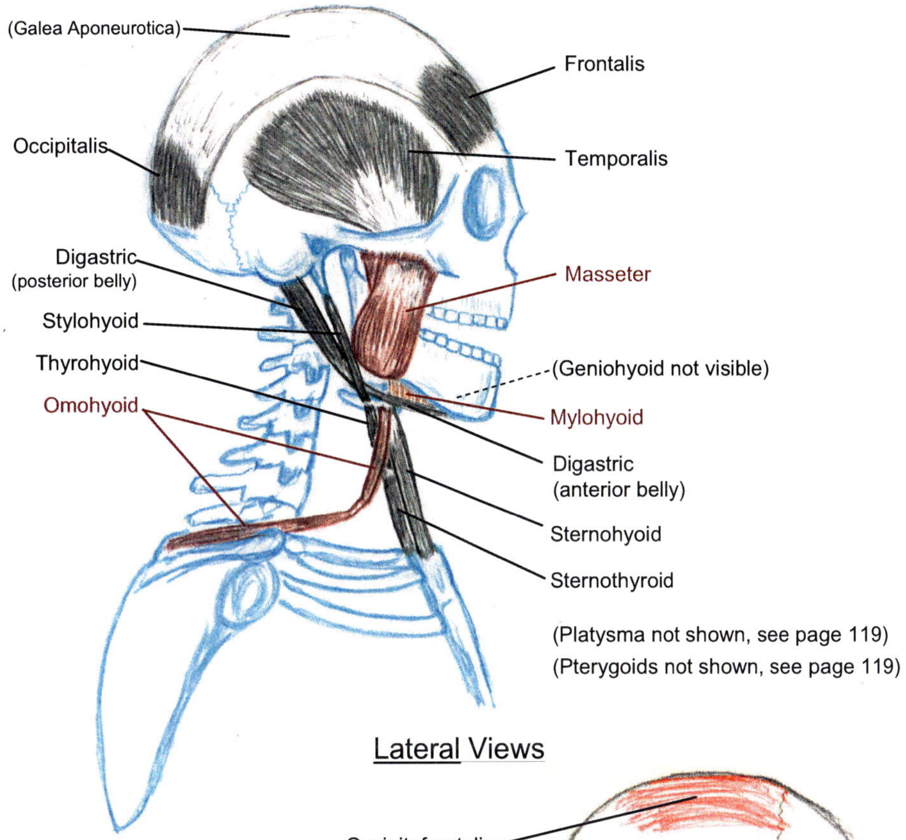

Lateral Views

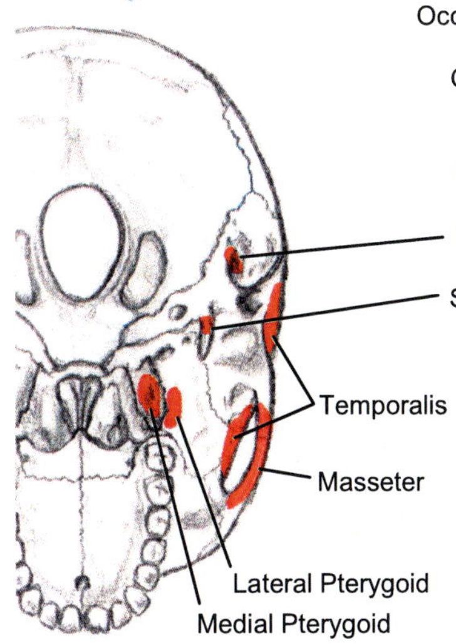

Inferior View

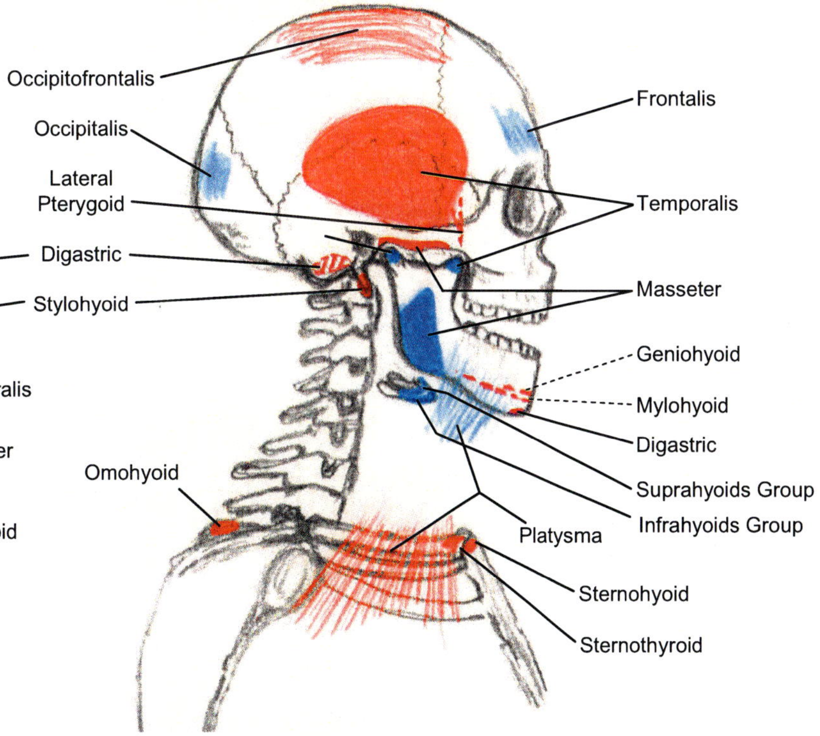

Origins are red
Insertions are blue

Attachment sites for all muscles in Group 6

Mastering Muscles & Movement © 2007   Chapter 5 – Muscles That Move the Axial Skeleton   115

Face, Jaw

Muscles Acting On Face, Jaw

Mandible moves at the TMJ=Temporomandibular joint, UL=Unilateral action, BL=Bilateral action

Face, Jaw	Origin	Insertion	Action
Masseter — moves the mandible	Zygomatic arch	Angle and ramus of the mandible (Has 2 bellies: superficial & deep)	BL: Elevation of the mandible
Temporalis — moves the mandible	Temporal fossa and fascia	Coronoid process of the mandible	BL: Elevation and retraction of mandible.
Lateral Pterygoid — moves the mandible	<u>Lower head</u>: *Lateral* surface of lateral pterygoid plate of sphenoid bone. <u>Upper Head</u>: Greater wing of sphenoid	Neck of mandible and articular disc of TMJ	BL: Protrusion of mandible, assists depression of mandible. UL: Lateral deviation of mandible to the opposite side.
Medial Pterygoid — moves the mandible	<u>Deep head</u>: *Medial* surface of lateral pterygoid plate of sphenoid bone. <u>Superficial head</u>: Tuberosity of maxilla	Medial surface of angle of mandible (forms a "sling" with the masseter muscle on the outside)	BL: Assists elevation of mandible. UL: Lateral deviation of mandible to the opposite side.
Occipitofrontalis — moves the brows	Galea aponeurotica	<u>Frontalis belly</u>: Skin and fascia above the eyebrows. <u>Occipitalis belly</u>: Superior nuchal line of occiput	Raises the eyebrows, wrinkles the forehead
Platysma — moves the mandible and skin	Fascia of the pectoralis major	Lower mandible, subcutaneous tissue of lower face & angle of mouth	Pulls angle of mouth downward and thoracic skin upward. Assists depression of the mandible
Suprahyoids group *Geniohyoid, Mylohyoid, Stylohyoid, Digastric*	Underside of mandible, styloid process, mastoid process (Digastric has 2-bellies, origins at mastoid process and mandible)	Hyoid bone	*When the mandible is stable*: Elevation of the hyoid when swallowing. *When the hyoid is stable*: Depression of the mandible.
Infrahyoids group *Sternohyoid, Sternothyroid, Omohyoid, Thyrohyoid*	Manubrium of sternum, medial clavicle, superior border of scapula (omohyoid)	Hyoid bone and thyroid cartilage	Depression and stabilization of the hyoid bone and thyroid cartilage.

Table 6 (A) - Face, Jaw - Origin, Insertion, Action

(larger illustrations on page 119)

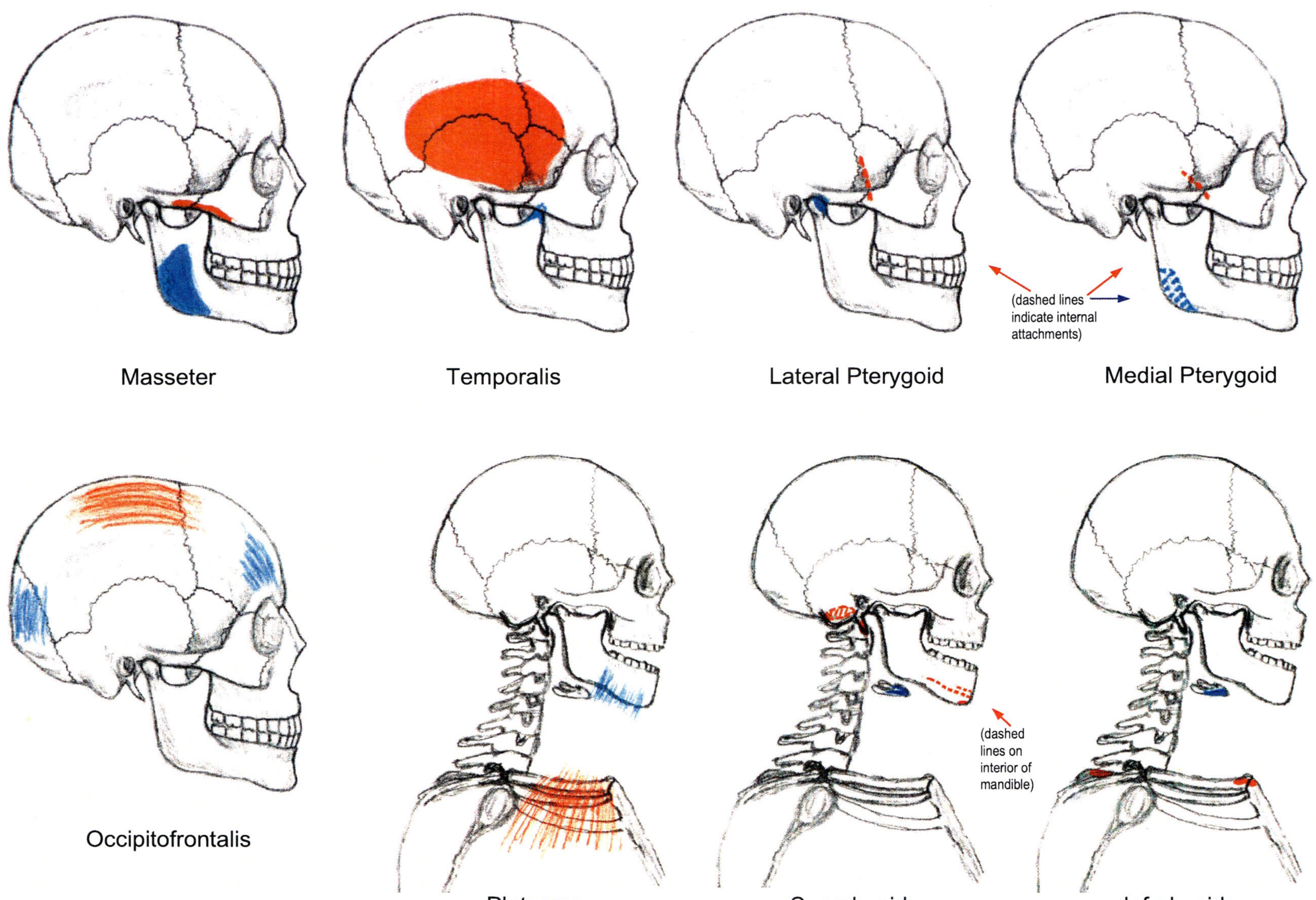

Face, Jaw

Masseter	Occipitofrontalis
Temporalis	Platysma
Lateral Pterygoid	Suprahyoids (dashed lines on interior of mandible)
Medial Pterygoid (dashed lines indicate internal attachments)	Infrahyoids

Figure 6 (A) - Face, Jaw - Muscle Attachments

Origins are red
Insertions are blue

Lift page to see muscle pictures

Mastering Muscles & Movement © 2007

Chapter 5 - Muscles That Move the Axial Skeleton   117

Face, Jaw

Muscles Acting On Face, Jaw — Mandible moves at the two TMJs (Temporomandibular Joints), UL=Unilateral action, BL=Bilateral action, X=Both UL & BL action

Face, Jaw	Elevation of mandible	Depression of mandible	Protrusion/ Protraction of mandible	Retrusion/ Retraction of mandible	Lateral Deviation of mandible	Other	Innervation	Cr. V	Cr. VII	Cr. XII	C1	C2	C3
1. Masseter	X			assist (deep belly)			Trigeminal N. (Cranial N. V)	N					
2. Temporalis	X			X (posterior fibers)		Many movements due to multiple fiber directions: Anterior = vertical Middle = diagonal Posterior = horizontal	Trigeminal N. (Cranial N. V)	N					
3. Lateral Pterygoid		assist	X		UL to opposite side	Pulls artcular disc forward when opening mouth	Trigeminal N. (Cranial N. V)	N					
4. Medial Pterygoid	assist		may assist		UL to opposite side	This and masseter create a "sling" around angle of the mandible	Trigeminal N. (Cranial N. V)	N					
5. Occipitofrontalis: Frontalis belly Occipitalis belly						Raises eyebrows, wrinkles forehead	Frontalis: Temporal branch Facial N. (Cr.VII) Occip: Posterior auricular br. Facial N. (Cr.VII)		N				
6. Platysma		assist				Pulls angle of mouth downward and thoracic skin upward	Facial N. (Cr. VII)		N				
7. Suprahyoids Group (Geniohyoid, Mylohyoid, Stylohyoid, Digastric)		X		assist		Elevates hyoid (when mandible is stable), or depresses mandible (when hyoid is stable)	G: Hypoglossal XII,C1 M: Trigeminal N. (V) S: Facial N. (VII) D: Trgiem.V,Facial VII	N	N	N	N		
8. Infrahyoids Group (Sternohyoid, Sternothyroid, Omohyoid, Thyrohyoid)		assist				Depresses and stabilizes the hyoid bone	Ansa cervicalis C1,C2,C3				N	N	N
(More muscles for the action) --->							Innervation						

Table 6 (B) - Face, Jaw - Synergists & Antagonists

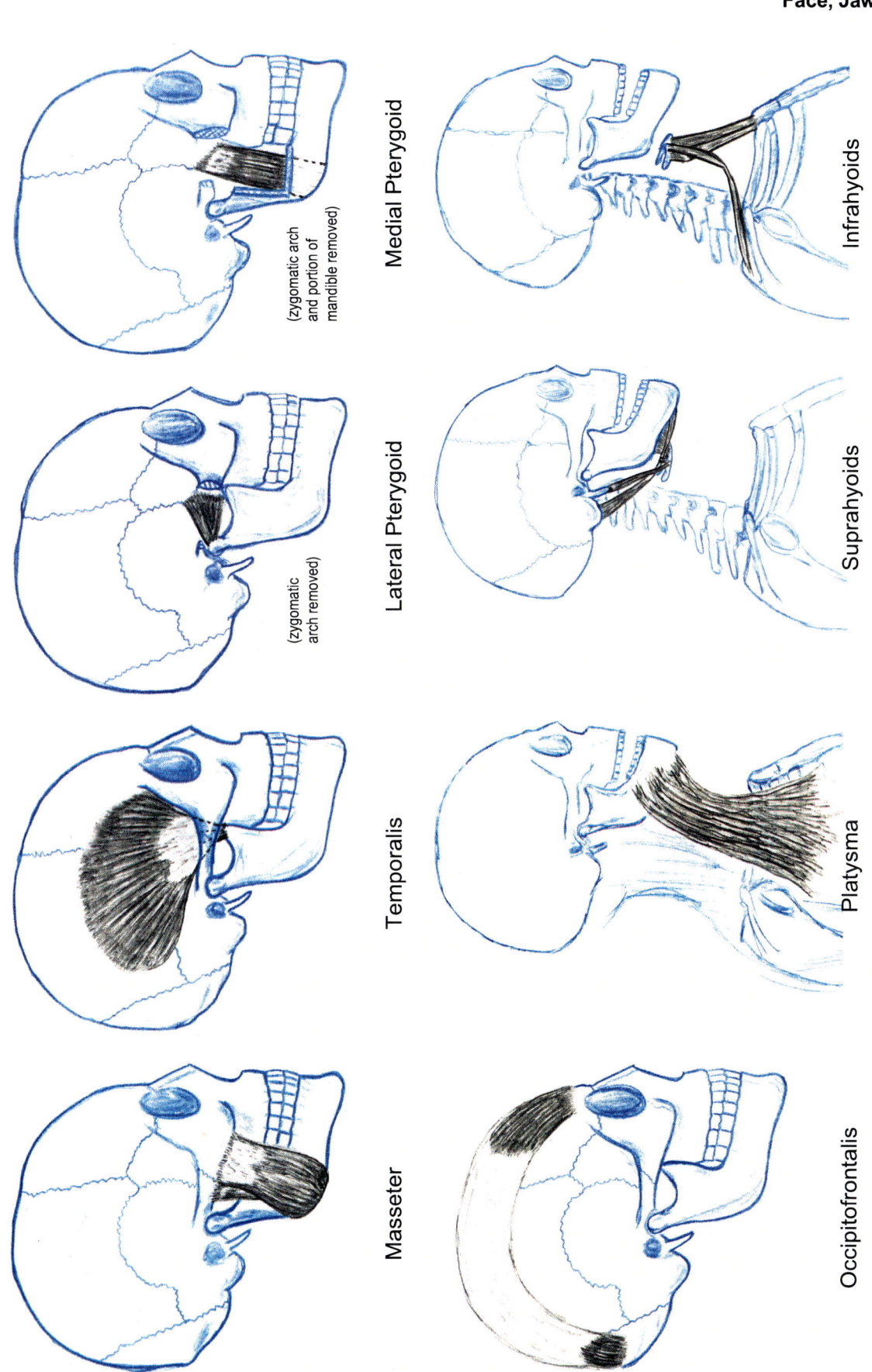

Figure 6 (B) - Face, Jaw - Muscle Pictures

Muscles Acting On

Face, Jaw ~ Notes ~ (palpation, fiber arrangement, lengthen/shorten, common uses, pathologies, cautions, etc.)

#	Muscle		#	Muscle	Notes
1.	Masseter		5.	Occipitofrontalis *(Occipitalis & Frontalis)*	
2.	Temporalis		6.	Platysma	
3.	Lateral Pterygoid		7.	Suprahyoids Group	*Geniohyoid, Mylohyoid, Stylohyoid, Digastric*
4.	Medial Pterygoid		8.	Infrahyoids Group	*Sternohyoid, Sternothyroid, Omohyoid, Thyrohyoid*

Muscle Group 6 - Face, Jaw - (Notes)

Neck, Head

Movement of the Neck and Head

Muscle Group 7:

Sternocleidomastoid (SCM)	Splenius capitis
Scalenes group	Splenius cervicis
(anterior, middle, posterior)	Semispinalis capitis
Longus capitis & longus colli	Trapezius, upper fibers (reversed O/I)
Suboccipital group	Levator scapula (reversed O/I)

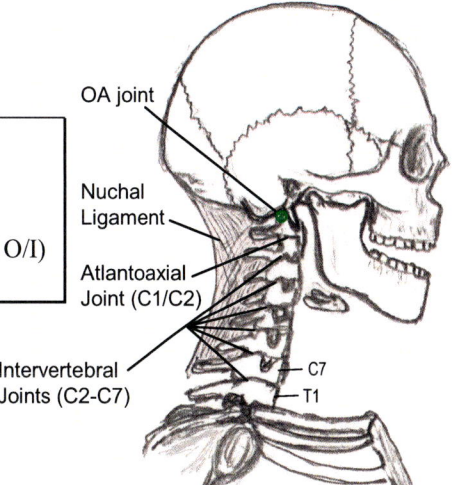

Joints

The muscles in this group primarily move the neck and head. Some are also used in the inhalation phase of breathing (reversed O/I actions). Two of the muscles also move the scapula, as described previously in Chapter 4, Muscle Group 1.

Occipitoatlantal Joint (OA)
    Occipital condyles ◄► Superior facets of C1 (atlas)
    Ellipsoid joint
        Flexion & Extension   (quite a lot, head rocks forward
                                        and backward -"stiff necked yes")
        Lateral flexion   (limited amount, tilt side to side – "duh")

(the OA joint is also called the **Atlantooccipital (AO)** joint)

Atlantoaxial Joint (Atlas/Axis = C1/C2, has 2 parts):

- **Atlantoodontoid Joint**
    Inner anterior arch of **C1** ◄► Odontoid process of **C2**
    Pivot joint (allows rotation)

- **Atlantoaxial Facet Joints (2)**
    <u>Pairs</u>: **C1** inferior facets ◄► **C2** superior facets
    Gliding joints (they glide opposite directions during rotation)

    C1/C2 has <u>more</u> rotation than any other spinal segment (true rotation occurs at the atlantoodontoid pivot joint).

Intervertebral Joints (C2–C7, have 2 parts):
(please review page 111)

- **Intervertebral Discs**
    Between vertebral bodies
    Amphiarthrotic/Cartilaginous joint

- **Intervertebral Facets (2)**
    <u>Pairs</u>: Inferior facets of vertebra above ◄► Superior
        facets of vertebra below
    Gliding joints

(C2-C7 intervertebral joints allow all six actions shown in box)

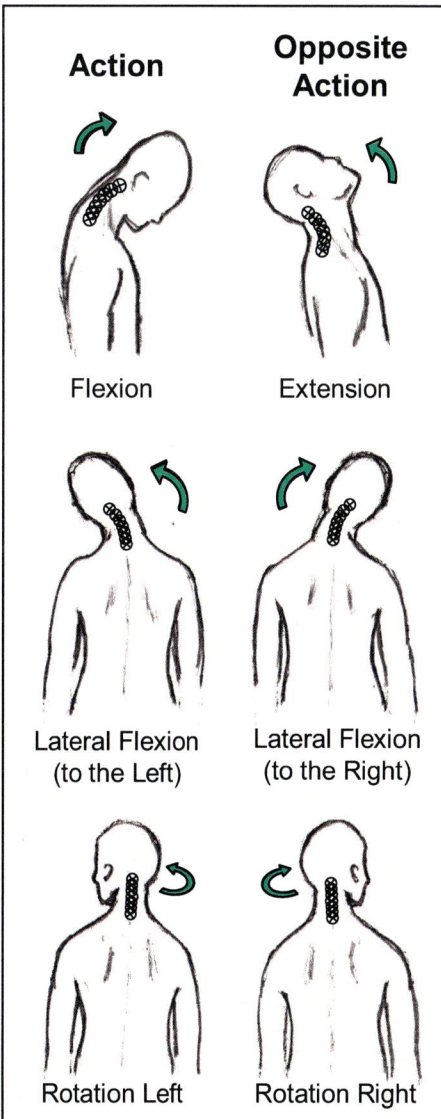

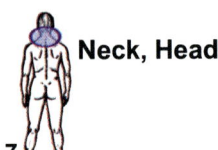

Neck, Head

Movements available

To move the head and neck, muscles anchor below and pull on structures above. Movement dynamics can be divided into three categories:

Head Movement: Head sitting on Atlas (C1) moves without the neck moving.
• Muscles have a short span – from C1 and C2 to the occiput.

Neck-Initiated Movement: Neck is pulled by a muscle, and the head goes along with it.
• Muscles insert on the cervical spine, and do not cross over to the skull.

Head-Initiated Movement: Skull is pulled by a muscle, and the neck "usually" goes along with it.
• Muscles span several vertebrae and insert on the occiput, creating a complex dynamic between head and neck.
• The head may tilt or rotate along with moving the neck.

Bones and bony landmarks

Muscles that move the neck and head have attachments on the skull, vertebrae, upper thorax, and shoulder girdle. Review the bony landmarks and other structures listed below, referring to the drawings in Chapter 2, pages 40-46.

Skull (p. 43)
    Occiput
        External occipital protuberance
        Superior nuchal line, Inferior nuchal line
        Occipital condyles
        Foramen magnum
    Temporal bone (mastoid process)

Sternum, upper 2 ribs, scapula, and clavicle (p. 40, 46)

Focus: Cervical spine (p. 45)

Cervical vertebrae have:
    Transverse foramen in TVPs (for blood vessels)
    Bifid spinous processes (except C1 & C7)
    Nearly horizontal facets

Atlas (C1)
    No SP, broader TVPs
    No Body (anterior arch instead)
    Superior facets are where the occiput sits

Axis (C2)
    Odontoid process (dens) above,
        Dens fits into anterior arch of C1
    "Body" articulates to C3 below

Middle and lower cervicals (C3 – C7)
    "Normal" cervical vertebrae

Nuchal Ligament
    Also called Ligamentum Nuchae
    Attachment site for the more superficial muscles

Deep Anterior Neck Muscles
(Prevertebral Muscles)

There are four muscles that lie deep in the anterior neck. These *prevertebral* muscles are not included in the Group 7 tables, but are summarized here for completeness. Three of the muscles insert on the base of the occiput, just anterior or lateral to the foramen magnum, and the fourth spans all cervical and the upper 3 thoracic vertebrae. They assist flexion of the neck/head, and the RCL assists lateral tilting of the head on C1.

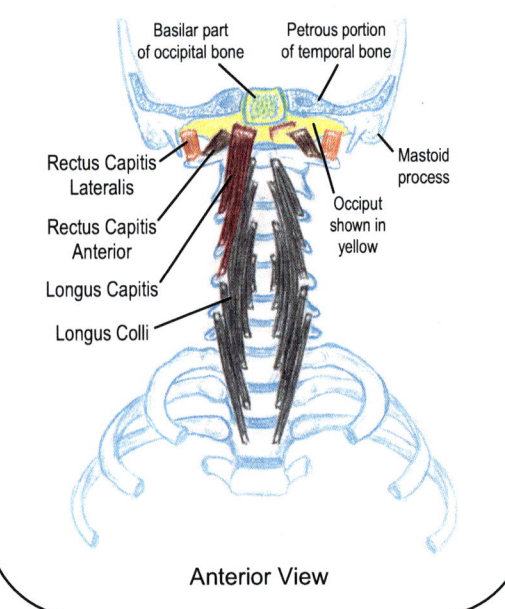

Anterior View

122 Chapter 5 – Muscles That Move the Axial Skeleton

Neck, Head

Muscle Group 7 – Muscles that move the neck and head are illustrated as a group on this page. The following four pages have tables and figures that describe each muscle individually, and provide many ways of comparing and contrasting the muscles to each other.

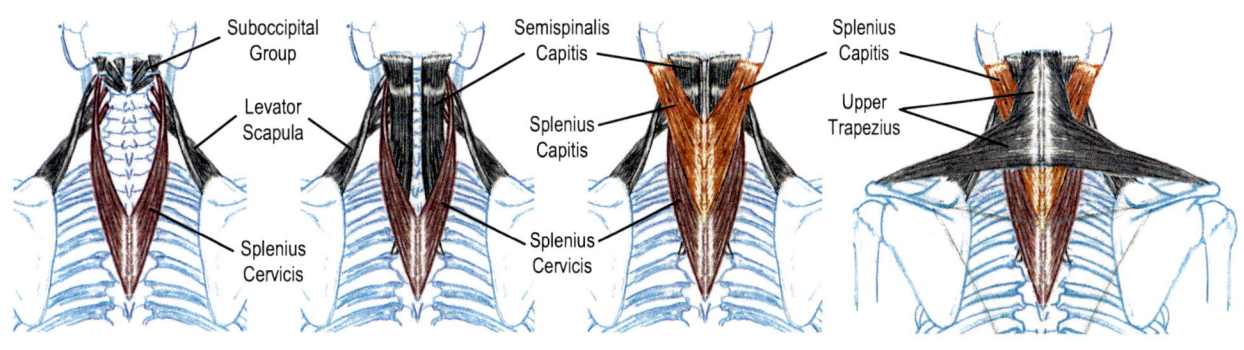

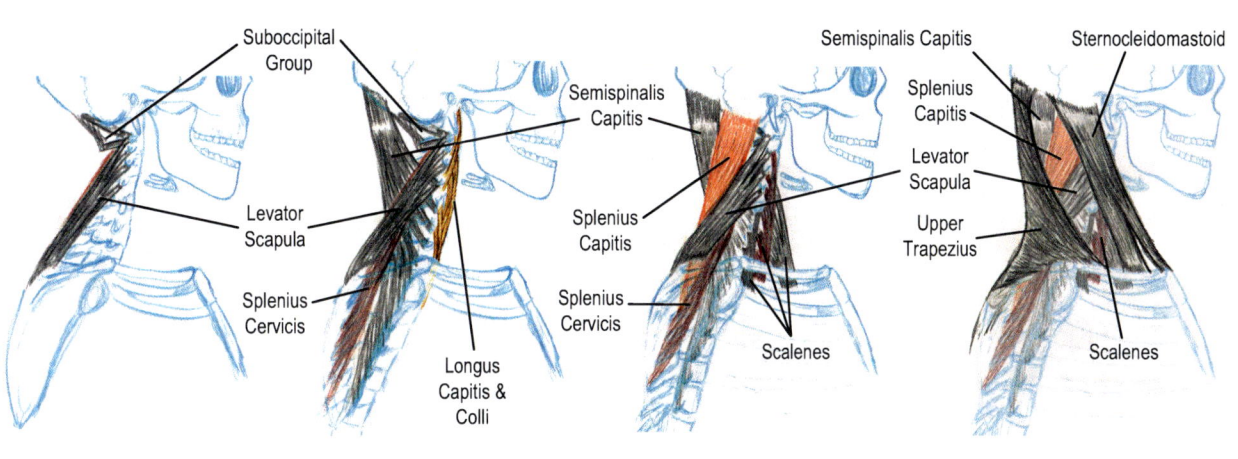

Note: Some of the muscles in this group use reversed O/I actions.

Base of Skull – Inferior View

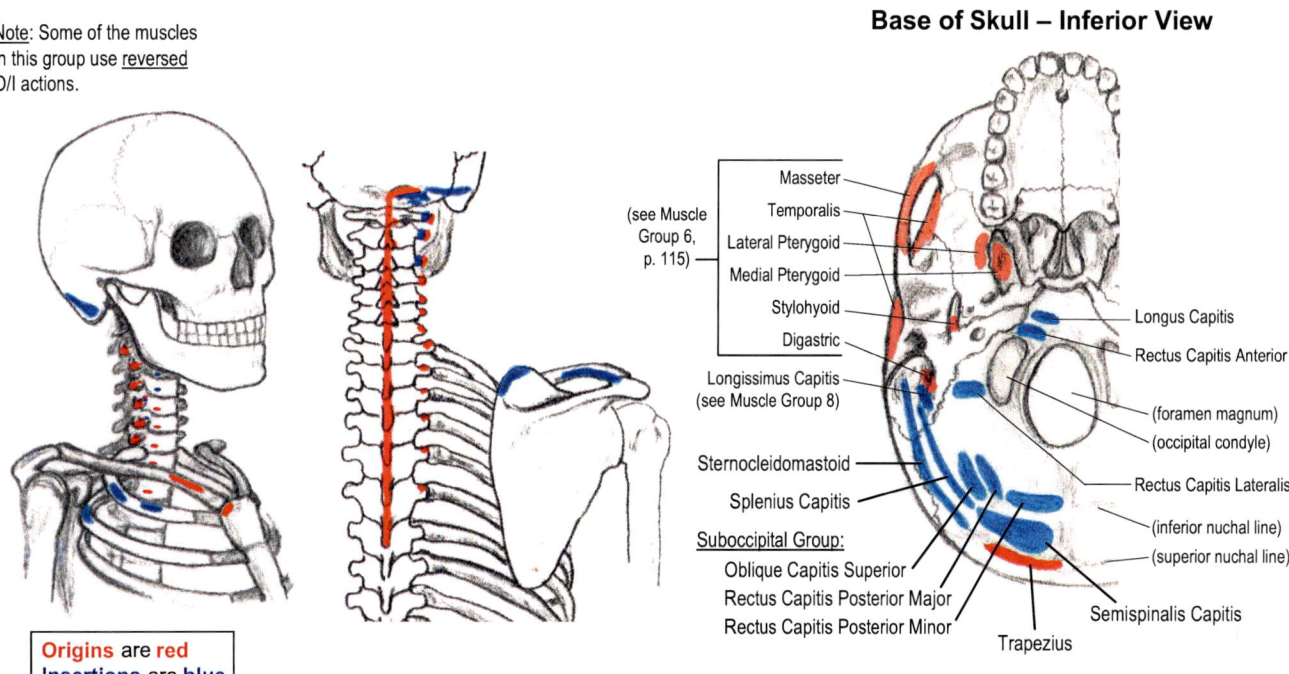

Origins are red
Insertions are blue

Attachment sites for all muscles in Group 7

Mastering Muscles & Movement © 2007   Chapter 5 – Muscles That Move the Axial Skeleton   **123**

Neck, Head

TVP=Transverse process of vertebra, SP=Spinous process of vertebra, UL=Unilateral action, BL=Bilateral action

Muscles Acting On Neck, Head	Origin	Insertion	Action
Sternocleidomastoid (SCM) moves the head and neck	Sternal head: Manubrium of the sternum Clavicular head: Medial ⅓ of the clavicle <u>Overall Description</u>: Medial clavicle, and manubrium of the sternum	Mastoid process of the temporal bone, (and lateral part of superior nuchal line of occiput)	BL: Flexion of the neck. (also assist inhalation by lifting ribcage) UL: Lateral flexion of head/neck, rotation of head to opposite side. (rotating also makes face turn upward)
Scalenes: Anterior Scalene / Middle Scalene / Posterior Scalene move the neck	Anterior: TVP's C3-C6 Middle: TVP's C2-C7 Posterior: TVP's C5-C7 <u>Overall Description</u>: TVP's of C2-C7	Anterior: Anterior-medial Rib 1 Middle: Lateral Rib 1 Posterior: Lateral Rib 2 <u>Overall Description</u>: Ribs 1 & 2	BL: Raise ribs #1 & 2 for inhalation, assist flexion of the neck (ant.&mid.). UL: Lateral flexion of neck (ant.&mid.–assist rotation to opp. side)
Longus Capitis, Longus Colli move the head and neck	Capitis: TVP's of C3-C5 Colli: TVP's of C3-C5, and Anterior bodies of C5-T3	Capitis: Inferior surface of occiput (anterior to the foramen magnum) Colli: Anterior arch/bodies C1-C4, & TVP's of C5-C6	BL: Flexion of the head/neck. (UL: May assist lateral flexion and rotation of the head/neck)
Suboccipitals Group move the head • Rectus Capitis Posterior Major • Rectus Capitis Posterior Minor • Oblique Capitis Superior • Oblique Capitis Inferior	RCP Maj: SP of C2 RCP Min: Posterior tubercle of C1 OCS: TVP of C1 OCI: SP of C2 <u>Overall Description</u>: SP's & TVP's of C1 and C2 (atlas and axis)	RCP Maj: ⎫ RCP Min: ⎬ Inferior nuchal line of occiput OCS: ⎭ OCI: TVP of C1 <u>Overall Description</u>: Lower occipital bone & TVP's of C1	BL: Rock the head back into extension (all) UL: Lateral flexion of the head (RCPM, RCPm, and OCS) Rotation to the same side (OCI)
Splenius Capitis moves the head and neck	Lower half of nuchal ligament (from C3-C6), and SP's of C7-T3	Mastoid process of the temporal bone, (and lateral part of superior nuchal line of occiput - deep to SCM insertion)	BL: Extension of the head/neck. UL: Rotation of head/neck to the same side, lateral flexion of the head/neck
Splenius Cervicis moves the neck	SP's of T3-T6	TVP's of C1-C3	BL: Extension of the neck. UL: Rotation of the neck to the same side, lateral flexion of the neck
Semispinalis Capitis moves the head and neck	TVP's of C4-C7 and T1-T6	Occipital bone (between the superior and inferior nuchal lines)	BL: Extension of the head/neck. (see Muscle Group 8 for description of the complete semispinalis muscle)
Levator Scapula This is reversed O/I action, i.e., scapula is held fixed, so moves the neck (See also Muscle Group 1)	TVP's of C1-C4	Superior angle of scapula	With scapula held fixed: BL: Extension of neck. UL: Lateral flexion of neck, rotation of neck to same side
Trapezius, upper fibers This is reversed O/I action, i.e., scapula is held fixed, so moves the head & neck (See also Muscle Group 1)	Upper fibers: Occiput, nuchal ligament, and SP of C7	Upper fibers: Lateral clavicle	With scapula held fixed: BL: Extension of head/neck. UL: Lateral flexion of head/neck, and rotation to opposite side.

Table 7 (A) - Neck, Head - Origin, Insertion, Action

(larger illustrations on page 127)

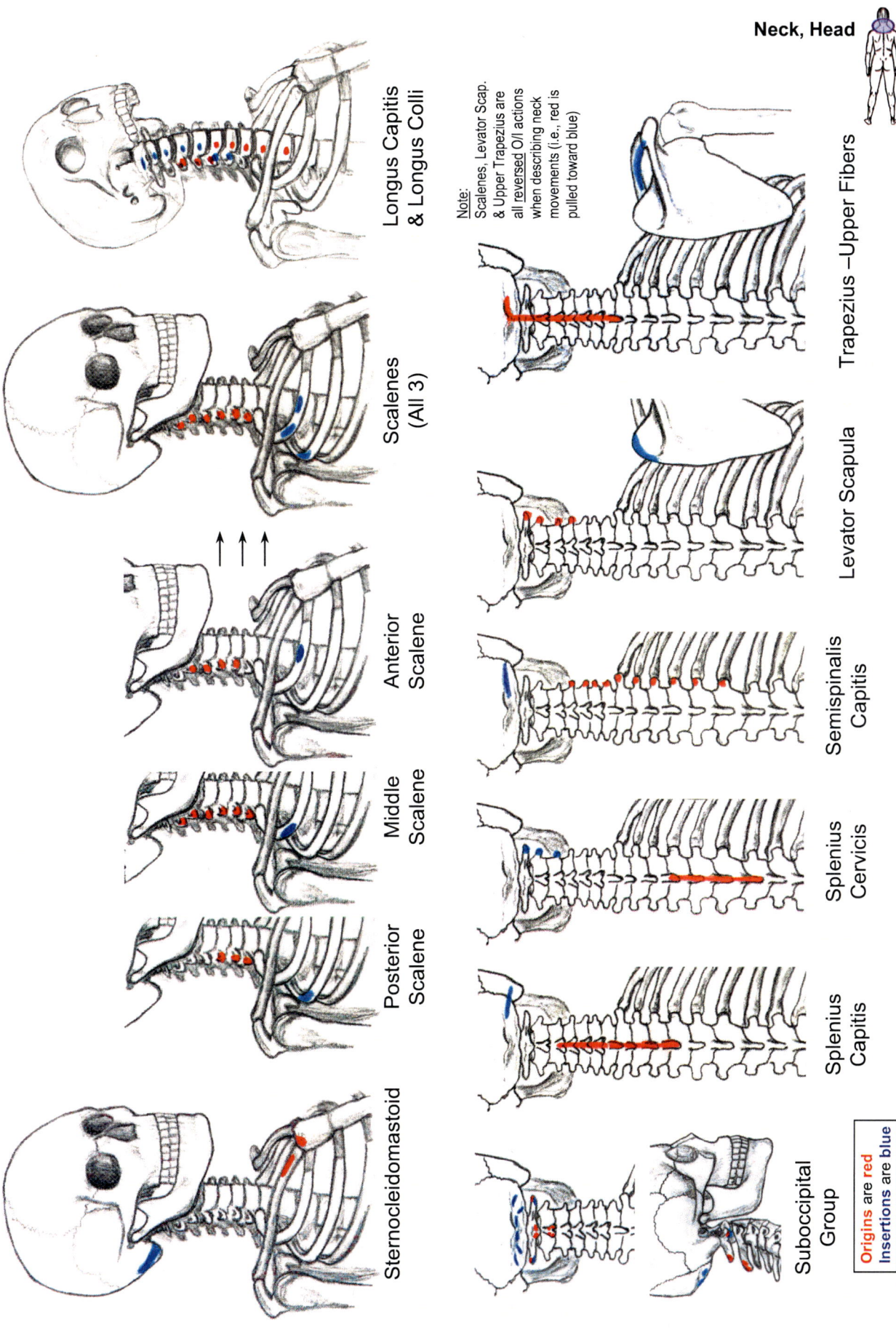

Figure 7 (A) - Neck, Head - Muscle Attachments

Neck, Head

UL=Unilateral action, BL=Bilateral action

Muscles Acting On: Neck, Head

Muscles Acting On	Flexion	Extension	Lateral Flexion	Rotation to same side	Rotation to opposite side	Inhalation	(other)	Innervation	Cr. XI	C1	C2	C3	C4	C5	C6	C7	C8
1. Sternocleidomastoid	BL head/neck		UL head/neck		UL head/neck	BL assist (lifts upper rib cage)	Helps stabilize head	Spinal Accessory N. (Cranial N. XI), and C2, C3	N		N	N					
2. Scalenes: Anterior Scalene / Middle Scalene / Posterior Scalene	BL neck (Ant. & Mid.)		UL neck (all)		UL assist (Ant. & Mid.)	BL (raises ribs 1 & 2)	Helps stabilize neck	Anterior rami of Ant. C5, C6 / Mid. C3-C8 / Post. C6-C8				N N	N N	N N N	N N N	N N N	N N
3. Longus Capitis	BL head/neck							Ventral rami of C1-C6		N	N	N	N	N	N		
4. Longus Colli	BL head/neck																
5. Suboccipitals Group — Rectus capitis posterior major / Rectus capitis posterior minor / Oblique capitis superior / Oblique capitis inferior								Suboccipital N. (C1)		N							
6. Splenius Capitis		BL head	UL head	UL head				Dorsal rami of middle cervicals			N	N	N	N	N		
7. Splenius Cervicis		BL neck	UL neck	UL head/neck				Dorsal rami of lower cervicals						N	N	N	
8. Semispinalis Capitis		BL head/neck		UL neck			This is the top portion of a spinal muscle (see Group 8)	Dorsal rami of C1-C5		N	N	N	N	N			
9. Levator Scapula (reversed O/I action, i.e., scapula held fixed)		BL neck	UL neck	UL neck			This muscle is also in Group 1 (moves scapula)	Dorsal scapular N. (C5), and C3, C4				N	N	N			
10. Trapezius, Upr fibers (reversed O/I action, i.e., scapula held fixed)		BL head/neck	UL head/neck		UL head/neck		This muscle is also in Group 1 (moves scapula)	Spinal Accessory N. (Cranial N. XI), and C3, C4	N			N	N				
(More muscles for the action) --->		See also Table 8	See also Table 8	See also Table 8		See also Table 9		Innervation									

Table 7 (B) - Neck, Head - Synergists & Antagonists

Chapter 5 – Muscles That Move the Axial Skeleton

Neck, Head

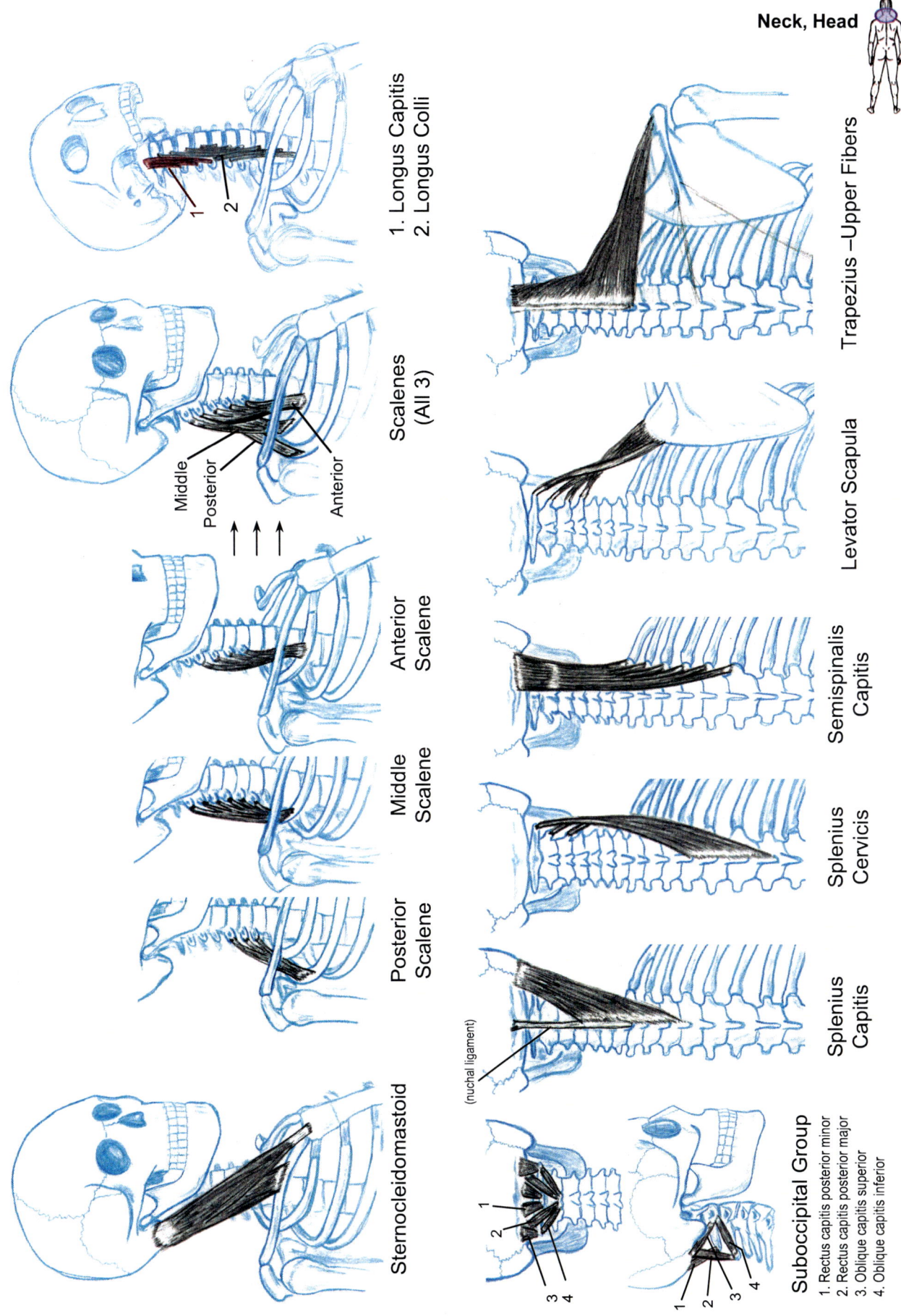

Figure 7 (B) - Neck, Head - Muscle Pictures

Chapter 5 - Muscles That Move the Axial Skeleton

Neck, Head

Muscles Acting On

Neck, Head ~ Notes ~ *(palpation, fiber arrangement, lengthen/shorten, common uses, pathologies, cautions, etc.)*

1. Sternocleidomastoid

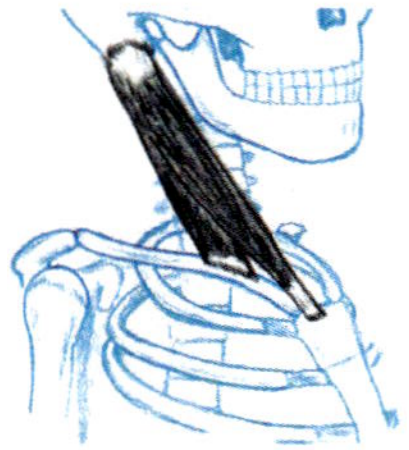

2. Scalenes
   - Middle
   - Posterior
   - Anterior

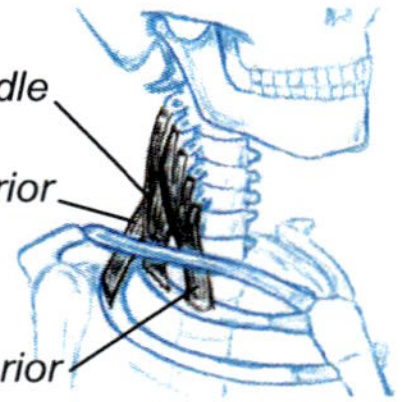

3. Longus Capitis
4. Longus Colli

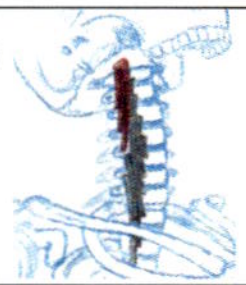

5. Suboccipitals Group

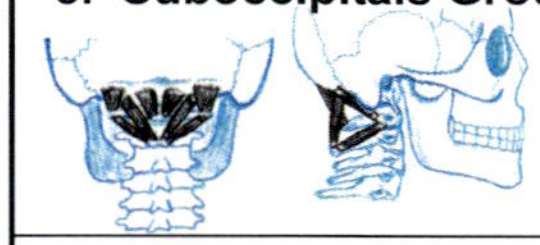

6. Splenius Capitis

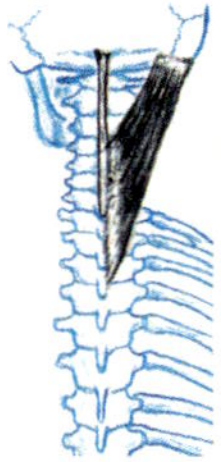

7. Splenius Cervicis

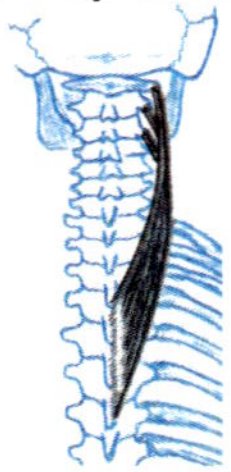

8. Semispinalis Capitis

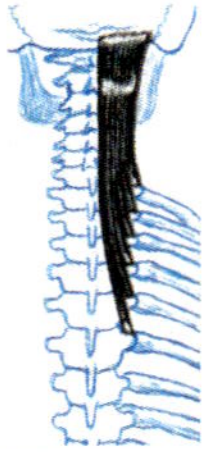

9. Levator Scapula — *Reversed O/I action (i.e., scapula held fixed)*

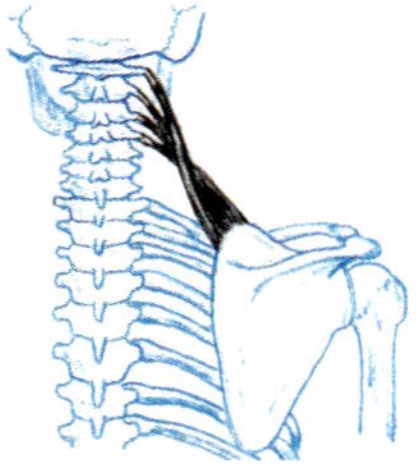

10. Trapezius, Upper fibers — *Reversed O/I action (i.e., scapula held fixed)*

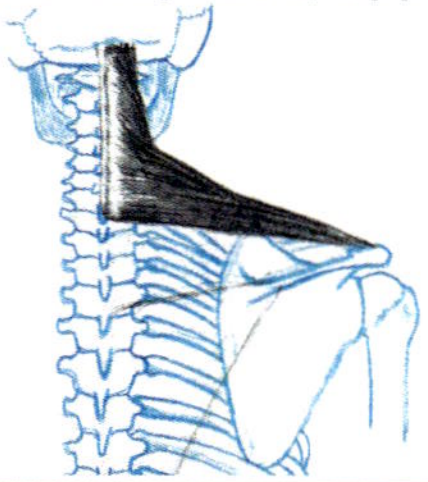

Muscle Group 7 - Neck, Head - (Notes)

128 Chapter 5 - Muscles That Move the Axial Skeleton

Movement of the Spine

Muscle Group 8:

Erector Spinae Group:	Transversospinal Group:
Spinalis	Semispinalis
Longissimus	Multifidis
Iliocostalis	Rotatores
	Other: Quadratus lumborum

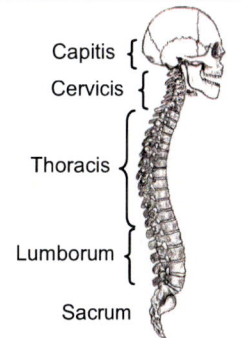

Joints

This group is generally called simply "the back muscles" or "the paraspinal muscles". The muscles are all located on the posterior axial skeleton, and move the axial body in extension, rotation, and lateral bending.

Occipitoatlantal Joint (OA)
Atlantoaxial Joint (C1/C2)
(above 2 joints: see Group 7 – Movement of the Neck, page 121)

Intervertebral Joints (have 2 components):
(please review page 111)

- **Intervertebral Discs (all vertebrae C2 through S1)**
    Between vertebral bodies
    Amphiarthrotic/Cartilaginous joints
    Movements: Compress, rotate, tilt in all directions

- **Intervertebral Facets (all vertebrae C1 through S1)**
    Pairs: Inferior facets of vertebra above ◄► Superior facets of vertebra below
    Gliding joints

(C2-S1 intervertebral joints allow all six actions shown in box)

Sacroiliac Joints
    Lateral **sacrum** ◄► Medial **Ilium**
    Part gliding, part fibrous/synarthrotic

Costovertebral & Costotransverse Joints
    Posterior **Ribs** ◄► Thoracic **Vertebrae**
    (Will be covered in the Thorax/Breathing section, page 137)

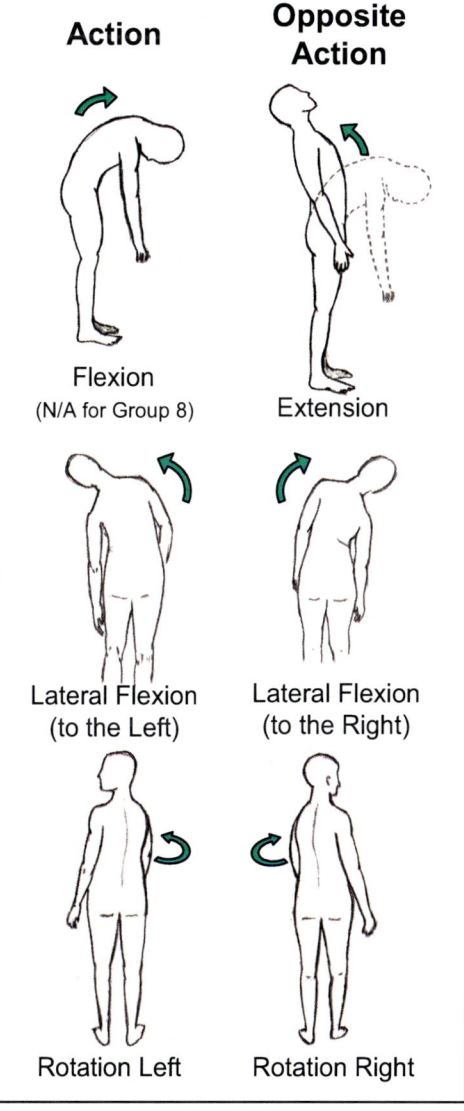

Movement

Movements of the spine are: Flexion, extension, right & left lateral flexion, right & left rotation

The cervical, thoracic, and lumbar sections of the spine have different amounts and types of mobility. These differences are due to the facet shapes & slants, lengths/slopes of spinous and transverse processes, disc thicknesses, ribs, ligaments, and other factors. The following table summarizes the structure and mobility of the spinal segments.

Spine

Summary of Vertebral Column by Section

Type	Vert.	Curvature	Structure of Vertebrae	Movement / Mobility
Cervical	7	Lordotic (secondary)	Small oval body, SPs short & horizontal, SPs bifid at tip, facets nearly horizontal (~30°)	Good mobility in flexion/extension & rotation, sidebending limited somewhat by TVPs and body shape, nuchal ligament resists over-flexion.
Thoracic	12	Kyphotic (primary)	Thin disc, long sloped SPs, facets nearly vertical facing front/back (in coronal plane, on a slight arc), ribs attach	Generally limited, some rotation is allowed, facets and sloped SPs limit extension, ribs and facets restrict sidebending
Lumbar	5	Lordotic (secondary)	Thick disc, short SPs, vertical facets facing side to side (in sagittal plane)	Mobile in flexion/extension and lateral flexion, rotation limited by facets
Sacrum	5 (fused)	Kyphotic (primary)	Fused	No movement within, but some mobility exists at L5/S1 joint and at sacroiliac joints
Coccyx	2-4 (fused)	Kyphotic (primary)	Fused	Joint between coccyx and sacrum allows movement, and can sublux with a fall on it

Comments

- T11/T12 is the first intervertebral joint above the lumbars that allows good rotation, so is stressed a lot.

- L5/S1 joint is curved forward to match into the tilt of the sacrum – so an anterior pelvic tilt is especially stressful on L5/S1 disc, and can cause L5 to slide forward on S1.

- Be careful to not confuse *spine* flexion with flexion at the *hip joint* (e.g., when bending to touch toes).

Interspinales and Intertransversarii

There are two groups of tiny paired muscles that interconnect vertebrae in the cervical and lumbar spine. They are not included in the Group 8 tables, but are illustrated here for completeness.

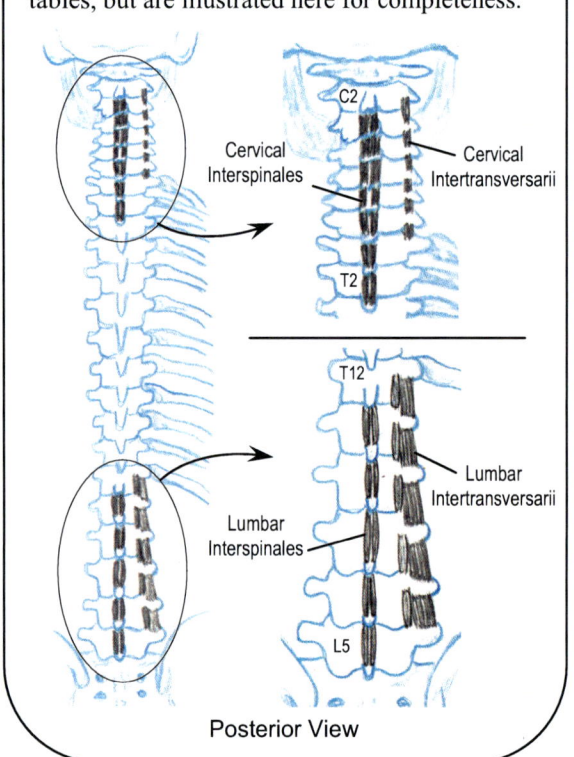

Posterior View

Bones and bony landmarks

Group 8 muscles move the spine, and have attachments on the posterior axial skeleton and the hip bone. Review the bony landmarks and other structures listed below, referring to Chapter 2, pages 43-47.

Occiput (p. 43)
Temporal Bone (p. 43) – Mastoid process

Spine (p. 44)
- Cervical, thoracic, and lumbar vertebrae
- Sacrum – L5/S1 joint (with disc and facets)
  – Articulation with hip bones (sacroiliac jt)

Ribs (p. 46)
- 12 ribs each side (7 true ribs, 5 false ribs)
- Bottom two false ribs are "floating ribs"
- *Costo* = Rib, *Chondro* = cartilage
- Costal angle (on posterior aspect of ribs)

Pelvis (p. 47)
- Coxal bones (hip bones)
    Made up of 3 fused bones:
    1. Ilium
        – Iliac crest (posterior medial portion)
    2. Ischium
    3. Pubis

Tendinous Structures
- Thoracolumbar Aponeurosis (TLA)
    Also called Thoracolumbar Fascia
    Also Lumbar Fascia, or Common Tendon
- Iliolumbar ligament

Spine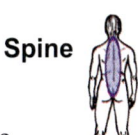

Muscle Group 8 – The back muscles that move the spine are illustrated as a group on this page. The next four pages have tables and figures that describe each muscle individually, and provide many ways of comparing and contrasting the muscles to each other.

Erector Spinae Group (ESG)
1 Layer: Side-by-Side

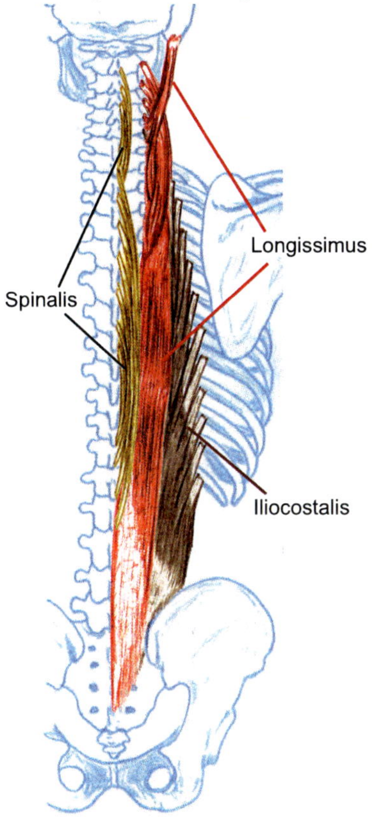

- Spinalis
- Longissimus
- Iliocostalis

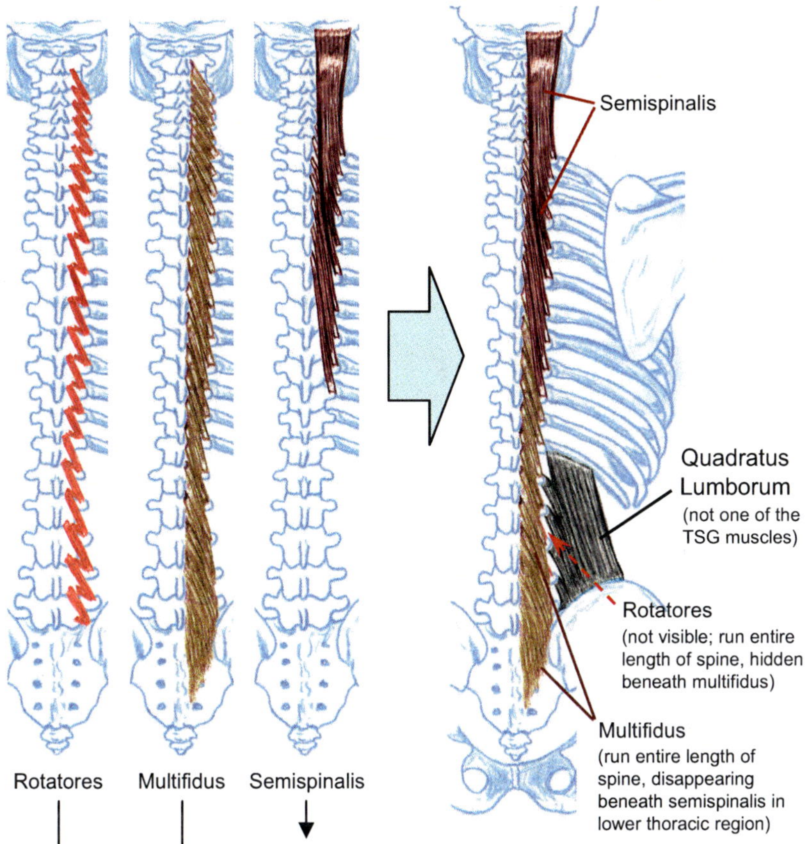

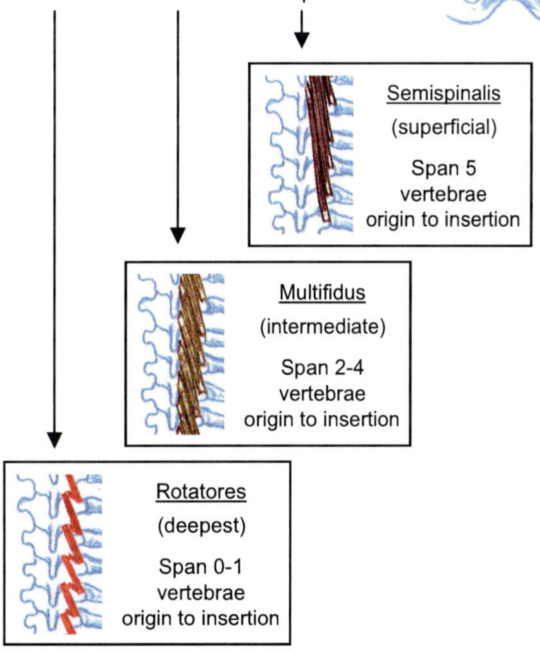

All Posterior Views

Origins are red
Insertions are blue

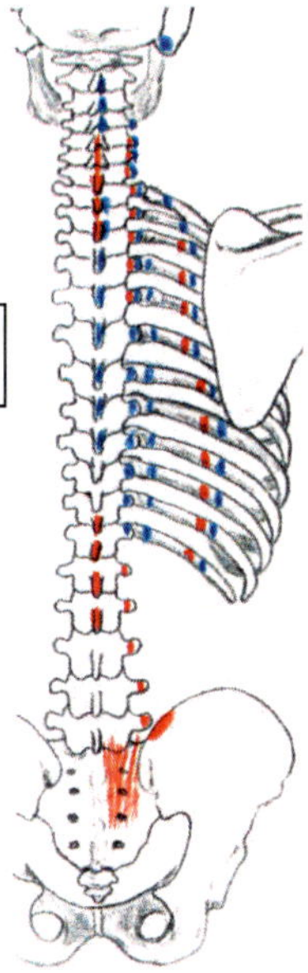

Attachment Sites for ESG

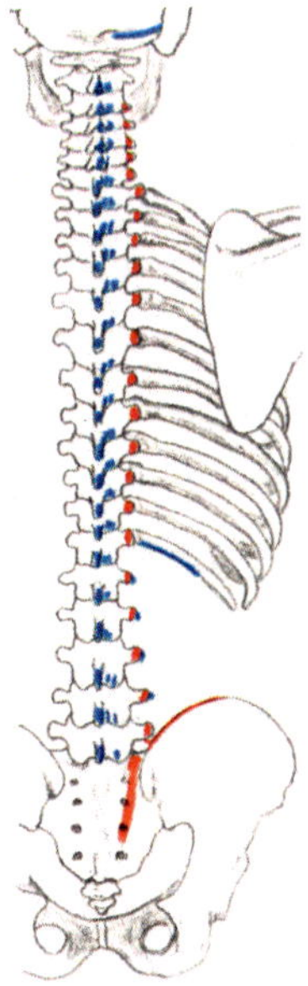

Transversospinal Group (TSG)
3 Layers

Rotatores | Multifidus | Semispinalis

- Semispinalis
- Quadratus Lumborum (not one of the TSG muscles)
- Rotatores (not visible; run entire length of spine, hidden beneath multifidus)
- Multifidus (run entire length of spine, disappearing beneath semispinalis in lower thoracic region)

Semispinalis (superficial)
Span 5 vertebrae origin to insertion

Multifidus (intermediate)
Span 2-4 vertebrae origin to insertion

Rotatores (deepest)
Span 0-1 vertebrae origin to insertion

Attachment Sites for TSG and Quadratus Lumborum

Mastering Muscles & Movement © 2007 — Chapter 5 – Muscles That Move the Axial Skeleton

Spine

TVP=Transverse process of vertebra, SP=Spinous process of vertebra, UL=Unilateral action, BL=Bilateral action

Muscles Acting On Spine — Origin, Insertion, Action

Spine	Origin	Insertion	Action
Erector Spinae Group – ESG			
Spinalis — Cervicis, Thoracis *(most medial of 3 columns of ESG)*	*Overall description:* SP's of upper lumbar & lower thoracic vertebrae, and SP's of upper thoracic & lower cervical vertebrae	*Overall description:* SP's of upper thoracic vertebrae, and SP's of upper cervical vertebrae	BL & UL: Extension of the spine
Longissimus — Capitis, Cervicis, Thoracis *(intermediate of 3 columns of ESG)*	*Overall description:* Lumbar fascia /common tendon, TVP's of all lumbar vertebrae, TVP's of upper thoracic & lower cervical vertebrae	*Overall description:* Ribs & TVP's of the thoracic and cervical vertebrae, and mastoid process of temporal bone	BL: Extension of the spine UL: Lateral flexion of the spine (Also, some fibers assist rotation of spine to the same side)
Iliocostalis — Cervicis, Thoracis, Lumborum *(most lateral of 3 columns of ESG)*	*Overall description:* Lumbar fascia /common tendon, and the posterior ribs #3 – 12 (at the angles of the ribs)	*Overall description:* Posterior aspect of all the ribs, and TVP's of the lower cervical vertebrae (at the angles of the ribs)	BL: Extension of the spine UL: Lateral flexion of the spine (Also, some fibers assist rotation of spine to the same side)
Transversospinal Group – TSG			
Semispinalis — Capitis, Cervicis, Thoracis *(most superficial of 3 layers of TSG)*	TVP's of thoracic vertebrae T1-T10 and cervical vertebrae C4-C7	SP's of vertebrae above – each spans 5 vertebrae, *and* to the occipital bone (capitis) (occiput attachment is between superior and inferior nuchal lines)	BL: Extension of the spine (Cervicis,Thor.) Extension of the head (Capitis) UL: Rotation of spine to opposite side (Cervicis and Thoracis)
Multifidus *(intermediate of 3 layers of TSG)*	Sacrum and TVP's of all vertebrae of the spine (except C1-C3)	SP's of vertebrae above – each spans 2 - 4 vertebrae above before inserting	BL: Extension of the spine UL: Rotation of spine to opposite side All: Stabilize vertebrae on each other
Rotatores (Longus and Brevis) *(deepest of 3 layers of TSG)*	Sacrum and TVP's of all vertebrae of the spine (except C1-C2)	Lamina of vertebrae above – Brevis: Inserts next vertebra up before inserting Longus: Spans 1 vertebra before inserting	BL: Extension of the spine UL: Rotation of spine to opposite side All: Stabilize vertebrae on each other
Other –			
Quadratus lumborum — moves the spine or the hip, or stabilizes rib #12	Iliac crest, iliolumbar ligament	Rib #12, TVP's of L1 - L4	BL: Extension of the lumbar spine UL: Lateral flexion of the lumbar spine *If the spine is held in place:* UL: Raises the ilium (hip hike) BL: Creates anterior pelvic tilt (*Also:* Stabilizes rib #12 during inhalation)

Table 8 (A) - Spine - Origin, Insertion, Action

(larger illustrations on page 135)

Chapter 5 – Muscles That Move the Axial Skeleton

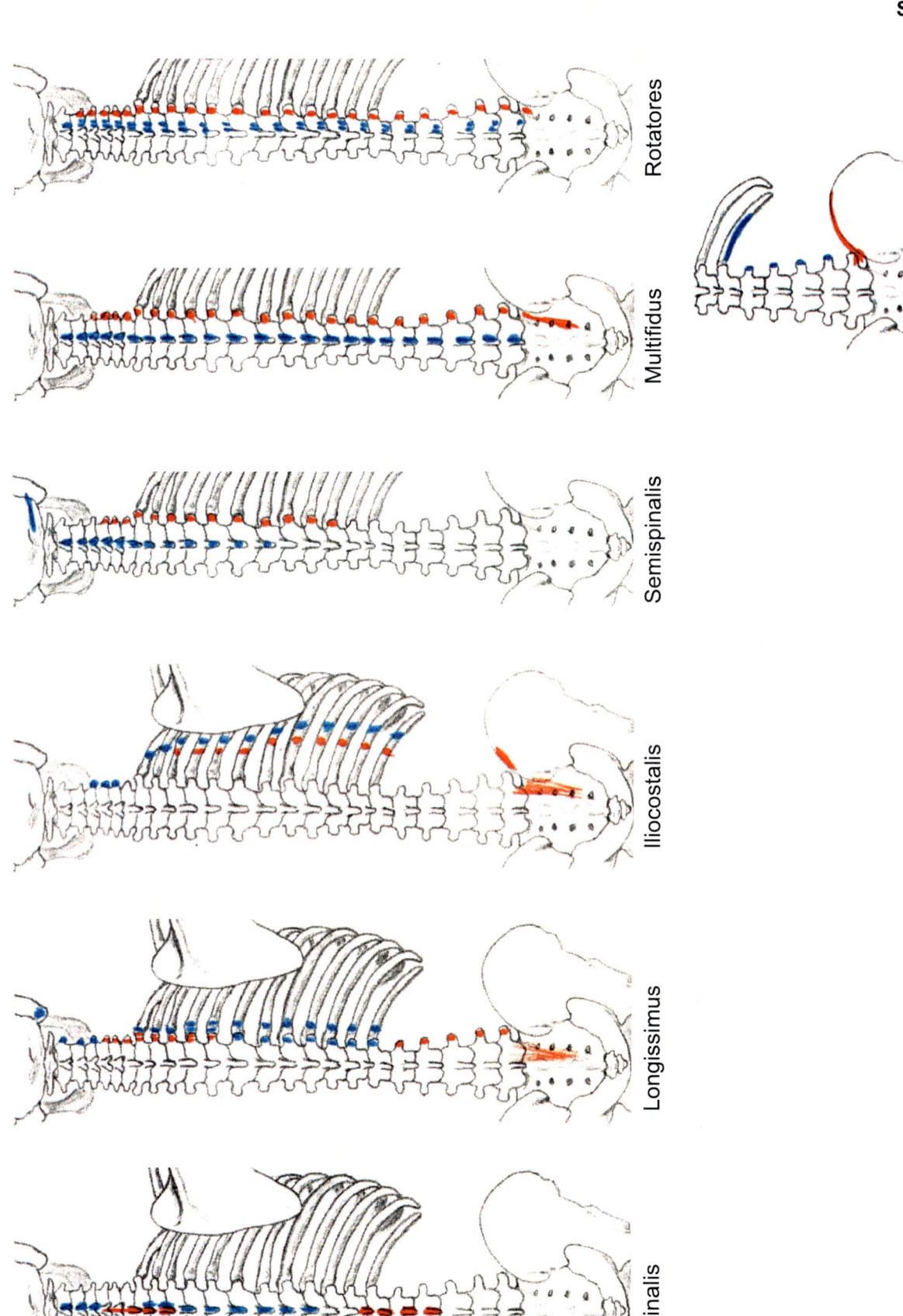

Figure 8 (A) - Spine - Muscle Attachments

Spine

UL=Unilateral action, BL=Bilateral action, SP=Spinous process of vertebra

Muscles Acting On Spine	Flexion	Extension	Lateral Flexion	Rotation	Other	(Notes)	Innervation	T12	L1	L2	L3	(all spinal segments)
(1-3) --- ERECTOR SPINAE GROUP --- ESG :												
1. Spinalis cervicis " thoracis		X				Most medial column of ESG. (goes SP to SP)	Dorsal primary divisions of spinal nerves					N
2. Longissimus capitis " cervicis " thoracis		BL	UL	UL (may assist) to same side	Upper fibers move head/neck, lower fibers move trunk	Intermediate column of ESG.	Dorsal primary divisions of spinal nerves					N
3. Iliocostalis cervicis " thoracis " lumborum		BL	UL	UL (may assist) to same side		Most lateral column of ESG.	Dorsal primary divisions of spinal nerves					N
(4-6) --- TRANSVERSOSPINAL GROUP --- TSG: *(deep to ESG)*												
4. Semispinalis (capitis, cervicis, thoracis)		BL		UL to opposite side	A major function of TSG group is to stabilize the spine, one vertebra on another.	Most superficial of TSG group	Dorsal primary divisions of spinal nerves					N
5. Multifidus		BL		UL to opposite side		Next deepest	Dorsal primary divisions of spinal nerves					N
6. Rotatores (longus and brevis)		BL		UL to opposite side		Deepest	Dorsal primary divisions of spinal nerves					N
OTHER :												
7. Quadratus Lumborum		BL (lumbar)	UL (lumbar)		Reverse O/I: UL: "hip hike" BL: anter. pelvic tilt	Used in breathing: Stabilizes rib #12 during inhalation	Lumbar plexus (T12, L1-L3)	N	N	N	N	
(More muscles for the action) --->			See also Table 9	See also Table 9			**Innervation**					

Table 8 (B) - Spine - Synergists & Antagonists

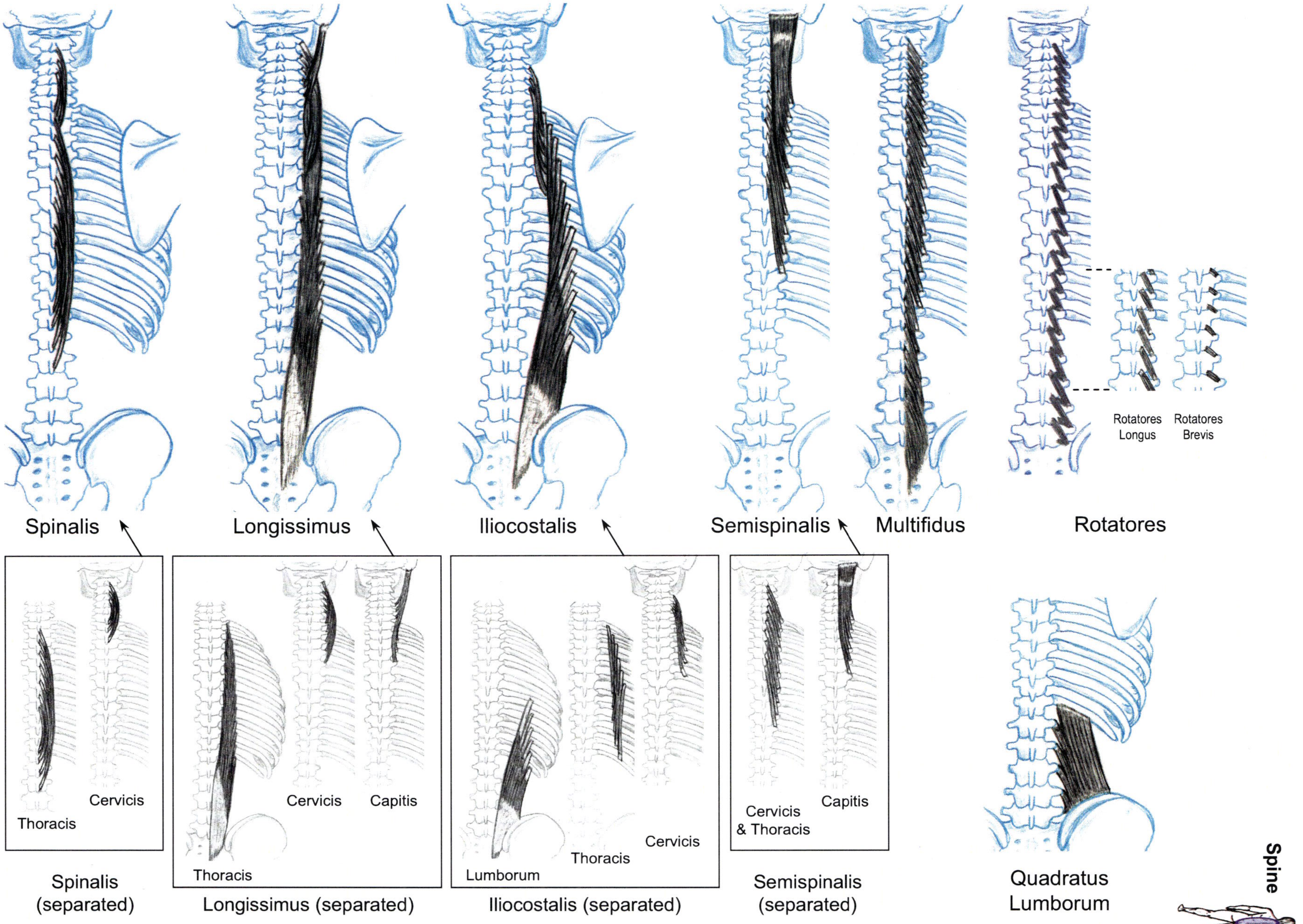

Spine

Figure 8 (B) - Spine - Muscle Pictures

Rotatores
- Rotatores Brevis
- Rotatores Longus

Quadratus Lumborum

Multifidus

Semispinalis

Semispinalis (separated)
- Capitis
- Cervicis & Thoracis

Iliocostalis

Iliocostalis (separated)
- Cervicis
- Thoracis
- Lumborum

Longissimus

Longissimus (separated)
- Capitis
- Cervicis
- Thoracis

Spinalis

Spinalis (separated)
- Cervicis
- Thoracis

Mastering Muscles & Movement © 2007 — Chapter 5 - Muscles That Move the Axial Skeleton

Muscles Acting On

Spine — ~ Notes ~
(palpation, fiber arrangement, lengthen/shorten, common uses, pathologies, cautions, etc.)

1. Spinalis
  - Cervicis
  - Thoracis

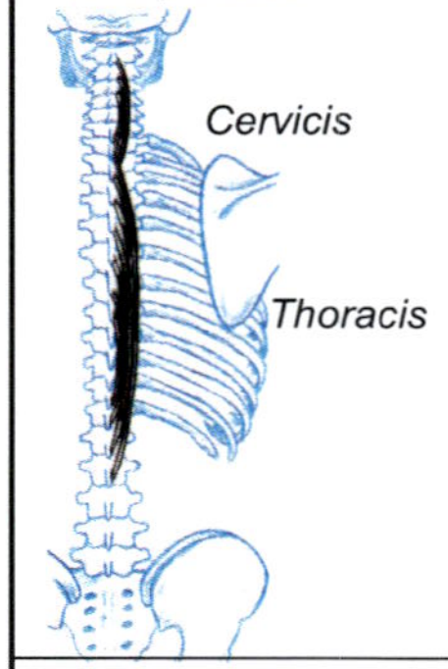

2. Longissimus — Capitis, Cervicis, Thoracis
  - Capitis
  - Cervicis
  - Thoracis

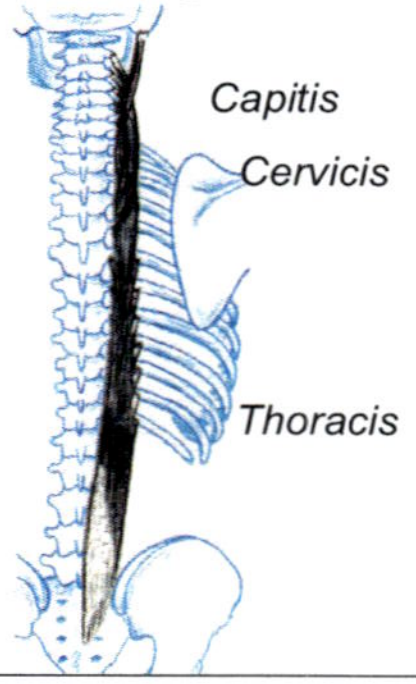

3. Iliocostalis
  - Cervicis
  - Thoracis
  - Lumborum

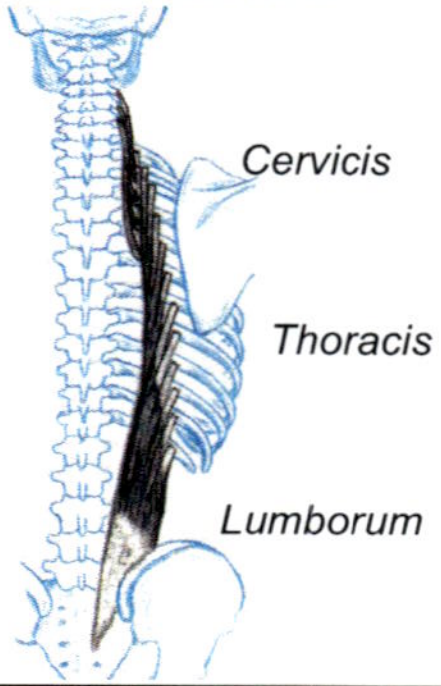

4. Semispinalis
  - Capitis
  - Cervicis
  - Thoracis

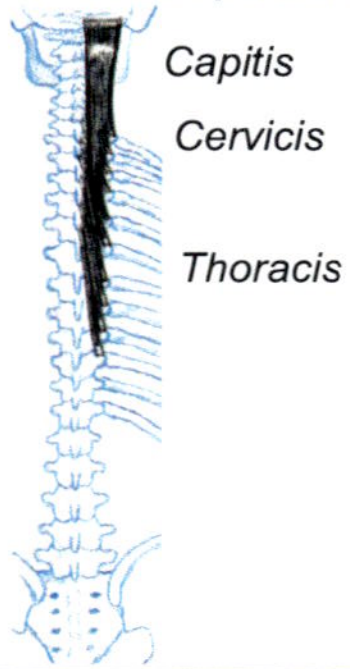

5. Multifidus

6. Rotatores
  - Longus
  - Brevis

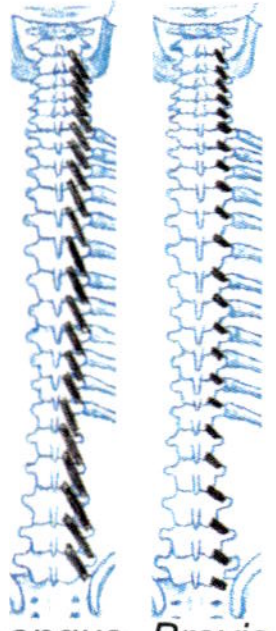

7. Quadratus Lumborum

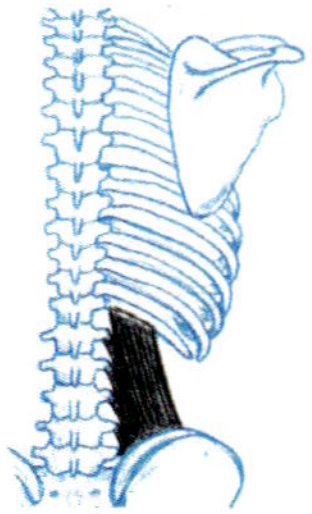

Comparison of Transversospinal Muscles
- Semispinalis 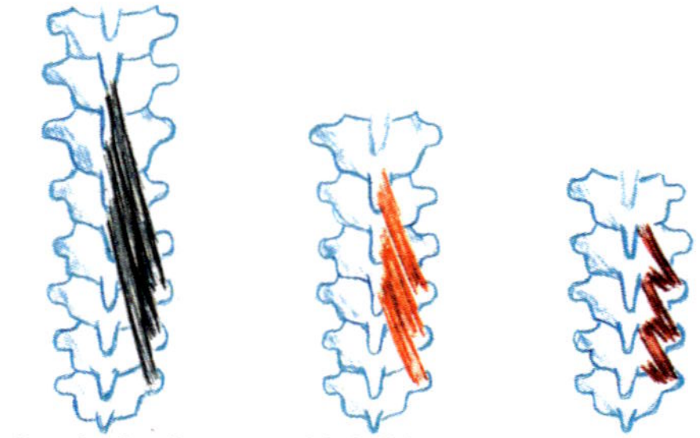
- Multifidus
- Rotatores Longus & Brevis

Muscle Group 8 - Spine - (Notes)

136   Chapter 5 - Muscles That Move the Axial Skeleton   Mastering Muscles & Movement © 2007

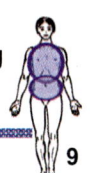

Thorax, Abdomen, Breathing

Movement of the Thorax, Abdomen, Breathing

Muscle Group 9:

Abdominal Group:	Other muscles:
Rectus abdominis	Diaphragm
External oblique	External intercostals
Internal oblique	Internal intercostals
Transverse abdominis	Serratus posterior superior
	Serratus posterior inferior
	Levator costae

Joints

The muscles in this group move the trunk of the body. This includes flexing and twisting the upper body, compressing the abdomen, and moving the ribs up and down while breathing.

Intervertebral Joints (Discs + Facets)
(see previous section: Group 8 – Movement of the Spine, page 129)

Rib Joints – Anterior: (for rib joints, please review page 46)

- **Sternocostal Joints (J1)**   (also called sterno*chondral* joints)

    Sternum ◀▶ Costal cartilages of **ribs** (Ribs 1 to 7)
    1: Fibrous Joint
    2-7: Gliding Joints

- **Costochondral Joints (J2)**

    Ribs ◀▶ Costal cartilages
    (Ribs 1 to 10: Fibrous junctions of bone to cartilage)

Rib Joints – Posterior:

- **Costovertebral Joints (J3)**
    Head of **rib ◀▶** Costal facets on **vertebral bodies**
    Gliding Joints

- **Costotransverse Joints (J4)**
    Tubercle of **rib ◀▶** Costal facet on **TVP of vertebra**
    Gliding Joints

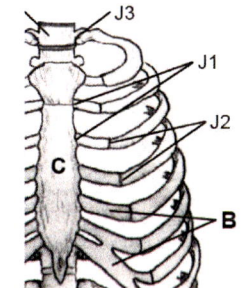

Movements available

Spine/Trunk:
    Flexion, Extension, R&L Lateral Flexion, R&L Rotation

Ribs – Elevation, Depression

Abdomen – Compression of abdominal contents

Lungs – Breathing (see next page)

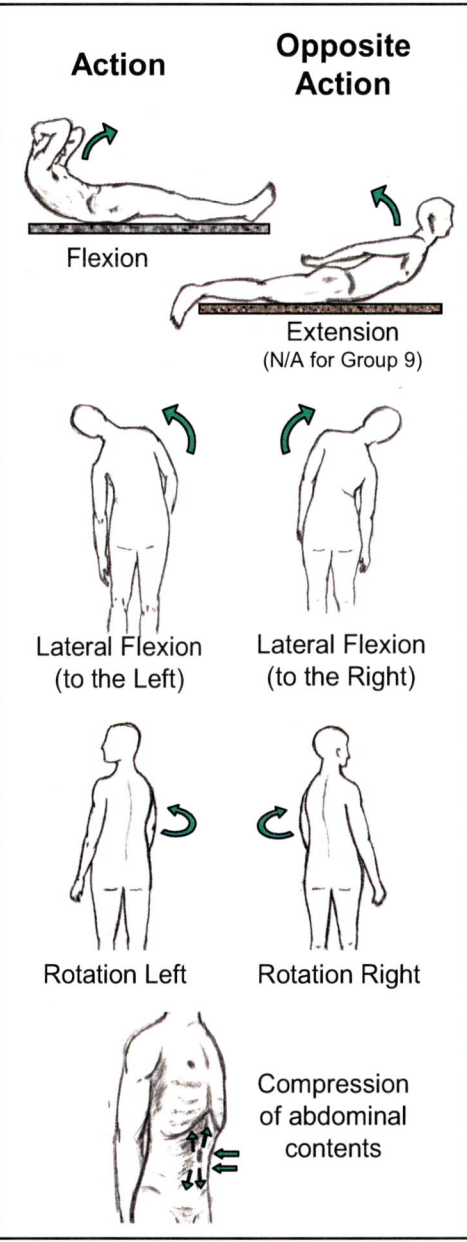

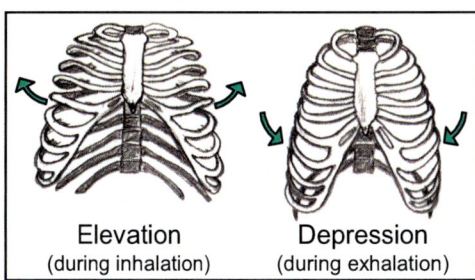

Elevation (during inhalation)   Depression (during exhalation)

Chapter 5 – Muscles That Move the Axial Skeleton

Thorax, Abdomen, Breathing

Respiration

Inhalation Phase (also called inspiration)

Quiet Inhalation – The diaphragm muscle contracts, pulling its central tendon downward. This spreads and elevates the lower ribs and pulls the floor of the thoracic cavity downward, causing air to suck into the lungs. The abdominal contents are pushed down, causing the belly to protrude.

Deep and Forced Inhalation – Rib-lifter muscles kick in as needed to draw more air into the lungs. A common recruitment order is: Scalenes, external intercostals, levator costae, sternocleidomastoid, pectoralis minor, serratus posterior superior, pectoralis major (esp. with arms raised or braced on hips). Actual order of muscle recruitment varies greatly depending on habit, emotion, posture, etc.

Note: The ribcage expands in all directions as the ribs elevate:
- Lateral – Ribs swing out and up (bucket handles) – The most movement
- Anterior – Sternum raises and swings out (pump handle) – Medium movement
- Posterior – Back expands as angles of ribs rise – Slight movement

Exhalation Phase (also called expiration)

Quiet Exhalation – A passive process: relaxation of the previously active diaphragm. Air is expelled due to elastic recoil of lungs and ribs being allowed back down and in.

Forced Exhalation – Muscles are recruited to compress the abdomen (abdominal muscles) and pull depression of the ribs (internal intercostals, serratus posterior inferior, quadratus lumborum, iliocostalis). Forced exhalation muscles are also used for speech, singing, coughing, sneezing, vomiting, defecation.

Bones and bony landmarks

Muscles that move the thorax and abdomen have attachments on the spine, ribs, pelvis, sternum, and fascia. Review the bony landmarks and other structures listed below, referring to Chapter 2, pages 44-47.

Spine (p. 44)
    Thoracic vertebrae
    Lumbar vertebrae

Pelvis (p. 47)

    Landmarks on ilium & pubis:
        Iliac crest
        Anterior Superior Iliac Spine (ASIS)
        Pubic crest
        Pubic symphysis
        Pubic tubercle

Sternum (p. 46)
    Xiphoid process

Ribs (p. 46)
    12 ribs each side (7 true ribs, 5 false ribs)
    Bottom two false ribs are "floating ribs"

    Articulation with sternum in front
    Articulation with thoracic vertebrae in back
    Costal cartilage
    Costal margin / lower thoracic "aperture"
    Costal angle (posterior angle of ribs)

Tendinous and Other Structures
    Lumbar fascia (thoracolumbar aponeurosis)
    Abdominal aponeurosis
    Rectus sheath
    Linea alba
    Inguinal ligament
        - Spans from ASIS to the pubic tubercle
    Thoracic, abdominal and pelvic organs

Thorax, Abdomen, Breathing

Muscle Group 9 – Muscles that move the thorax and abdomen and facilitate breathing are illustrated as a group on this page. The next four pages have tables and figures that describe each muscle individually, and provide many ways of comparing and contrasting the muscles to each other.

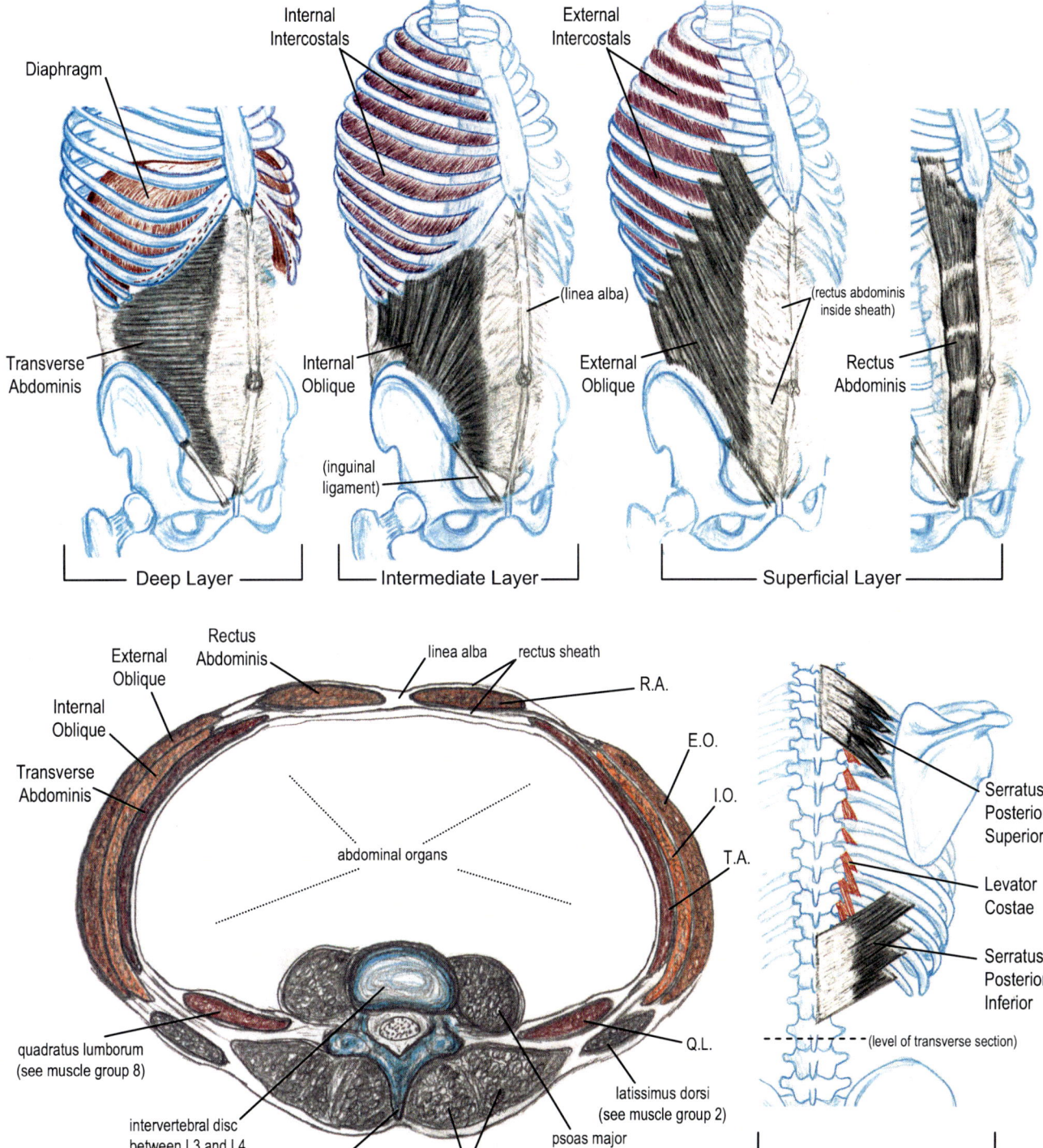

Transverse Section at the Level of 3rd Lumbar Vertebra (L3)

Chapter 5 – Muscles That Move the Axial Skeleton

Thorax, Abdomen, Breathing

SP=Spinous process of vertebra, TVP=Transverse process of vertebra, UL=Unilateral action, BL=Bilateral action

Muscles Acting On Thorax, Abdomen, Breathing	Origin	Insertion	Action
Rectus Abdominis moves the trunk	Pubic crest and symphysis pubis	Costal cartilages of ribs #5-7, Xiphoid process of sternum	Flexion of the spine, Assists compression of abdominal contents *When spine is stabilized*: Pulls public bone up (posterior pelvic tilt)
External Oblique moves the trunk	Lower 8 ribs, lateral aspect. *(Interdigitates with serratus anterior)*	Anterior iliac crest, inguinal ligament, abdominal aponeurosis to the linea alba	BL: Flexion of the spine, Compression of abdominal contents UL: Lateral flexion of the spine, Rotation to the *opposite* side
Internal Oblique moves the trunk	Lumbar fascia, iliac crest, Lateral inguinal ligament	Last 3 or 4 ribs (lower margins), Abdominal aponeurosis to the linea alba	BL: Flexion of the spine, Compression of abdominal contents UL: Lateral flexion of the spine, Rotation to the *same* side
Transverse Abdominis compresses the abdomen	Lower 6 ribs (interior surface), Lumbar fascia, iliac crest, Lateral inguinal ligament	Abdominal aponeurosis to the linea alba	Compression of abdominal contents
Diaphragm muscle of respiration	Bodies of upper lumbar vertebrae, interior surface of lower ribs & xiphoid process	Central tendon of the diaphragm *(The diaphragm is shaped like a dome or umbrella. The central tendon is a large clover-shaped aponeurosis at the top of the dome that is not attached to any bones.)*	Draws down the central tendon, creating inhalation. (The movement increases the volume inside the thoracic cavity. It also expands the *lower* ribs in all directions, and compresses down on the abdominal contents.)
External Intercostals move the ribs	Inferior border of rib above *Between all ribs.*	Superior border of rib below *(angle of fibers follows the external oblique muscle fibers)*	Draw ribs upward during inhalation (Assists rotation of thoracic spine to opposite side)
Internal Intercostals move the ribs	Superior border of rib below *Between all ribs.*	Inferior border of rib above *(angle of fibers follows the internal oblique muscle fibers)*	Draw ribs downward during exhalation (Assists rotation of thoracic spine to same side)
Serratus Posterior Superior moves the ribs	SP's of lower cervical & upper thoracic vertebrae (C6 - T2)	Posterior upper ribs (ribs #2 - 5)	Draws ribs upward during inhalation
Serratus Posterior Inferior moves or stabilizes the ribs	SP's of lower thoracic & upper lumbar vertebrae (T11 - L2)	Posterior lower ribs (ribs #9 - 12)	Draws ribs downward and backward. Stabilizes lower ribs against the upward pull of the diaphragm.
Levator Costae move the ribs	TVPs of vertebrae C7 and T1 - T11	Posterior ribs 1 - 12 (Brevis: Attach to rib below (ribs 1-12) Longus: Second slip of muscle attaches to 2nd rib below, ribs 10-12 only)	Draw ribs upward during inhalation

Table 9 (A) - Thorax, Abdomen, Breathing - Origin, Insertion, Action

(larger illustrations on page 143)

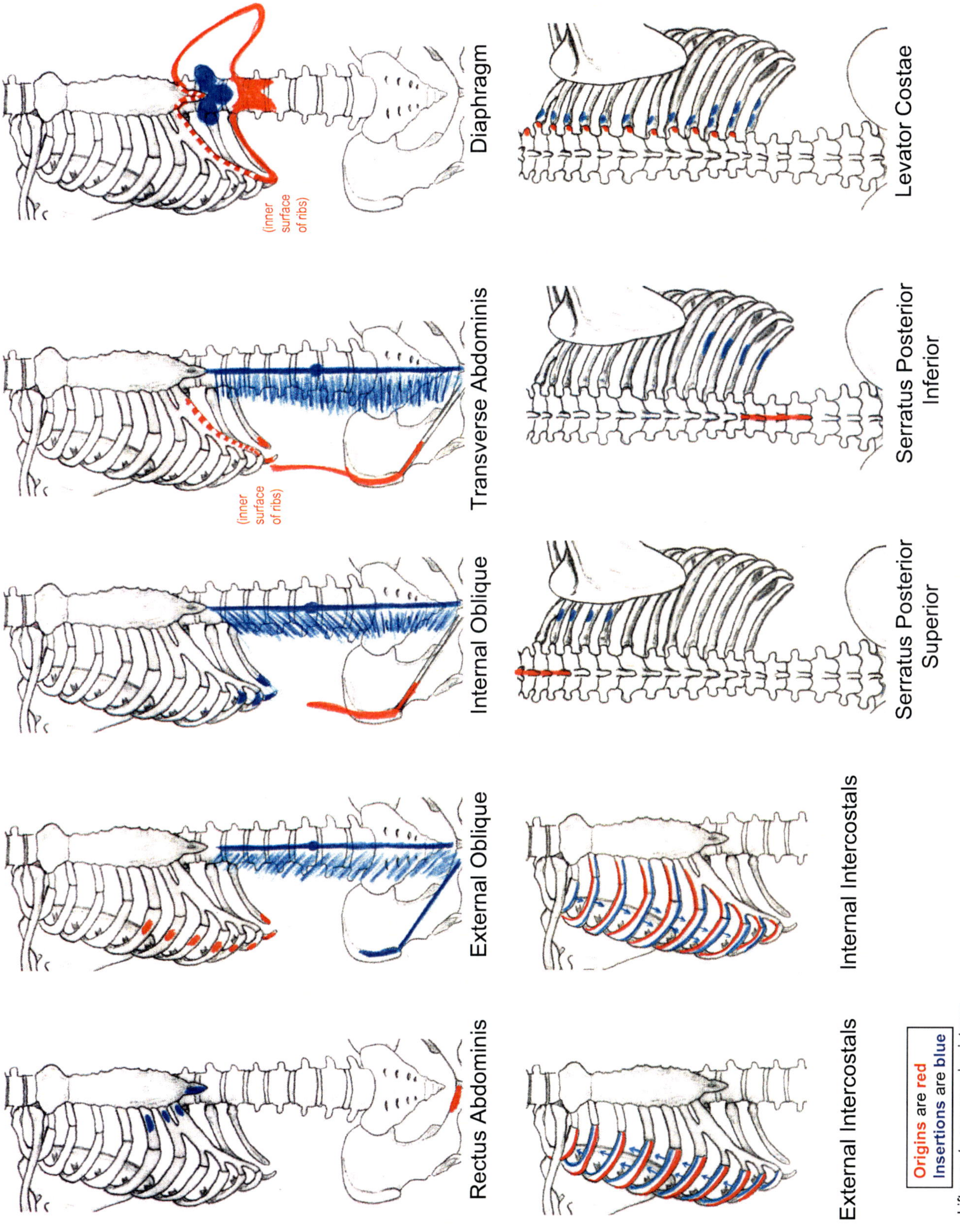

Thorax, Abdomen, Breathing

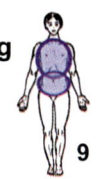

Figure 9 (A) - Thorax, Abdomen, Breathing - Muscle Attachments

Origins are red
Insertions are blue

Lift page to see muscle pictures

Chapter 5 - Muscles That Move the Axial Skeleton

Thorax, Abdomen, Breathing

UL=Unilateral action, BL=Bilateral action

Muscles Acting On Thorax, Abdomen, Breathing	Flexion	Extension (none)	Lateral Flexion	Rotation	Compress Abdominal Contents	Inhalation/ Inspiration	Exhalation/ Expiration	Other	Innervation	C3 to C5	T1 to T4	T5	T6	T7	T8	T9	T10	T11	T12	L1
1. Rectus Abdominis	X				assist		X		Intercostal nerves (T6-T12)				N	N	N	N	N	N	N	
2. External Oblique	BL		UL	UL to opposite side	X		X	Interdigitates w/ Serratus Anterior & Lat. Dorsi	Intercostal (T8-12)						N	N	N	N	N	
3. Internal Oblique	BL		UL	UL to same side	X		X		Intercostal (T8-12) iliohypogastric N. (T12,L1) & ilioinguinal N. (L1)								N	N	N	N
4. Transverse Abdominis					X		X	Interdigitates w/ Diaphragm	Intercostal (T7-12) iliohypogastric N. (T12,L1) & ilioinguinal N. (L1)					N	N	N	N	N	N	N
5. Diaphragm					(X) ↓ (downward, pushes out abdominals)	X	X forced exhalation, cough, etc.	Draws central tendon down, and expands lower ribs	Phrenic nerve (C3, C4, C5)	N										
6. External Intercostals						X Draw ribs upward			Intercostal nerves (segmental)			N	N	N	N	N	N	N	N	
7. Internal Intercostals							X Draw ribs downward		Intercostal nerves (segmental)			N	N	N	N	N	N	N	N	
8. Serratus Post. Superior						X Raises ribs			Anterior primary rami of spinal nerves (T1-T4)		N									
9. Serratus Post. Inferior							X Draws ribs down	Counters pull of diaphragm	Anterior primary rami of spinal nerves (T9-T12)							N	N	N	N	
10. Levator Costae						X			Intercostal nerves (segmental)			N	N	N	N	N	N	N	N	
11. Other "Rib Lifters": Scalenes, Pectoralis Minor, Sternocleidomastoid, Pec Major (More muscles for the action) --->	see Table 8	see also Table 8	see also Table 8	see also Table 8				See Tables 1, 2, & 7 for muscle details												

Table 9 (B) - Thorax, Abdomen, Breathing - Synergists & Antagonists

Chapter 5 – Muscles That Move the Axial Skeleton

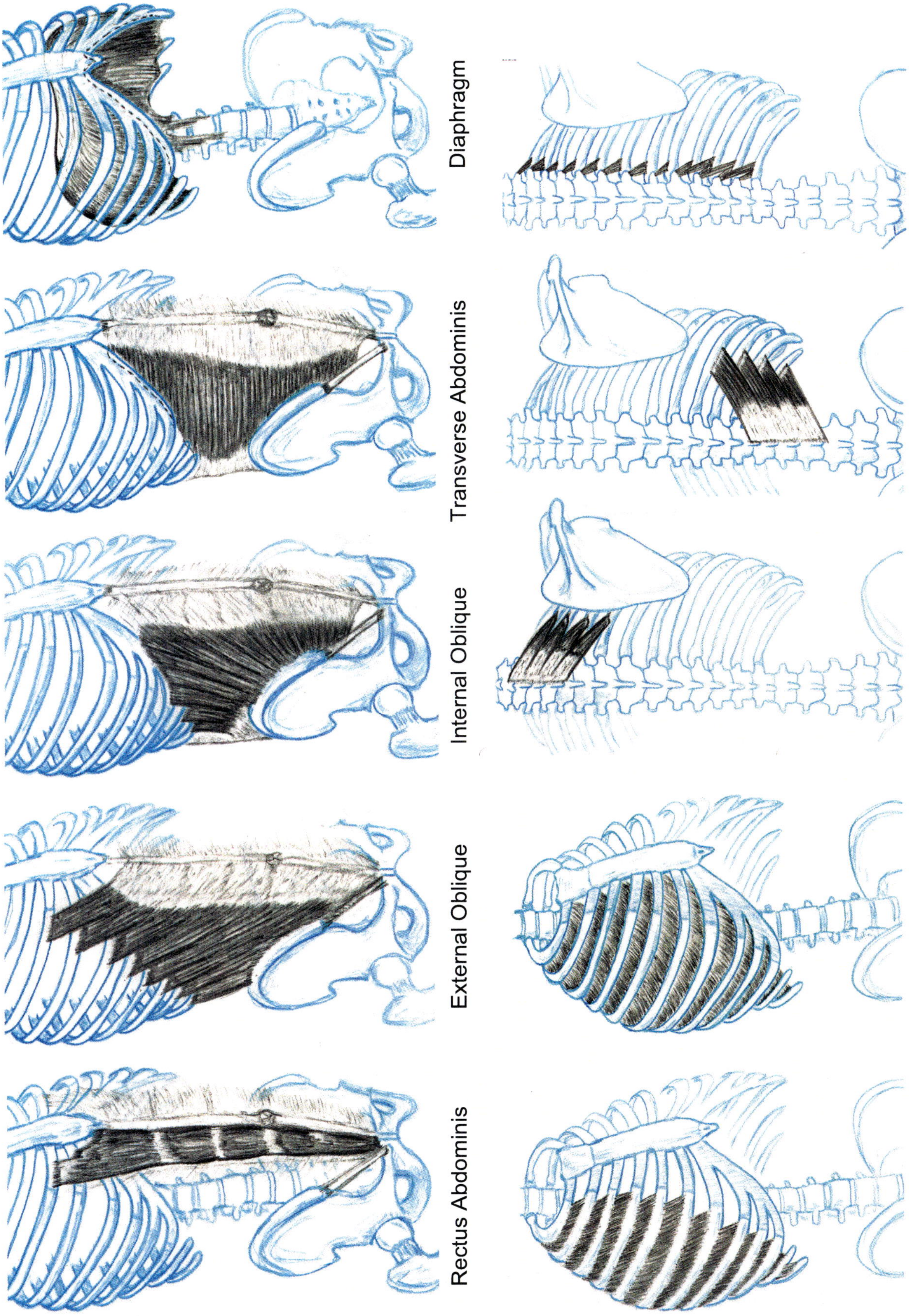

Thorax, Abdomen, Breathing

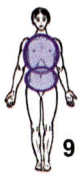

Figure 9 (B) - Thorax, Abdomen, Breathing - Muscle Pictures

Chapter 5 - Muscles That Move the Axial Skeleton 143

Muscles Acting On
Thorax, Abdomen, Breathing

(palpation, fiber arrangement, lengthen/shorten, common uses, pathologies, cautions, etc.)

1. Rectus Abdominis	4. Transverse Abdominis	7. Internal Intercostals
2. External Oblique	5. Diaphragm	8. Serratus Posterior Superior 9. Serratus Posterior Inferior
3. Internal Oblique	6. External Intercostals	10. Levator Costae

Muscle Group 9 - Thorax, Abdomen, Breathing - (Notes)

144   Chapter 5 - Muscles That Move the Axial Skeleton

(Pages 145 - 150 reserved for expansion of Chapter 5)

Chapter 6
Muscles That Move the Lower Extremity

Introduction 152
Movement of the Hip Joint (Part 1) .... (Muscle Group 10) 153
Movement of the Hip Joint (Part 2) .... (Muscle Group 11) 161
Movement of the Knee (& Hip Joint, Part 3) .... (Muscle Group 12) 169
Movement of the Ankle, Foot, and Toes .... (Muscle Group 13) 177
Intrinsic Muscles of the Foot.. 185
Muscles of the Leg – by Compartment 186

Group 10 – Hip Joint (Part 1)

p. 153

Gluteus maximus
Gluteus medius
Gluteus minimus
Piriformis
The other 5 lateral rotators
    (GOGOQ)
Iliopsoas
    (Iliacus & Psoas major)

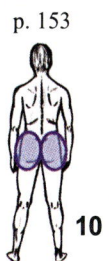

Group 11 – Hip Joint (Part 2)

p. 161

Sartorius
Tensor fascia latae
Pectineus
Adductor brevis
Adductor longus
Adductor magnus
Gracilis

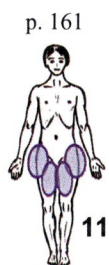

Group 12 – Knee (& Hip Joint, Part 3)

p. 169

Rectus femoris
Vastus lateralis
Vastus intermedius
Vastus medialis
Biceps femoris
Semitendinosus
Semimembranosus
Popliteus

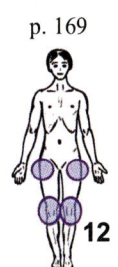

Group 13 – Ankle, Foot, Toes

p. 177

Gastrocnemius
Plantaris
Soleus
Tibialis posterior
Flexor digitorum longus
Flexor hallucis longus
Peroneus longus
Peroneus brevis
Tibialis anterior
Extensor digitorum longus
Extensor hallucis longus

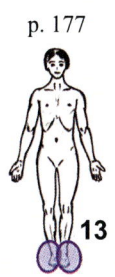

Introduction

The **lower extremity** is the lower-body portion of the appendicular skeleton (see page 34), and includes the hip bone, thigh, leg, and foot. The sacroiliac joints, where the hip bones articulate with the sacrum, are the joints connecting the lower extremities to the trunk.

This chapter describes the muscles that move the various joints within the lower extremity. The muscles are separated into four functional groups, with some overlap of function between groups for muscles that cross multiple joints:

Group 10 – Movement of the hip – part 1, which are the "shorter" muscles that move the femur at the hip joint
Group 11 – Movement of the hip – part 2, which are the "longer" muscles that move the femur at the hip joint
Group 12 – Movement of the knee (which includes multiple-joint muscles that also move the femur at the hip)
Group 13 – Movement of the ankle, foot, and toes

At the end of the chapter, additional illustrations present the intrinsic muscles of the foot.

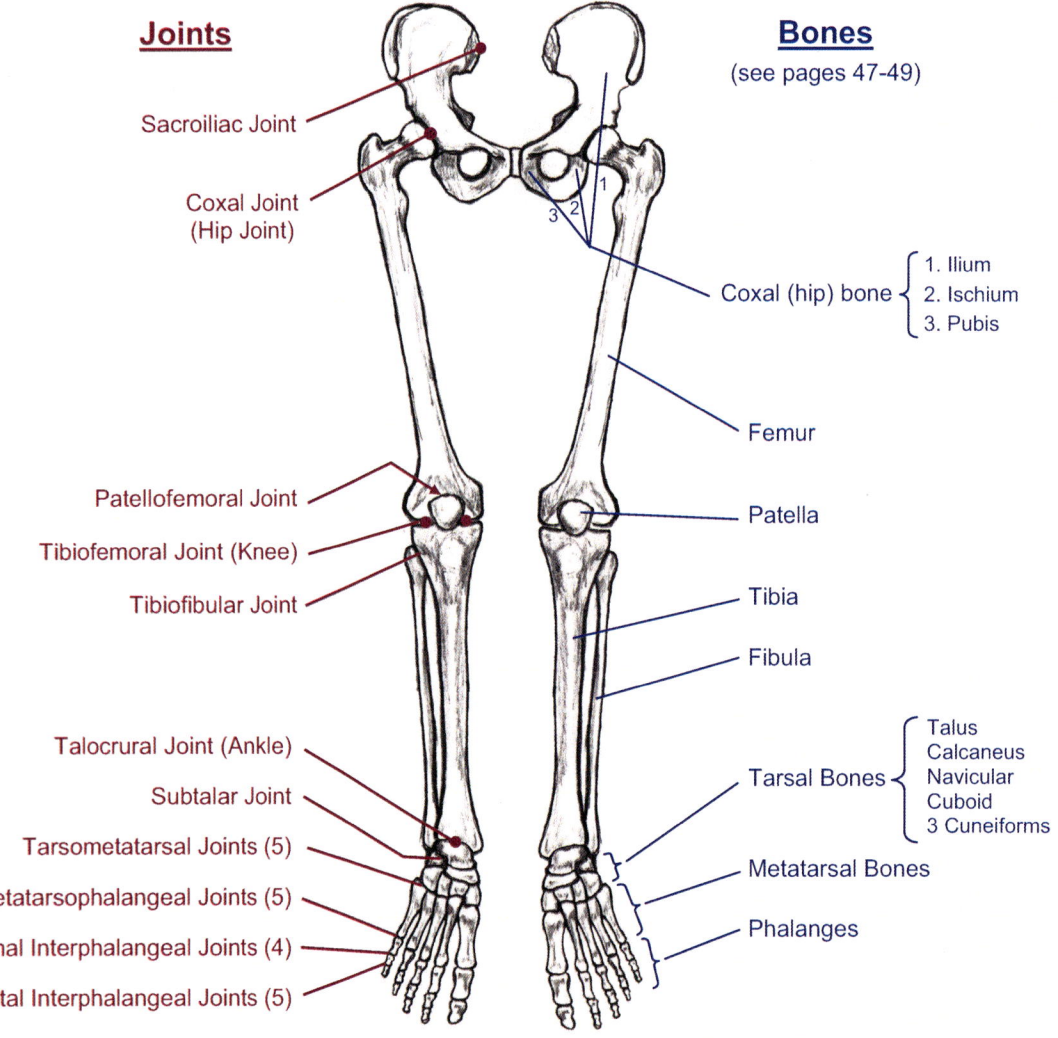

Movement of the Hip Joint (Part 1)

Hip Joint (Part 1)

Muscle Group 10:

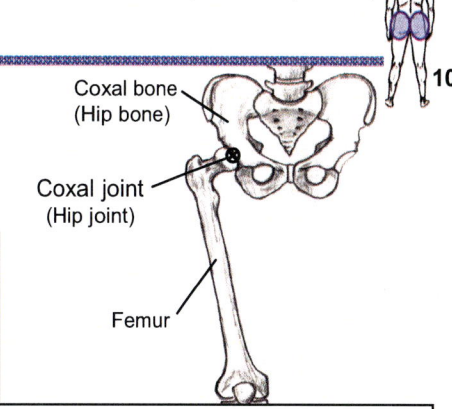

Gluteus maximus	Piriformis (deep lateral rotator #1 of 6)
Gluteus medius	The other 5 lateral rotators:
Gluteus minimus	Gemellus superior, Obturator internus
Iliopsoas { Iliacus & Psoas Major	Gemellus inferior, Obturator externus Quadratus femoris

Joints

This is the first of three groups of muscles that primarily move the femur at the hip joint (coxal joint). This group contains the "shorter" length muscles that mainly originate on the front or back of the ilium bone of the pelvis, and insert on the greater or lesser trochanter of the femur.

Coxal Joint (Hip Joint)

Head of **Femur** ◄► Acetabulum of the **Coxal Bone**

Ball and Socket Joint

Movements Available:

  Flexion
  Extension
  Abduction
  Adduction
  Lateral Rotation
  Medial Rotation

Other Joints

Small movements of the following joints are also created by the muscles in this group:

  Sacroiliac Joint
    Lateral **Sacrum** ◄► Posterior **Ilium**
    Part gliding, part fibrocartilagenous

  Intervertebral Joints of lumbar vertebrae
    Facets and discs
    (moved by the psoas major in this muscle group)

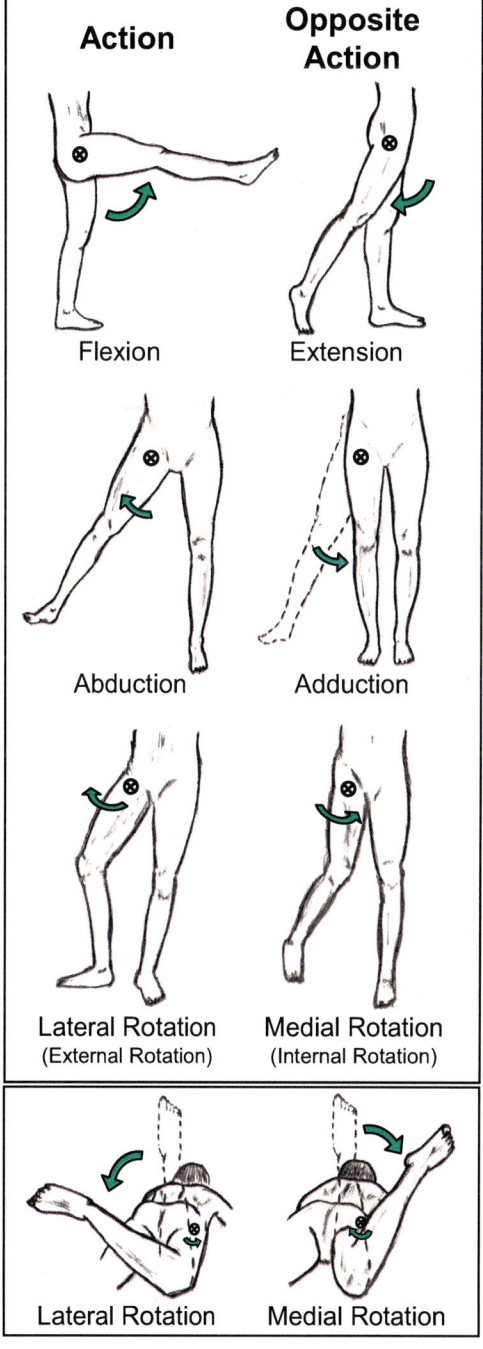

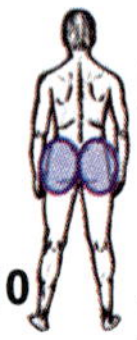

Hip Joint (Part 1)

Bones and bony landmarks

The "short" muscles that move the hip joint mainly have attachments on the pelvis and the femur. Review the bony landmarks and other structures listed below, referring to the drawings in Chapter 2, pages 47-48.

Coxal Bone (p. 47)
(Hip Bone, Os Coxae)

(Made up of 3 bones fused: Ilium, Ischium, Pubis)

    Landmarks on the Ilium:
        Iliac Fossa
            (anterior-medial surface of wing)
        Iliac Crest
        Posterior Superior Iliac Spine (PSIS)
        Gluteal surface
            (posterior-lateral surface of wing)
        Anterior Gluteal Line

    Acetabulum
        All 3 hip bones (ilium, ischium, pubis)
        intersect in the cavity of this socket

    Obturator foramen
        Hole encircled by pubis and ischium

Femur (p. 48)

    Head
    Neck
    Greater trochanter
    Lesser trochanter
    Gluteal tuberosity

Sacrum (p. 45)
    Muscles attach on both the
    posterior and anterior surfaces.

Lumbar vertebrae L1-L5, and thoracic T12

    Anterior bodies and TVP's – (for psoas major)

<u>Tendinous Structures</u>

    Sacrotuberous Ligament
    Iliotibial Tract / Iliotibial Band (ITB)
    Inguinal Ligament

Notes

Hip Joint (Part 1)

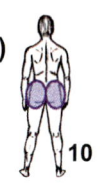

Muscle Group 10 – The first of three groups of muscles that move the hip joint are illustrated as a group on this page. The following four pages have tables and figures that describe each muscle individually, and provide many ways of comparing and contrasting the muscles to each other.

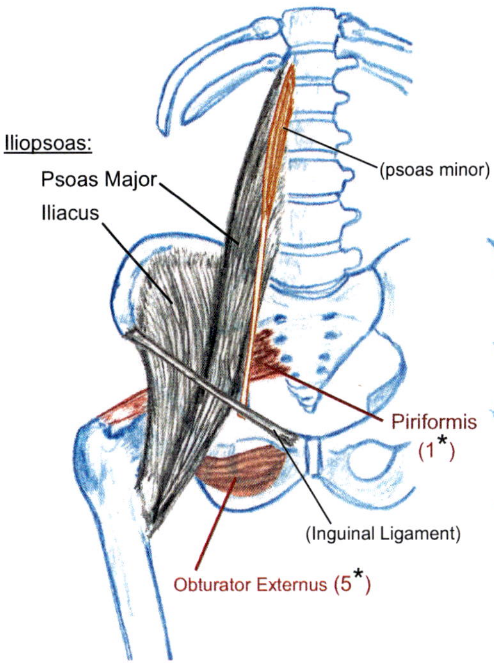

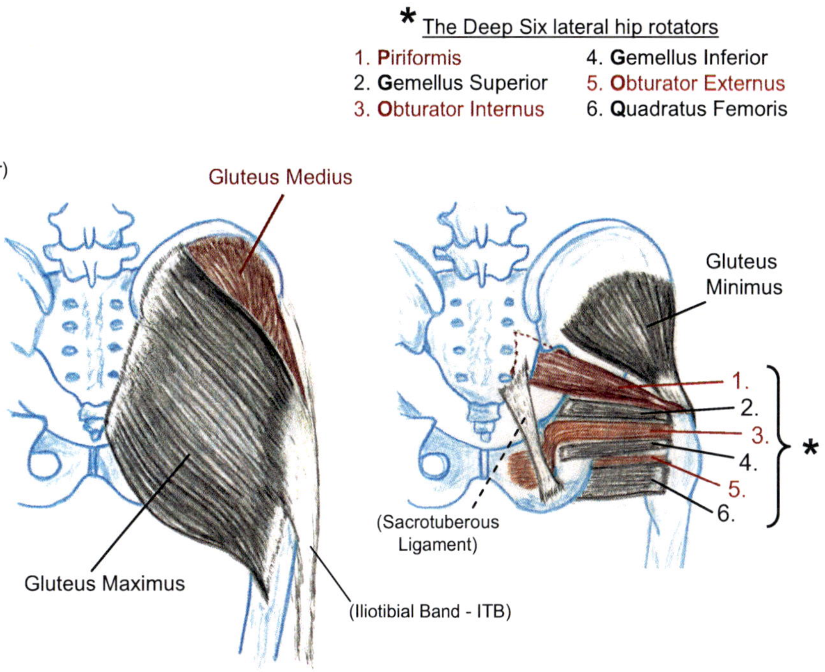

* The Deep Six lateral hip rotators
1. **P**iriformis   4. **G**emellus Inferior
2. **G**emellus Superior   5. **O**bturator Externus
3. **O**bturator Internus 6. **Q**uadratus Femoris

Iliopsoas:
- Psoas Major
- Iliacus
- (psoas minor)
- Gluteus Medius
- Gluteus Minimus
- Piriformis (1*)
- (Inguinal Ligament)
- Obturator Externus (5*)
- Gluteus Maximus
- (Sacrotuberous Ligament)
- (Iliotibial Band - ITB)
- 1. 2. 3. 4. 5. 6. *

Anterior View   **Posterior Views**

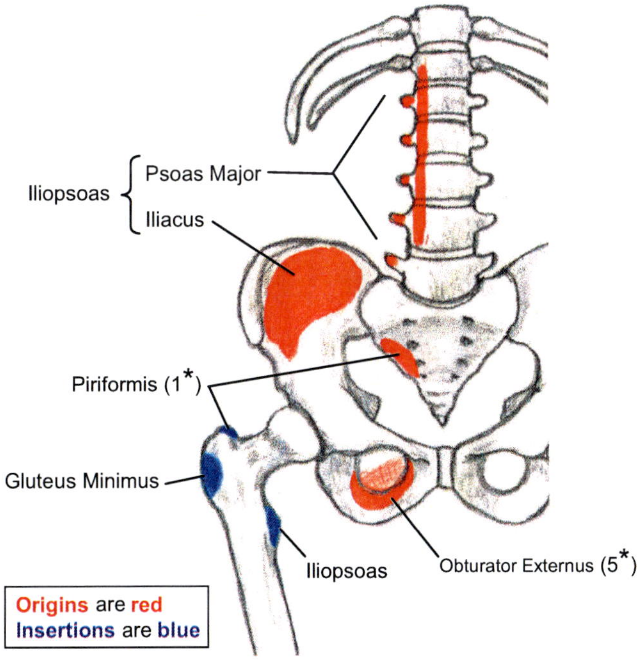

Iliopsoas { Psoas Major, Iliacus }
Piriformis (1*)
Gluteus Minimus
Iliopsoas
Obturator Externus (5*)

Gluteus Maximus
Gluteus Medius
Gluteus Minimus
Piriformis (1*)
Gluteus Medius
2, 3, 4, 5, 6
Deep Lateral Rotators *
5 (anterior)
Gluteus Maximus

Origins are red
Insertions are blue

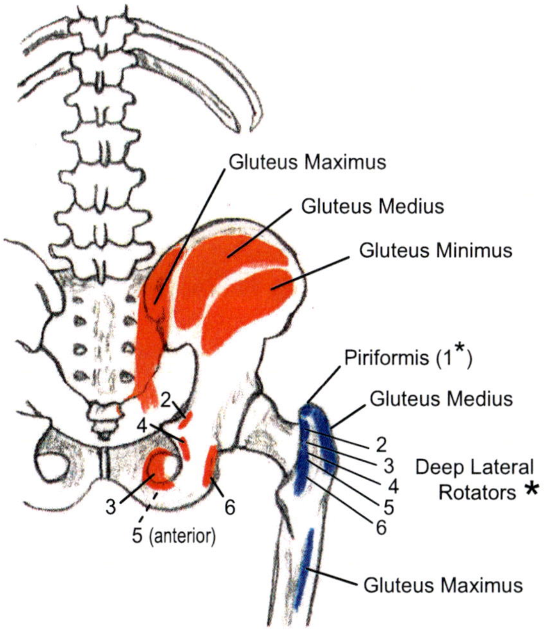

Attachment sites for all muscles in Group 10

Mastering Muscles & Movement © 2007   Chapter 6 – Muscles That Move the Lower Extremity

Hip Joint (Part 1)

TVP=Transverse process of vertebra, ITB=Iliotibial Band (another name for the iliotibial tract)

Muscles Acting On Hip Joint (Part 1)	Origin	Insertion	Action
Gluteus Maximus — moves the hip joint	Posterior iliac crest, ilium, and sacrum (also lateral coccyx and sacrotuberous ligament)	Gluteal tuberosity of femur, and the iliotibial tract (ITB)	Extension and lateral rotation at the hip joint (also lower fibers assist adduction, and upper fibers may assist abduction)
Gluteus Medius — moves the hip joint	Upper lateral surface of the ilium (upper half of the wing of the ilium, starting just below the iliac crest)	Greater trochanter of femur (lateral aspect)	All fibers: Abduction at the hip joint. Ant. fibers: Assist flexion and medial rotation Post. fibers: Assist extension and lateral rotation
Gluteus Minimus — moves the hip joint	Lower lateral surface of the ilium (lower half of the wing of the ilium, inferior to the origin of gluteus medius)	Greater trochanter of femur (anterior aspect)	Abduction and medial rotation at the hip joint. (Also may assist flexion)
Piriformis (Deep Lateral Rotator #1) — moves the hip joint	Anterior surface of sacrum	Greater trochanter of femur (superior aspect)	Lateral rotation at the hip joint
The Other 5 Deep Lateral Rotators (#2 - #6) Gemellus superior, Obturator internus, Gemellus inferior, Obturator externus, Quadratus femoris	Gemelli & Quad.Fem.: Ischium Obturators: Obturator foramen (ischium & pubis) All Deep 6 Collective: Sacrum, Ischium, and Pubis	Greater trochanter of femur (posterior-medial aspect)	Lateral rotation at the hip joint
Iliopsoas: Iliacus and Psoas Major — Moves the hip joint and the spine	Iliacus: Anterior iliac fossa Psoas Major: Bodies & TVP's of T12 and L1-L5	Both: Lesser trochanter of the femur	Flexion at the hip joint. (May assist lateral rotation at the hip joint) If the femur is fixed (in a standing position): Pulls on lumbar spine, increasing lordosis and anterior pelvic tilt.

(larger illustrations on page 159)

Table 10 (A) - Hip Joint (Part 1) - Origin, Insertion, Action

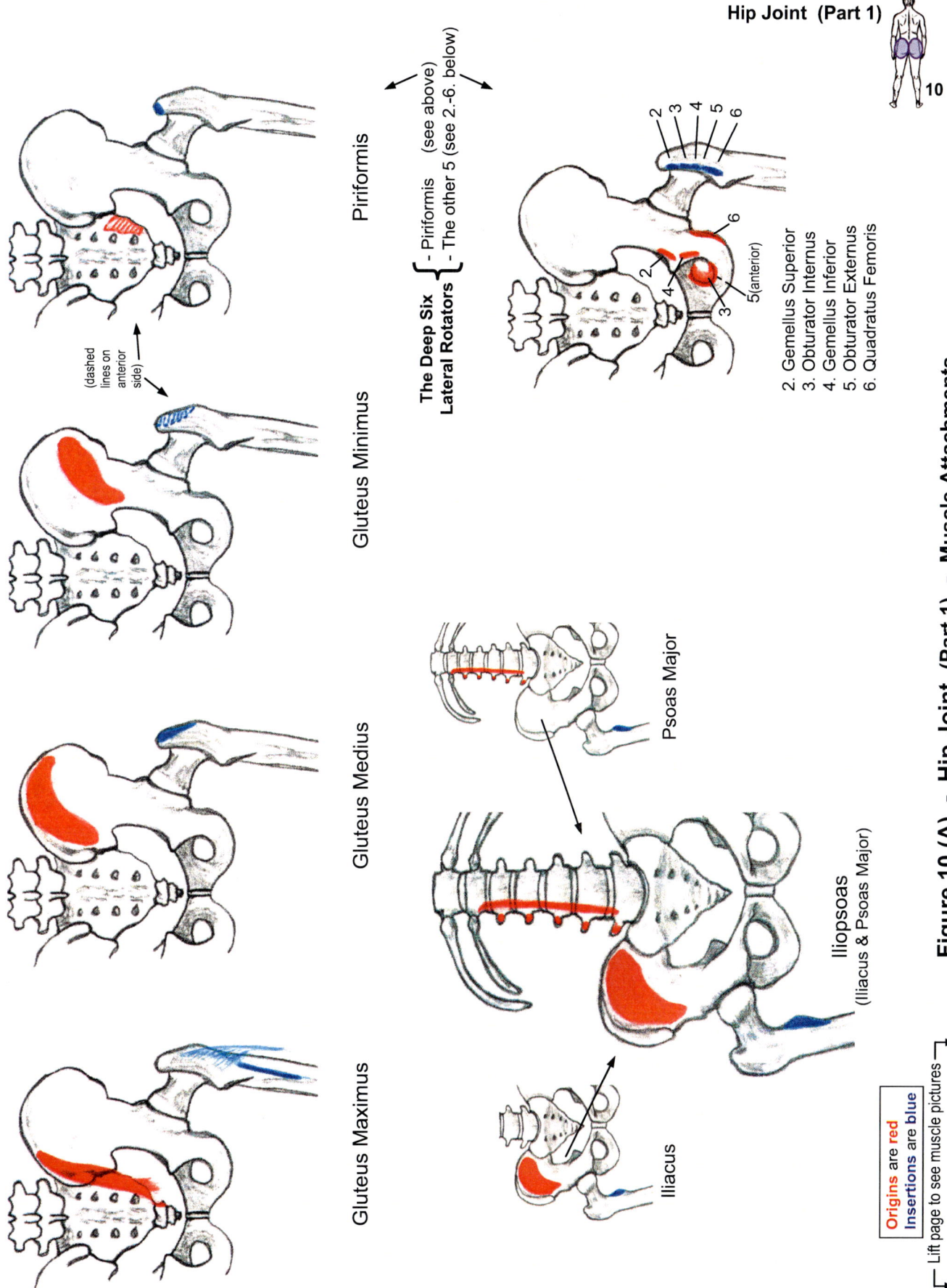

Figure 10 (A) - Hip Joint (Part 1) - Muscle Attachments

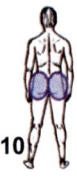

Hip Joint (Part 1)

Coxal joint = Hip joint = Head of femur seated in acetabulum of the coxal bone (hip bone)

Table 10 (B) - Hip Joint (Part 1) - Synergists & Antagonists

Muscles Acting On Hip Joint (Part 1)	Flexion @ Coxal jt.	Extension @ Coxal jt.	Abduction @ Coxal jt.	Adduction @ Coxal jt.	Medial Rotation @ Coxal jt.	Lateral Rotation @ Coxal jt.	Stabilization of Coxal jt.	Other	Innervation	L2	L3	L4	L5	S1	S2
1. Gluteus maximus		X	may assist (upper fibers)	assist (lower fibers)		X			Inferior gluteal N. (L5, S1, S2)				N	N	N
2. Gluteus medius	may assist (anterior fibers)	may assist (posterior fibers)	X (all fibers)		assist (anterior fibers)	assist (post. fibers) when hip is extended	X (main hip stabilizer)	This is the primary abductor	Superior gluteal N. (L4, L5, S1)			N	N	N	
3. Gluteus minimus	may assist		X		X		X		Superior gluteal N. (L4, L5, S1)			N	N	N	
4. Piriformis						X			Sacral Plexus (S1, S2)				N	N	N
5. The Other 5 Deep Lateral Rotators Gemellus superior Obturator internus Gemellus inferior Obturator externus Quadratus femoris						X			GS: SP- L5, S1, 2 OI: SP- L5, S1, 2 GI: SP- L4, 5, S1 OE: Obturator, L3,4 QF: SP- L4, 5, S1 (SP=Sacral Plexus)		N	N N N	N N N	N N N N	N N
Iliopsoas:															
6. Iliacus	X							may assist	Iliacus: Femoral nerve (L2, L3)	N	N				
7. Psoas major	X							Reverse O/I (femur fixed): increases lumbar lordosis, anterior pelvic tilt	Psoas Major: Lumbar plexus (L2-L4)	N	N	N			
(More muscles for the action) --->	see also Tables 11, 12	see also Tables 11, 12	see also Table 11	see also Table 11	see also Tables 11, 12	see also Tables 11, 12			**Innervation**						

Hip Joint (Part 1)

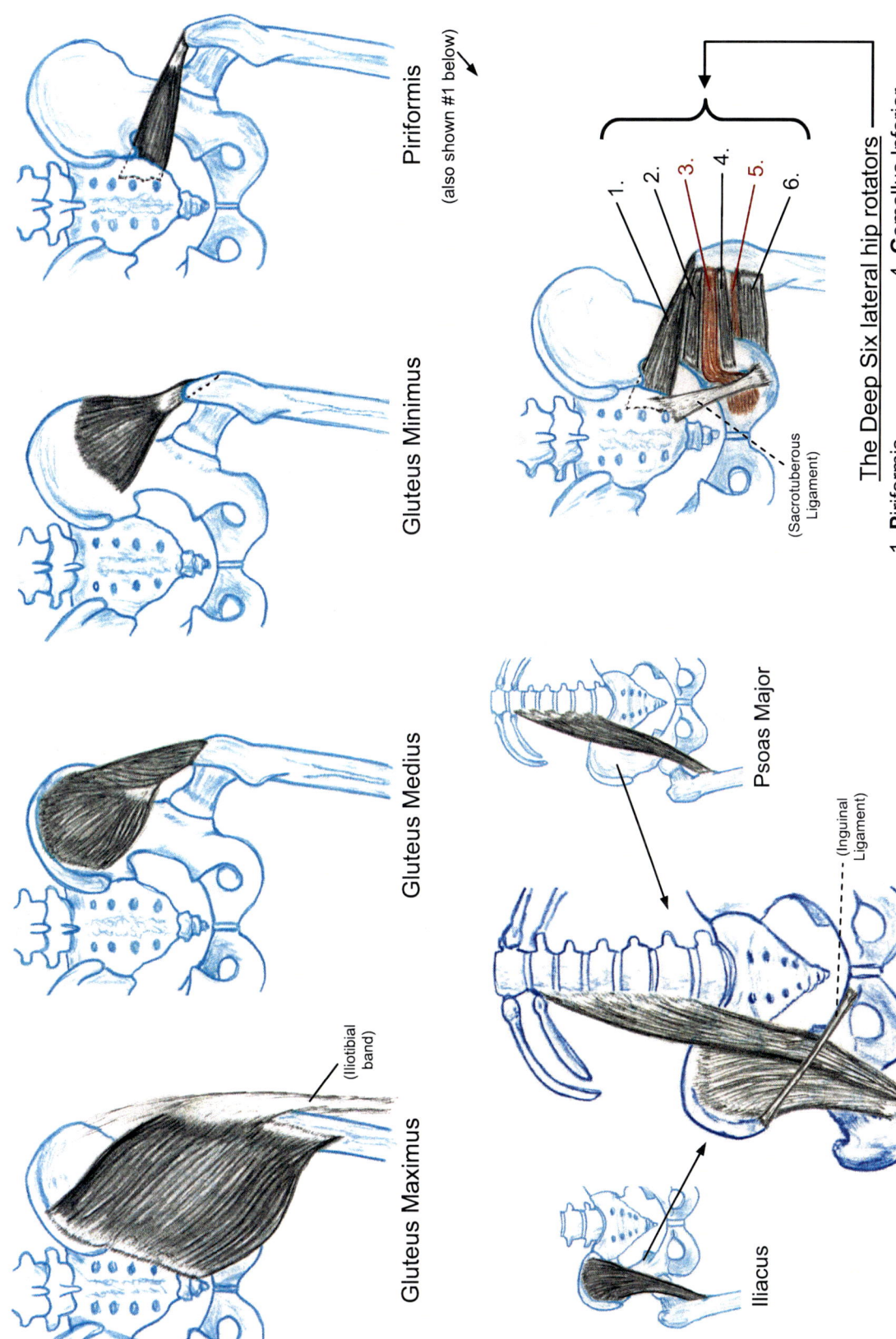

Figure 10 (B) - **Hip Joint (Part 1)** - **Muscle Pictures**

Hip Joint (Part 1)

Muscles Acting On Hip Joint (Part 1) — ~ Notes ~ *(palpation, fiber arrangement, lengthen/shorten, common uses, pathologies, cautions, etc.)*

1. Gluteus Maximus	**5. The Other 5 Deep Lateral Rotators** *#2 - #6* (*Piriformis is #1*) *Gemellus superior* *Obturator internus* *Gemellus inferior* *Obturator externus* *Quadratus femoris*
2. Gluteus Medius	**6. Iliacus**
3. Gluteus Minimus	**7. Psoas Major**
4. Piriformis *(Deep Lateral Rotator #1)*	**Iliopsoas** *(Iliacus & Psoas Major treated as one muscle)*

Muscle Group 10 - Hip Joint (Part 1) - (Notes)

Movement of the Hip Joint (Part 2)

Muscle Group 11:

Superficial Long Muscles:	The Adductor Group:
Sartorius	Pectineus
Tensor fascia latae (TFL)	Adductor brevis
	Adductor longus
	Adductor magnus
	Gracilis

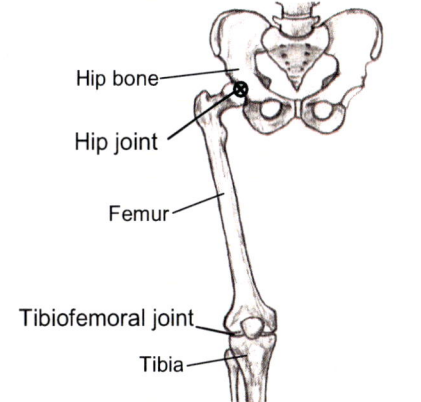

Joint

This is the second of three groups of muscles that move the femur at the hip joint (coxal joint). This group contains the "longer" length muscles that mainly originate on the iliac crest and pubic bone, and insert on the posterior shaft of the femur and the top of the tibia.

Coxal Joint (Hip Joint)

Acetabulum of the **Coxal Bone** ◄► Head of **Femur**

Ball and Socket Joint

Movements Available:

Flexion, Extension
Abduction, Adduction
Lateral Rotation, Medial Rotation

Other Joints

Three of the muscles in this group cross both the hip and the knee joints, and therefore also affect the knee (although the *main* knee movers are presented in the next section – Group 12: Movement of the Knee).

Tibiofemoral Joint (TF)

Condyles of **Femur** ◄► Condyles of **Tibia** (tibial plateau)
Modified Hinge Joint
Movements Available:
Flexion, Extension
Medial and Lateral Rotation (when the knee is flexed)

(Note: The TF joint is covered more fully
in the next section – Group 12: Movement of the Knee)

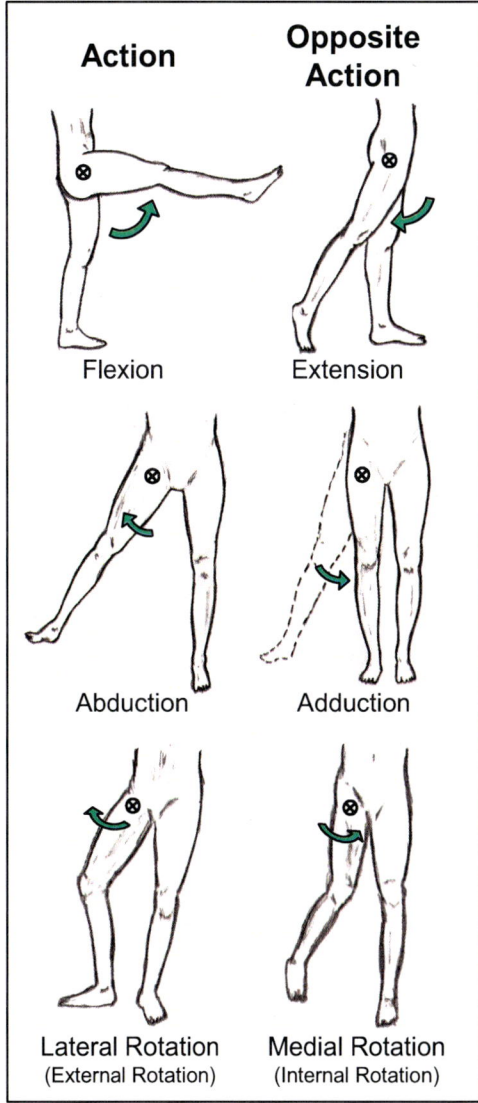

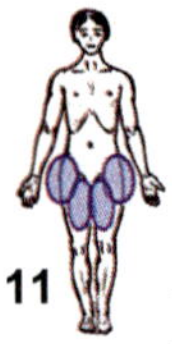

Hip Joint (Part 2)

Bones and bony landmarks

The "long" muscles that move the hip joint have attachments on the pelvis, femur and tibia. Review the bony landmarks and other structures listed below, referring to the drawings in Chapter 2, pages 47-48.

Coxal Bone (p. 47)
(Hip Bone, Os Coxae)

(Made up of 3 bones fused: Ilium, Ischium, Pubis)

    Ilium
        Anterior Superior Iliac Spine (ASIS)

    Ischium
        Ischial tuberosity
        Ramus of ischium

    Pubis
        Superior ramus of pubis
        Pubic crest
        Pubic tubercle
        Body of pubis
        Inferior ramus of pubis
        Symphysis pubis

Femur (p. 48)

    Condyles
    Linea aspera
        (medial lip, lateral lip)
    Pectineal line
    Adductor tubercle

Tibia (p. 48)

    Proximal Medial Shaft (PMS)
    Condyles

<u>Tendinous & Other Structures</u>

    Iliotibial tract / Iliotibial band (ITB)
    Pes anserinus
        (on proximal medial shaft (PMS) of tibia)
    Inguinal ligament
    Femoral triangle
    Adductor canal
    Adductor hiatus

Notes

Hip Joint (Part 2)

Muscle Group 11 – The second of three groups that move the hip joint are illustrated as a group on this page. The following four pages have tables and figures that describe each muscle individually, and provide many ways of comparing and contrasting the muscles to each other.

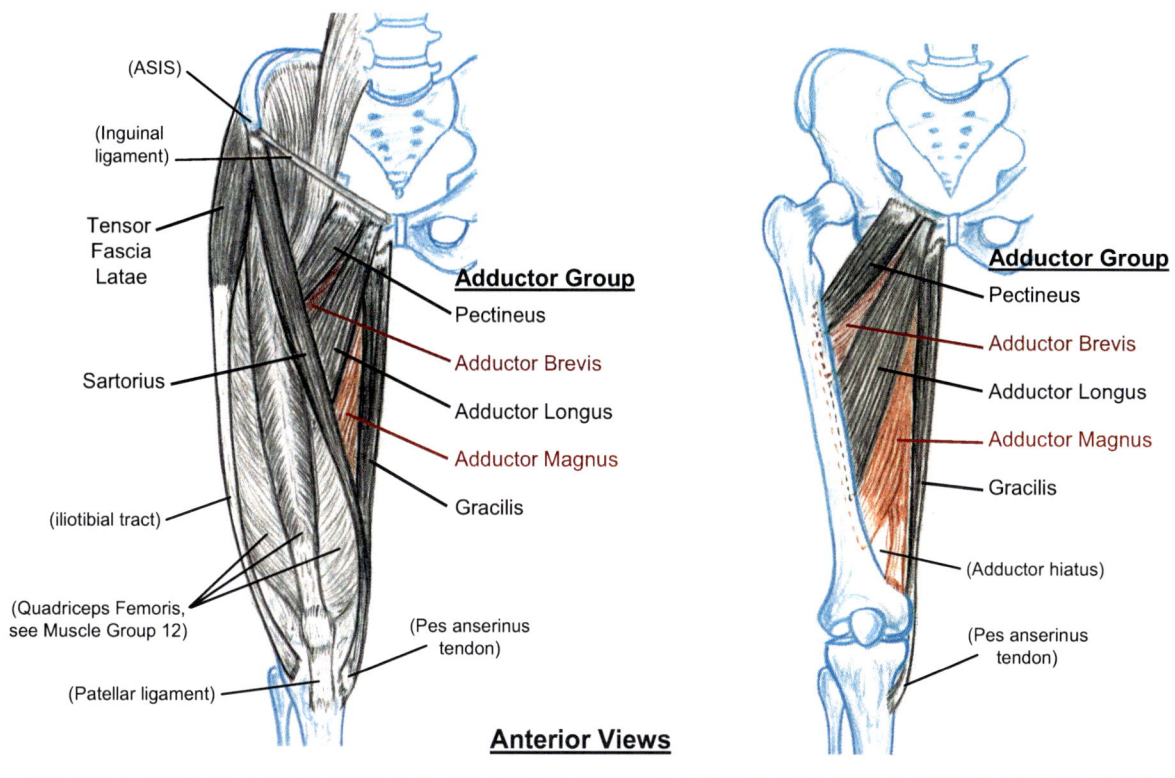

Anterior Views

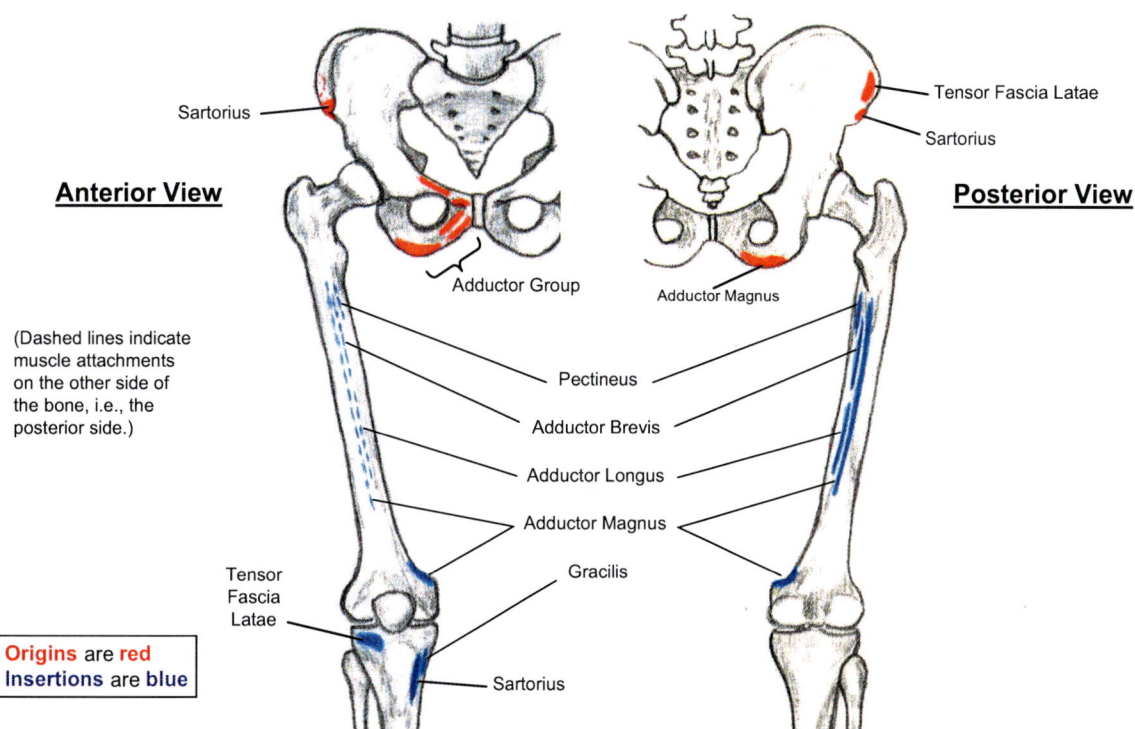

Attachment sites for all muscles in Group 11

Origins are red
Insertions are blue

Mastering Muscles & Movement © 2007   Chapter 6 – Muscles That Move the Lower Extremity   163

Hip Joint (Part 2)

Muscles Acting On Hip Joint (Part 2)

Muscle	Origin	Insertion	Action
Sartorius — moves the hip joint and knee	Anterior Superior Iliac Spine (ASIS) of the hip bone	Proximal medial shaft of tibia (by way of the pes anserinus tendon)	Flexion, abduction, and lateral rotation at the hip joint. Flexion of the knee and medial rotation of the tibia at the flexed knee.
Tensor Fascia Latae — moves the hip joint and stabilizes the knee	Iliac crest, just posterior to the ASIS (i.e., next to the sartorius origin)	Lateral condyle of the tibia via the iliotibial tract (anterior aspect of the lat. condyle)	Flexion, abduction, and medial rotation at the hip joint. Stabilizes the extended knee
Pectineus — moves the hip joint	Superior ramus of pubis	Pectineal line of femur (on posterior femur, proximal to linea aspera)	Adduction, flexion, and medial rotation at the hip joint
Adductor Brevis — moves the hip joint	Inferior ramus of pubis (near the obturator foramen, lateral to the gracilis attachment)	Proximal linea aspera of femur	Adduction, flexion, and medial rotation at the hip joint
Adductor Longus — moves the hip joint	Pubic tubercle	Mid linea aspera of femur	Adduction, flexion, and medial rotation at the hip joint
Adductor Magnus — moves the hip joint	Anterior: Inferior ramus of pubis, Posterior: Ramus of ischium, Ischiocondylar: Ischial tuberosity. *Overall Description*: Inferior pubic ramus, ischial ramus, & ischial tuberosity	Entire linea aspera, and adductor tubercle of femur (with hiatus in between for vessels to pass through)	All fibers: Adduction at the hip joint. Anterior fibers: Flexion and medial rotation at the hip joint. Posterior fibers: Extension at the hip joint
Gracilis — moves the hip joint and knee	Inferior ramus of pubis (medial edge of ramus, near the symphysis pubis)	Proximal medial shaft of tibia (by way of the pes anserinus tendon)	Adduction at the hip joint. Flexion of the knee and medial rotation of the tibia at the flexed knee (may assist flexion & medial rotation at the hip joint)

The Adductor Group (Pectineus, Adductor Brevis, Adductor Longus, Adductor Magnus, Gracilis)

(larger illustrations on page 167)

Table 11 (A) - Hip Joint (Part 2) - Origin, Insertion, Action

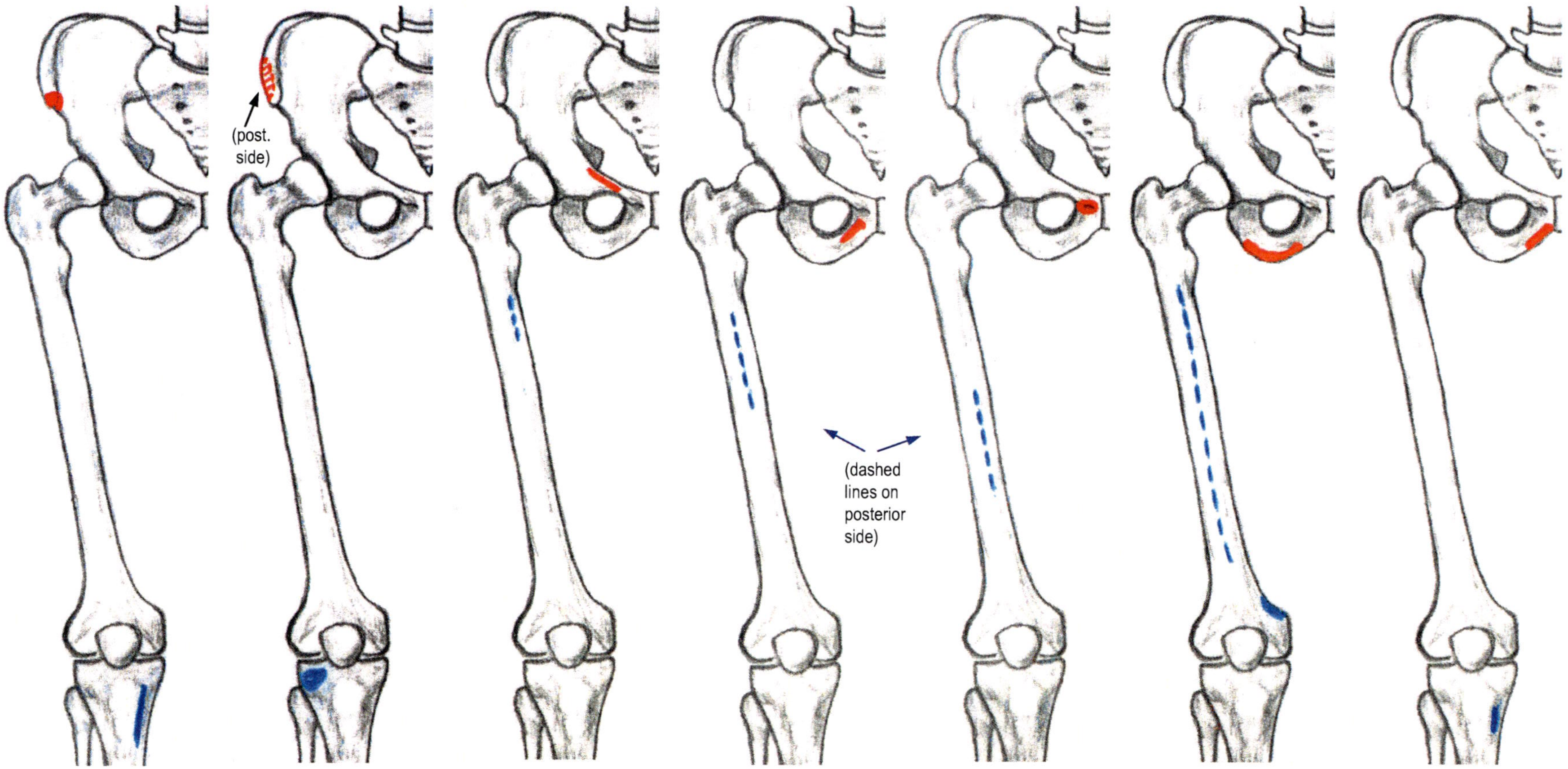

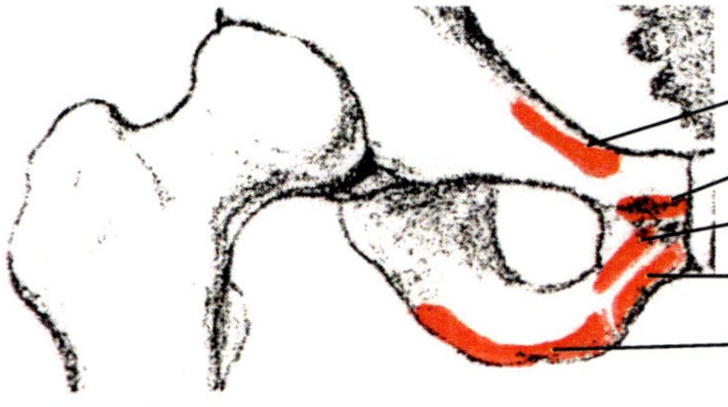

Hip Joint (Part 2)

Gracilis

Adductor Magnus

Adductor Longus

(dashed lines on posterior side)

Adductor Brevis

Pectineus

Tensor Fascia Latae
(post. side)

Sartorius

Adductor Origins (clockwise):

1. **Pectineus** (superior pubic ramus)
2. Add. **Longus** (pubic tubercle)
3. Add. **Brevis** (inferior pubic ramus – lateral to gracilis)
4. **Gracilis** (inferior pubic ramus – medial edge)
5. Add. **Magnus** (inferior pubic ramus, ischial ramus & ischial tuberosity)

Figure 11 (A) - Hip Joint (Part 2) - Muscle Attachments

Origins are **red**
Insertions are **blue**

Lift page to see muscle pictures

Mastering Muscles & Movement © 2007 — Chapter 6 - Muscles That Move the Lower Extremity

Hip Joint (Part 2)

Coxal jt. = Hip Joint, TF jt. = Tibiofemoral joint (knee)

Muscles Acting On: Hip Joint (Part 2)	Flexion @ Coxal jt.	Extension @ Coxal jt.	Abduction @ Coxal jt.	Adduction @ Coxal jt.	Medial Rotation @ Coxal jt.	Lateral Rotation @ Coxal jt.	Flexion @ Knee	Other	Innervation	L2	L3	L4	L5	S1	S2	S3
1. Sartorius	X		X			X	X	Medial rotation of tibia at flexed knee	Femoral N. (L2, L3)	N	N					
2. Tensor fascia latae	X		X		X			Stabilizes the extended knee	Superior Gluteal N. (L4, L5, S1)			N	N	N		
3. Pectineus	X			X	X				Femoral N. (L2, L3) (& sometimes Obturator N.)	N	N					
4. Adductor brevis	X			X	X			(deep to adductor longus)	Obturator N. (L2, L3, L4)	N	N	N				
5. Adductor longus	X			X	X				Obturator N. (L2, L3, L4)	N	N	N				
6. Adductor magnus	X Anterior fibers (which insert proximally)	X Posterior fibers (which insert distally)		X All fibers	X Anterior fibers			Can be an antagonist to itself (posterior vs. anterior fibers)	Anterior part: Obturator N. (L2,L3,L4) Posterior part: Sciatic N. (L4,L5, S1)	N	N	N	N	N		
7. Gracilis	may assist			X	may assist		X	Medial rotation of tibia at flexed knee	Obturator N. (L2, L3)	N	N					
(More muscles for the action) --->	see also Tables 10,12	see also Tables 10,12	see also Table 10	see also Table 10	see also Tables 10,12	see also Tables 10,12	see also Tables12,13		Innervation							

Table 11 (B) - Hip Joint (Part 2) - Synergists & Antagonists

Hip Joint (Part 2)

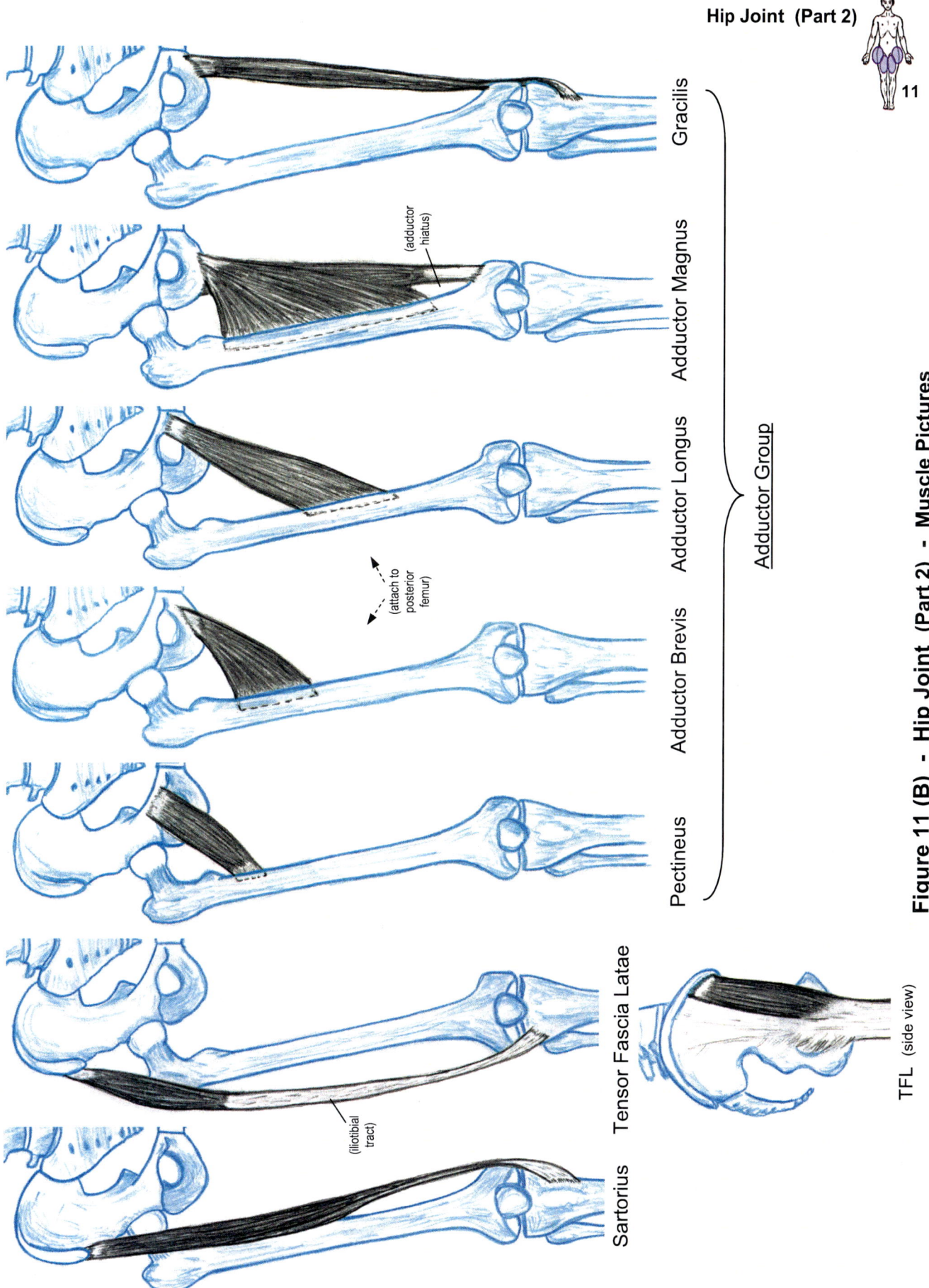

Figure 11 (B) - Hip Joint (Part 2) - Muscle Pictures

Hip Joint (Part 2)

Muscles Acting On Hip Joint (Part 2)

~ Notes ~ (palpation, fiber arrangement, lengthen/shorten, common uses, pathologies, cautions, etc.)

1. Sartorius	5. Adductor Longus
2. Tensor Fascia Latae	6. Adductor Magnus
3. Pectineus	7. Gracilis
4. Adductor Brevis	

Muscle Group 11 - Hip Joint (Part 2) - (Notes)

168 Chapter 6 - Muscles That Move the Lower Extremity

Knee (&Hip Joint, Part 3)

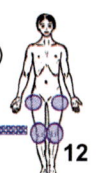

Movement of the Knee (& Hip Joint, Part 3)

Muscle Group 12:

Quadriceps Group:	Hamstrings Group:
Rectus femoris	Biceps femoris
Vastus lateralis	Semimembranosus
Vastus intermedius	Semitendinosus
Vastus medialis	
	Other:
	Popliteus

Joints

This group primarily moves the tibia/fibula at the knee. Many of the muscles are also strong movers of the femur at the hip joint, so this is *also* the third of the three groups of hip movers (with Groups 10 and 11).

Tibiofemoral Joint (TF) – The Knee

Condyles of **Femur** ◄► Condyles of **Tibia** (tibial plateau)

Modified Hinge joint
Movements Available:
    Flexion, Extension
    Medial and Lateral Rotation
    (when the knee is flexed)

Hip Joint
(See previous section:
Group 11: Hip Joint, Part 2)

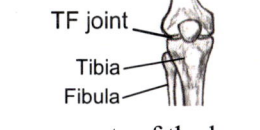

Other Joints

These joints are also involved with movements of the knee.

Patellofemoral
    Posterior **Patella** ◄► Patellar surface of **femur**
    Gliding joint
    Moves all directions, but primarily up & down

Proximal Tibiofibular
    Proximal lateral **Tibia** ◄► Head of **fibula**
    Gliding joint
    Slight movement

Distal Tibiofibular (@ distal tibia <> fibula)
    Fibrous joint (has almost no movement)

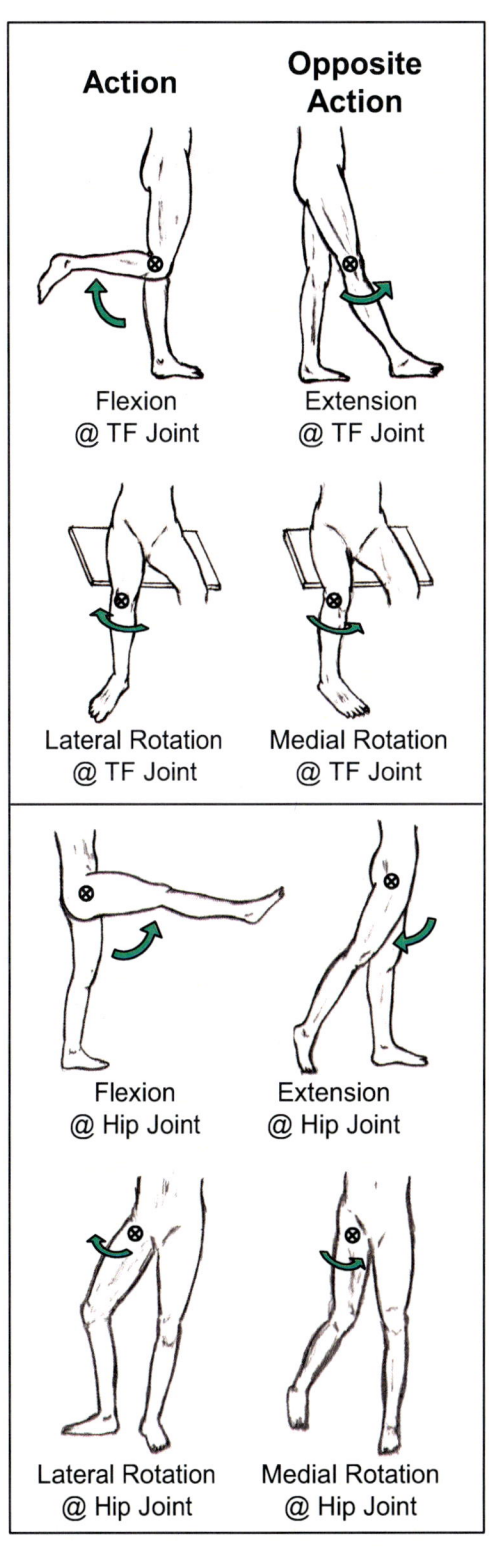

Action / Opposite Action:
- Flexion @ TF Joint / Extension @ TF Joint
- Lateral Rotation @ TF Joint / Medial Rotation @ TF Joint
- Flexion @ Hip Joint / Extension @ Hip Joint
- Lateral Rotation @ Hip Joint / Medial Rotation @ Hip Joint

Knee (&Hip Joint, Part 3)

Bones and bony landmarks

Muscles that move knee (and hip) have attachments on the pelvis, femur, patella, tibia and fibula. Review the bony landmarks and other structures listed below, referring to the drawings in Chapter 2, pages 47-48.

Pelvis (p. 47)

    Anterior Inferior Iliac Spine (AIIS)
    Ischial tuberosity

Femur (p. 48)

    Linea aspera
    Greater trochanter
    Lesser trochanter
    Lateral condyle
    Shaft
        Anterior, Posterior, Lateral, Medial
        Surfaces
    Patellar Surface

Patella

Tibia (p. 48)

    Tibial tuberosity
    Lateral & Medial Condyles (tibial plateau)
    PMS – Proximal Medial Shaft
    Pes Anserinus attachment

Fibula (p. 48)

    Head

<u>Tendinous and Other Structures:</u>

    Patellar Ligament (Patellar Tendon)
    Pes Anserinus tendon
    Popliteal Fossa
    Knee Ligaments & Menisci

Notes

Knee (& Hip Joint, Part 3)

Muscle Group 12 – Muscles that move the knee (and in many cases the hip) are illustrated as a group on this page. The next four pages have tables and figures that describe each muscle individually, and provide many ways of comparing and contrasting the muscles to each other.

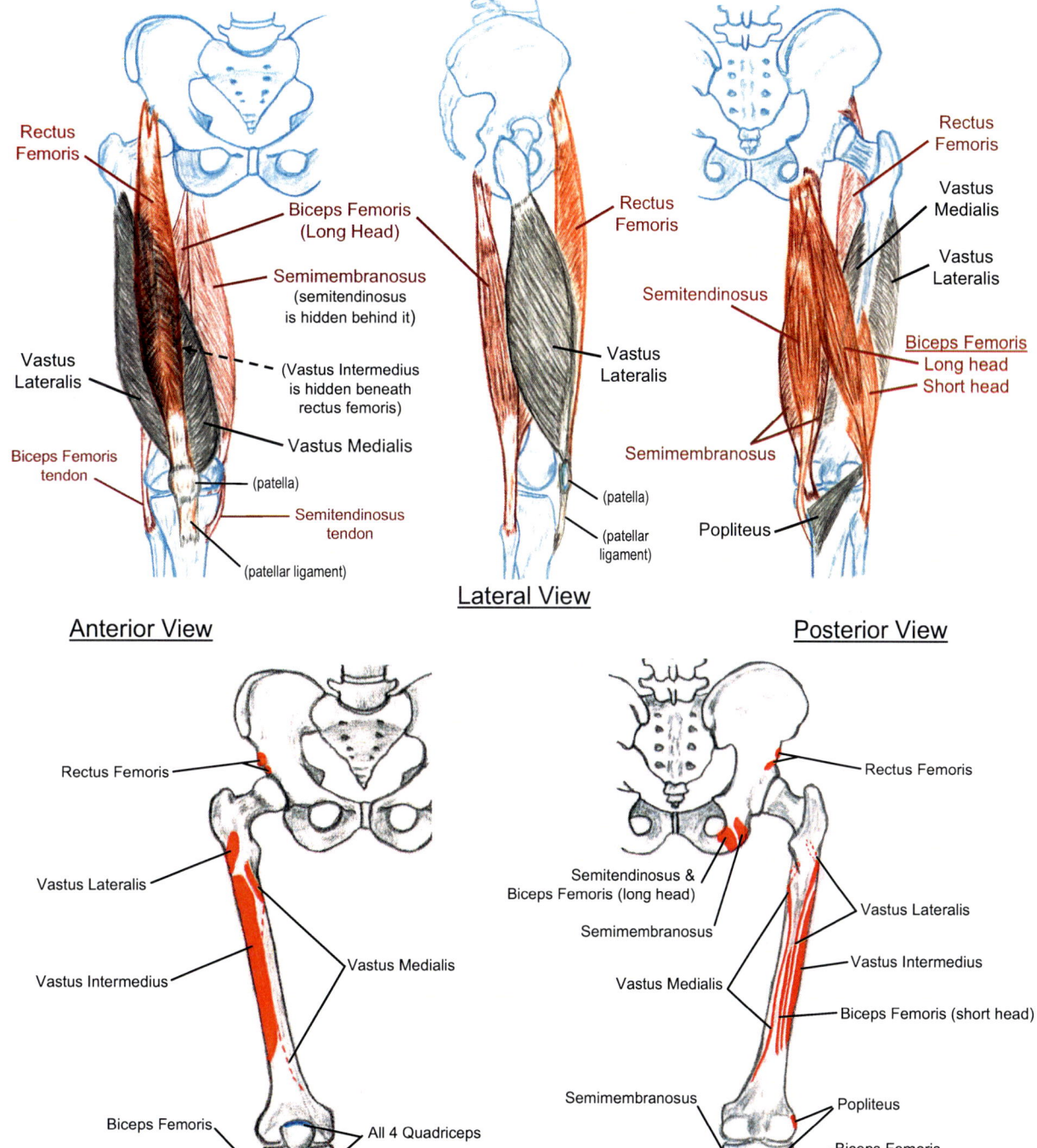

Attachment sites for all muscles in Group 12

Knee (& Hip Joint, Part 3)

Muscles Acting On Knee (& Hip Joint, Part 3)

*The tibia is only capable of rotation (at the tibiofemoral joint) when the knee is in a flexed position.

Muscle	Origin	Insertion	Action
Rectus Femoris (Quadricep) — moves the knee and the hip joint	Anterior Inferior Iliac Spine (AIIS) of the hip bone (and superior margin of the acetabulum just below the AIIS)	Tibial tuberosity via the patellar ligament	Extension at the knee, Flexion at the hip joint
Vastus Lateralis (Quadricep) — moves the knee	Posterior lateral femur, lateral lip of linea aspera (and wraps to anterior at the base of the greater trochanter)	Tibial tuberosity via the patellar ligament	Extension at the knee
Vastus Intermedius (Quadricep) — moves the knee	Anterior and lateral shaft of femur (upper 2/3 of the shaft)	Tibial tuberosity via the patellar ligament	Extension at the knee
Vastus Medialis (Quadricep) — moves the knee	Posterior medial femur, medial lip of linea aspera (and wraps to anterior at the base of the lesser trochanter)	Tibial tuberosity via the patellar ligament	Extension at the knee
Biceps Femoris (Hamstring) — moves the knee and the hip joint	Long head: Ischial tuberosity Short head: Lateral lip of linea aspera (distal half)	Head of fibula	*Both* heads: Flexion and lateral rotation* at the knee Long head: Extension and lateral rotation at the hip joint.
Semitendinosus (Hamstring) — moves the knee and the hip joint	Ischial tuberosity	Proximal medial shaft of tibia (by way of the pes anserinus tendon)	Flexion and medial rotation* at the knee, Extension and medial rotation at the hip joint
Semimembranosus (Hamstring) — moves the knee and the hip joint	Ischial tuberosity	Posterior medial condyle of tibia	Flexion and medial rotation* at the knee, Extension and medial rotation at the hip joint
Popliteus — moves the knee	Lateral condyle of the femur	Proximal posterio-medial tibia	Medial rotation* at the knee, May assist flexion at the knee When weight bearing: Lateral rotation of femur, to unlock straightened knee

Anterior Views

Posterior Views

(larger illustrations on page 175)

Table 12 (A) - Knee (& Hip Joint, Part 3) **- Origin, Insertion, Action**

Chapter 6 – Muscles That Move the Lower Extremity

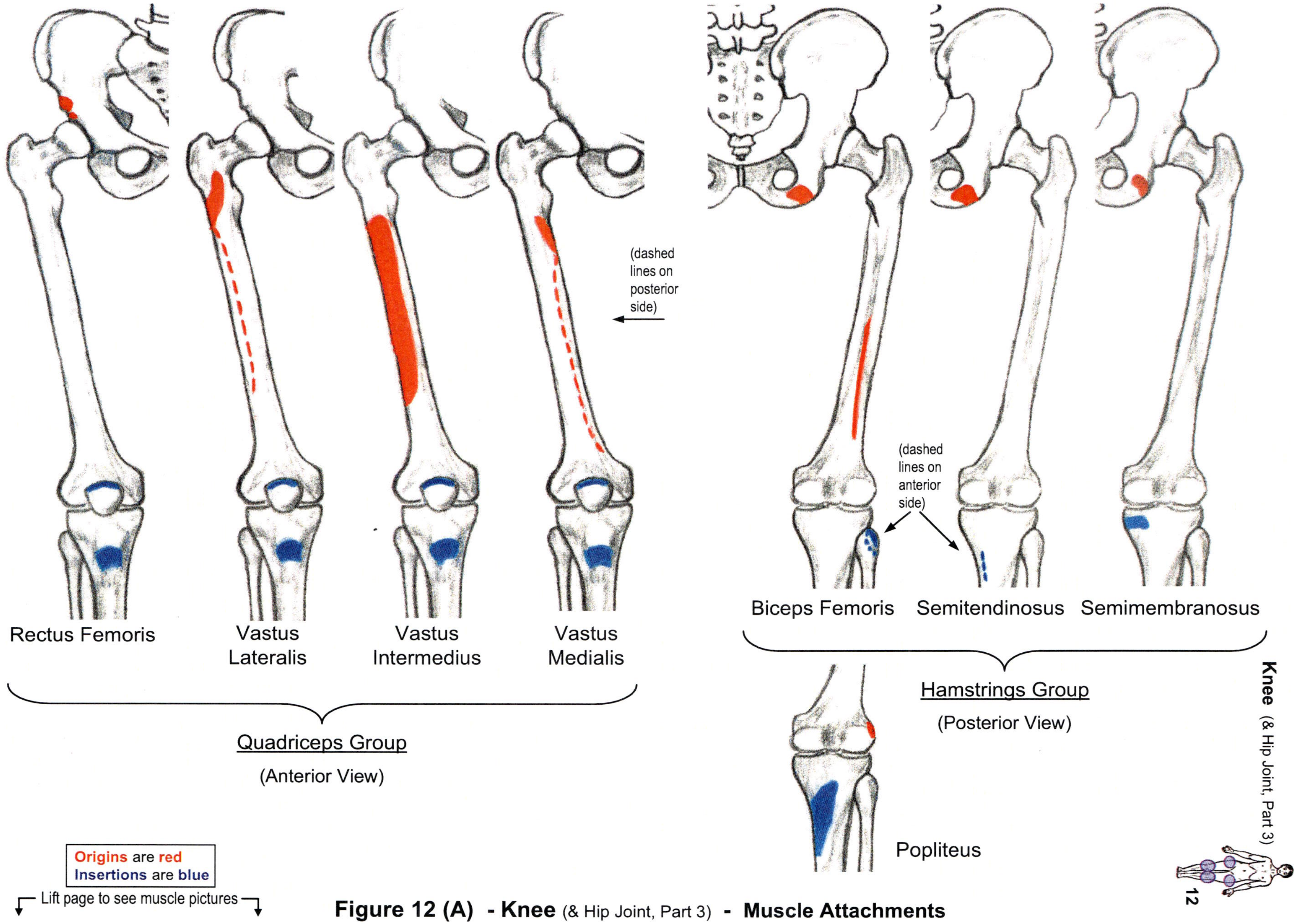

Knee (& Hip Joint, Part 3)

Hamstrings Group (Posterior View)

- Semimembranosus
- Semitendinosus
- Biceps Femoris

(dashed lines on anterior side)

Popliteus

Quadriceps Group (Anterior View)

- Vastus Medialis
- Vastus Intermedius
- Vastus Lateralis
- Rectus Femoris

(dashed lines on posterior side)

Origins are red
Insertions are blue

Lift page to see muscle pictures

Figure 12 (A) - Knee (& Hip Joint, Part 3) **- Muscle Attachments**

Mastering Muscles & Movement © 2007 — Chapter 6 - Muscles That Move the Lower Extremity

Knee (& Hip Joint, Part 3)

TF jt.=Tibiofemoral joint (knee)

Muscles Acting On Knee (& Hip Joint, Part 3)	Flexion @ knee	Extension @ knee	Rotation @ knee (if flexed)	Flexion @ Hip jt.	Extension @ Hip jt.	Medial Rotation @ Hip jt.	Lateral Rotation @ Hip jt.	Other	Innervation	L2	L3	L4	L5	S1	S2	S3
1. **Rectus femoris** (Quadricep)		X		X					Femoral N. (L2, L3, L4)	N	N	N				
2. **Vastus lateralis** (Quadricep)		X						Makes up all of the lateral thigh. It is deep to the iliotibial tract	Femoral N. (L2, L3, L4)	N	N	N				
3. **Vastus intermedius** (Quadricep)		X						It is deep to the other 3 quads	Femoral N. (L2, L3, L4)	N	N	N				
4. **Vastus medialis** (Quadricep)		X						Keeps patella pulled medially so it tracks properly	Femoral N. (L2, L3, L4)	N	N	N				
5. **Biceps femoris** (Hamstring)	X		X (lateral rotation)		X (long head)		X (long head)	This is the lateral hamstring	Long head: Tibial part of sciatic N. (S1, S2, S3) Short hd: Peroneal part of sciatic N. (L5, S1, S2)				N	N	N	N
6. **Semitendinosus** (Hamstring)	X		X (medial rotation)		X	X		Note: All hamstrings can tilt the pelvis posteriorly.	Tibial part of the sciatic nerve (L5, S1, S2)				N	N	N	
7. **Semimembranosus** (Hamstring)	X		X (medial rotation)		X	X		Semimemb. is broad, flat, bipennate, deep to semitendinosus	Tibial part of the sciatic nerve (L5, S1, S2)				N	N	N	
8. **Popliteus**	may assist		X (medial rotation)					When weight bearing: Lateral rotation of femur, to unlock knee	Tibial N. (L4, L5, S1)			N	N	N		
(More muscles for the action --->)	see also Tables 11.13			see also Tables 10,11	see also Tables 10,11	see also Tables 10,11	see also Tables 10,11		Innervation							

Table 12 (B) - Knee (& Hip Joint, Part 3) - Synergists & Antagonists

Knee (& Hip Joint, Part 3)

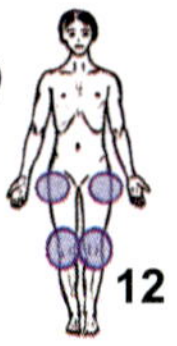

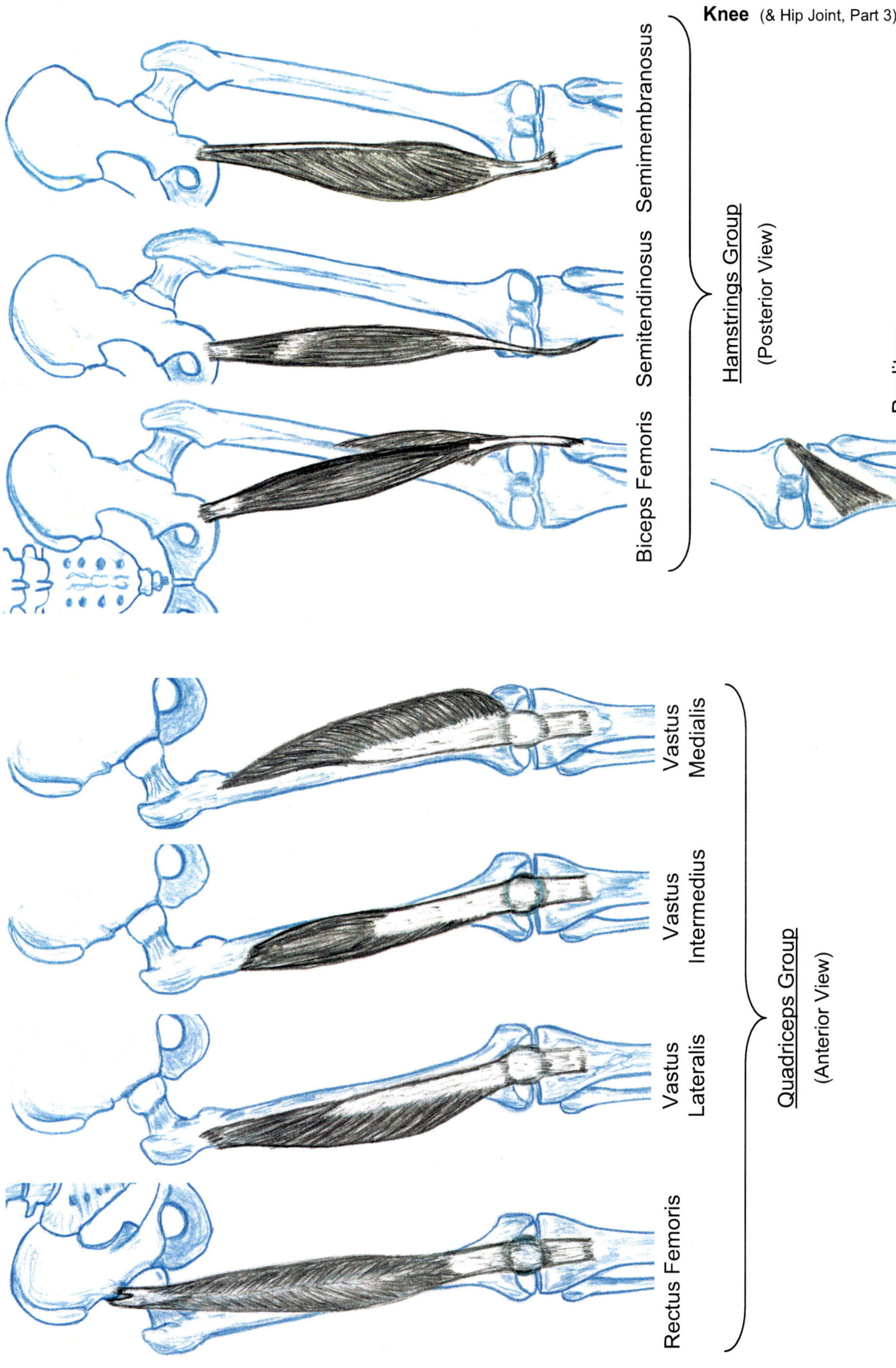

Semimembranosus

Semitendinosus

Biceps Femoris

Hamstrings Group
(Posterior View)

Popliteus

Vastus Medialis

Vastus Intermedius

Vastus Lateralis

Rectus Femoris

Quadriceps Group
(Anterior View)

Figure 12 (B) - **Knee** (& Hip Joint, Part 3) - **Muscle Pictures**

Mastering Muscles & Movement © 2007   Chapter 6 - Muscles That Move the Lower Extremity

Muscles Acting On
Knee (& Hip Joint, Part 3)

~ Notes ~ (palpation, fiber arrangement, lengthen/shorten, common uses, pathologies, cautions, etc.)

1. Rectus Femoris
2. Vastus Lateralis
3. Vastus Intermedius
4. Vastus Medialis
5. Biceps Femoris
6. Semitendinosus
7. Semimembranosus
8. Popliteus

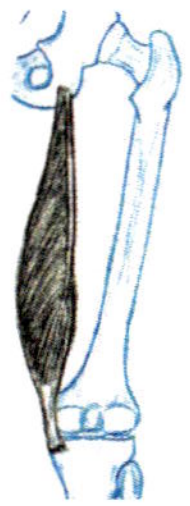

Muscle Group 12 - Knee (& Hip Joint, Part 3) - (Notes)

Movement of the Ankle, Foot, and Toes

Muscle Group 13:

Gastrocnemius	Peroneus brevis (fibularis b.)
Soleus	Peroneus longus (fibularis l.)
Plantaris	
Tibialis posterior	Tibialis anterior
Flexor digitorum longus	Extensor digitorum longus
Flexor hallucis longus	Extensor hallucis longus

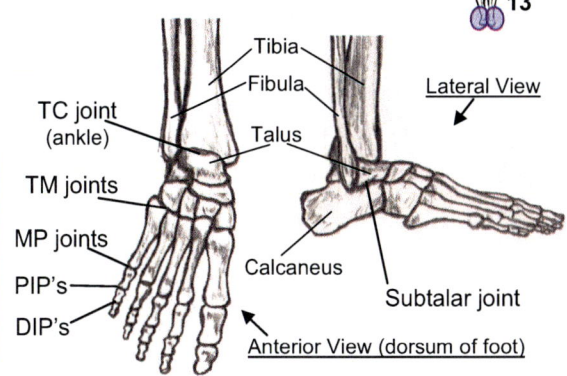

Joints

The muscles in this group move the ankle, foot, and toes. A couple of the muscles also cross the knee joint and therefore affect the knee. There are many joints involved, and it can be challenging to visualize which joints are in play with some of the more complex foot movements.

Talocrural Joint (TC) - Ankle
  Distal **tibia** & distal **fibula** ◄► **Talus**
  Hinge joint
  Movements available: Plantar flexion (Flexion)
  Dorsiflexion (Extension)

Subtalar Joint
  Inferior aspect of **Talus** ◄► Superior aspect of **Calcaneus**
  Gliding joint
  Movements available: Inversion (Supination)
  Eversion (Pronation)

Tarsometatarsal Joints (TM) (#1-#5)
  Distal row of **tarsals** ◄► Bases of **metatarsals**
  Gliding joints

Metatarsophalangeal Joints (MP) (#1-#5)
  Heads of **metatarsals** ◄► Bases of proximal **phalanges**
  Condyloid joints
  Movement of the toes: Flexion, Extension
  Abduction, Adduction

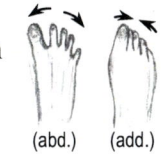
(abd.)   (add.)

Interphalangeal Joints (PIP and DIP) (#1-#5)
  Joints between the **phalanges** of the toes
  Hinge joints
  Movement of the toes: Flexion, Extension

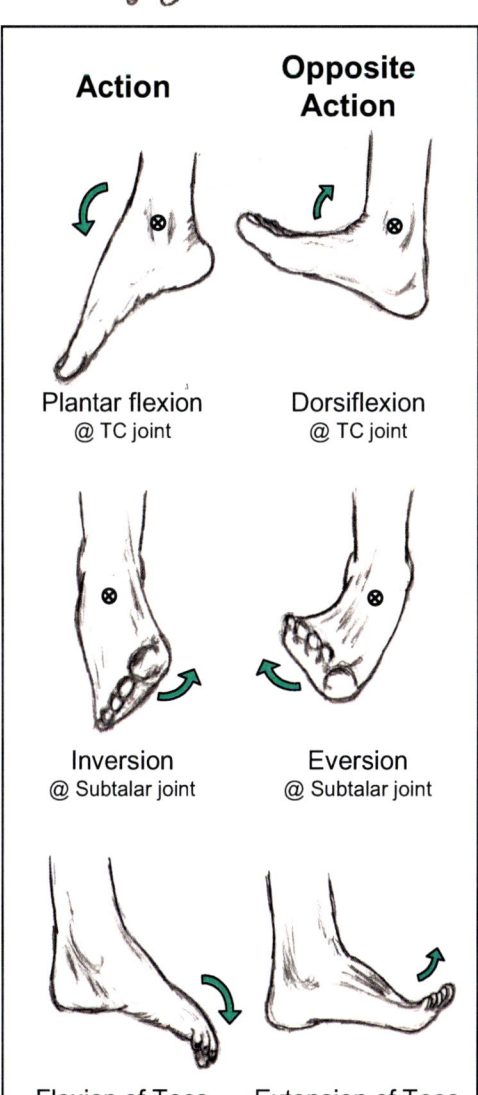

PIP = Proximal Interphalangeal (between the <u>proximal</u> phalanx and the <u>middle</u> phalanx)
DIP = Distal Interphalangeal   (between the <u>middle</u> phalanx and the <u>distal</u> phalanx)
Note: The big toe (hallux) has only 2 phalanges, so has only a DIP joint (no PIP)

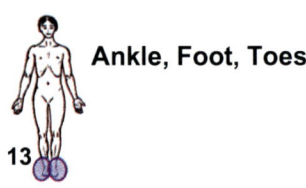

Ankle, Foot, Toes

Other Joints

The following joints are included here for completeness, but are not considered primary joints involved with ankle and foot movements. They are special combinations of intertarsal joints (interfaces between tarsal bones) that are important when studying the arches and flexibility of the foot.

    Transverse Tarsal Joint
    Talocalcaneonavicular Joint (TCN)
    Longitudinal and transverse <u>arches</u> of the foot

Tripod Arches of the Foot

Bones and Bony Landmarks

Muscles that move the ankle, foot and toes have attachments on the bones of the leg, foot, and posterior knee. Review the bony landmarks and other structures listed below, referring to Chapter 2, page 48-49.

Femur (p. 48)
    Medial and Lateral Condyles
    Medial and Lateral Epicondyles

Tibia (p. 48)
    Medial and Lateral Condyles
    Soleal Line
    Medial malleolus
    Shaft
        Posterior, Anterior, Medial, Lateral
        Combination aspects, e.g., posteriolateral
        Lengthwise positions, e.g.,
        "proximal two-thirds", "middle one-half"

Fibula (p. 48)
    Head
    Lateral malleolus
    Shaft
        Posterior, Anterior, combinations,
        lengthwise positions (as with tibia above)

Ankle & Foot Bones (p. 49)
    Tarsals
        Talus
        Calcaneus
        Cuboid
        Navicular
        3 Cuneiforms
            (1st: medial, 2nd: middle, 3rd: lateral)
    Metatarsals (5)
        5 bones: 1st=medial through 5th=lateral
    Phalanges (14)
        Digit #1 = Hallux (big toe)
            Proximal, Distal phalanges
        Digits #2-#5 = toes medial to lateral
            Proximal, Middle, Distal phalanges

<u>Tendinous Structures</u>
Interosseous membrane between tibia and fibula
Calcaneal Tendon/Achilles Tendon
Plantar Aponeurosis /Plantar Fascia
Retinacula: (Extensor, Flexor, Peroneal/Fibular)

Tendon Arrangements and Compartments of the Leg (see pages 179 and 186)

Medial malleolus tendons: Tibialis Posterior & two Flexor muscles (**TP, FDL, FHL**)
Dorsum of foot tendons: Tibialis Anterior & two Extensor muscles (TA, EHL, EDL)
Lateral malleolus tendons: Peroneus Longus & Peroneus Brevis (PL, PB)
Anatomical Stirrup: Tendons of per. long. and tib. ant. form "stirrup" under foot

Leg Compartments: 1. Anterior, 2. Lateral, 3. Deep posterior, 4. Superficial posterior

Intrinsic Muscles of the Foot (see pages 184-185)

There are twelve muscles that reside within the structure of the foot itself. These *intrinsic* muscles of the foot are not included in the Group 13 tables, but a separate table and illustrations are given on pages 184-185.

Plantar Muscles

Dorsal Muscles

Ankle, Foot, Toes

Muscle Group 13 – Muscles that move the ankle, foot, and toes are illustrated as a group on this page. The next four pages have tables and figures that describe each muscle individually, and provide many ways of comparing and contrasting the muscles to each other.

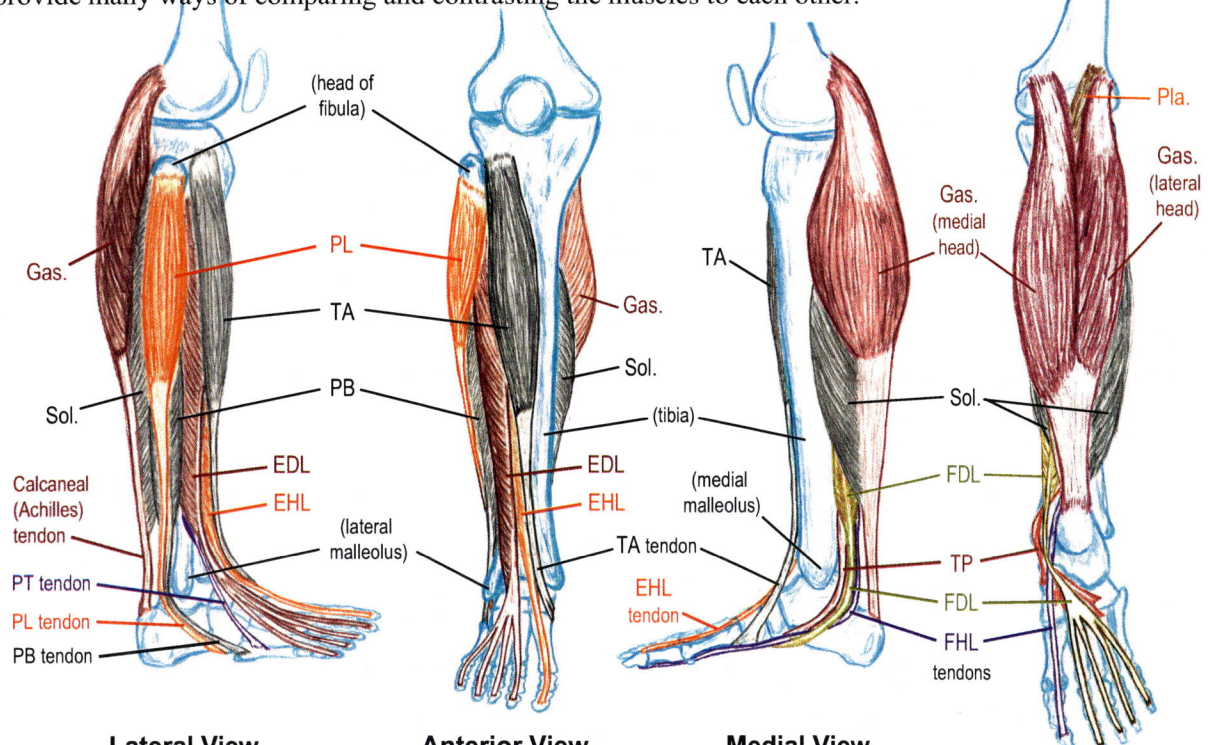

Lateral View   **Anterior View**   **Medial View**   **Posterior View**

Gas. - Gastrocnemius
Sol. - Soleus
Pla. - Plantaris

PL - Peroneus Longus
PB - Peroneus Brevis
PT - Peroneus Tertius

TA - Tibialis Anterior
EDL - Extensor Digitorum Longus
EHL - Extensor Hallucis Longus

TP - Tibialis Posterior
FDL - Flexor Digitorum Longus
FHL - Flexor Hallucis Longus

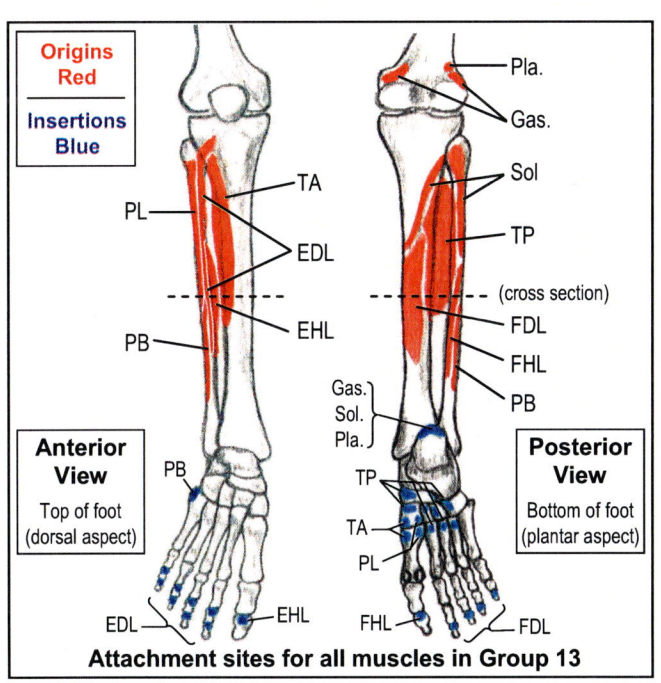

Attachment sites for all muscles in Group 13

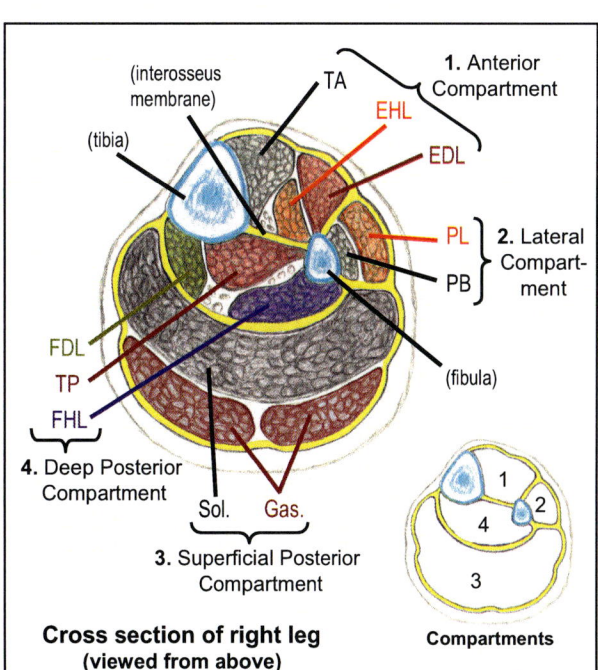

Cross section of right leg (viewed from above)

Compartments:
1. Anterior Compartment
2. Lateral Compartment
3. Superficial Posterior Compartment
4. Deep Posterior Compartment

Mastering Muscles & Movement © 2007   Chapter 6 – Muscles That Move the Lower Extremity

Ankle, Foot, Toes

DIP & PIP=Distal & Proximal Interphalangeal, MP=Metatarsophalangeal, Toes #1-#5: 1=big toe, 5=little toe

Muscles Acting On Ankle, Foot, Toes	Origin	Insertion	Action
Gastrocnemius — moves the ankle and knee	Posterior condyles of femur (lateral & medial)	Calcaneus via the achilles tendon	Plantar flexion of the ankle, Flexion of the knee (also stabilizes the knee in standing, walking, running)
Plantaris — moves the ankle and knee	Posterior lateral epicondyle of femur	Calcaneus via the achilles tendon (small spot on the medial side)	Weak plantar flexion of ankle, may assist with inversion of the foot and flexion of the knee
Soleus — moves the ankle	Proximal posterior shaft and head of fibula, Soleal line & middle medial edge of tibia	Calcaneus via the achilles tendon	Plantar flexion of the ankle
Tibialis Posterior — moves the foot	Posterior lateral tibia, Posterior medial fibula, and interosseus membrane.	Plantar aspect of all tarsals except talus, and bases of metatarsals #2-4 (Tarsal attachments: calcaneus, navicular, cuboid, 3 cuneiforms)	Inversion of the foot, Plantar flexion of the ankle
Flexor Digitorum Longus — moves toes #2-5 and the foot	Posterior tibia (starts 1/3 of the way down)	Base of distal phalanges #2-5 (plantar aspect)	Flexion of toes #2-5, Inversion of foot, Plantar flexion of ankle
Flexor Hallucis Longus — moves toe #1 and the foot	Posterior fibula (starts 1/3 of the way down)	Base of distal phalanx of hallux - big toe (plantar aspect)	Flexion of toe #1 (hallux), Inversion of foot, Plantar flexion of ankle
Peroneus Brevis (also called *Fibularis Brevis*) — moves the foot	Distal half of fibula (lateral aspect)	Tuberosity of the 5th metatarsal	Eversion of the foot, Assists plantar flexion of ankle
Peroneus Longus (also called *Fibularis Longus*) — moves the foot	Head and proximal two-thirds of fibula (lateral aspect)	Medial (1st) cuneiform and base of 1st metatarsal (plantar aspect)	Eversion of the foot, Assists plantar flexion of ankle
Tibialis Anterior — moves the foot	Lateral condyle and proximal half of tibia (lateral aspect) (and interosseus membrane)	Medial (1st) cuneiform and base of 1st metatarsal (medial edge of plantar aspect)	Dorsiflexion of the ankle, Inversion of the foot
Extensor Digitorum Longus — moves toes #2-5 and the foot	Lateral condyle of tibia, and proximal 2/3 of fibula (anterior aspect)	Middle & distal phalanges #2-5 (dorsal aspect)	Extension of toes #2-5, Dorsiflexion of the ankle, Eversion of the foot
Extensor Hallucis Longus — moves toe #1 and the foot	Middle portion of fibula (anterior medial aspect) (and interosseus membrane)	Base of distal phalanx of hallux - big toe (dorsal aspect)	Extension of toe #1 (hallux), Dorsiflexion of the ankle (May assist inversion of foot)

(larger illustrations on page 183)

Gas. Pla. Sol. TP FDL FHL PB PL TA EDL EHL

Table 13 (A) - Ankle, Foot, Toes - Origin, Insertion, Action

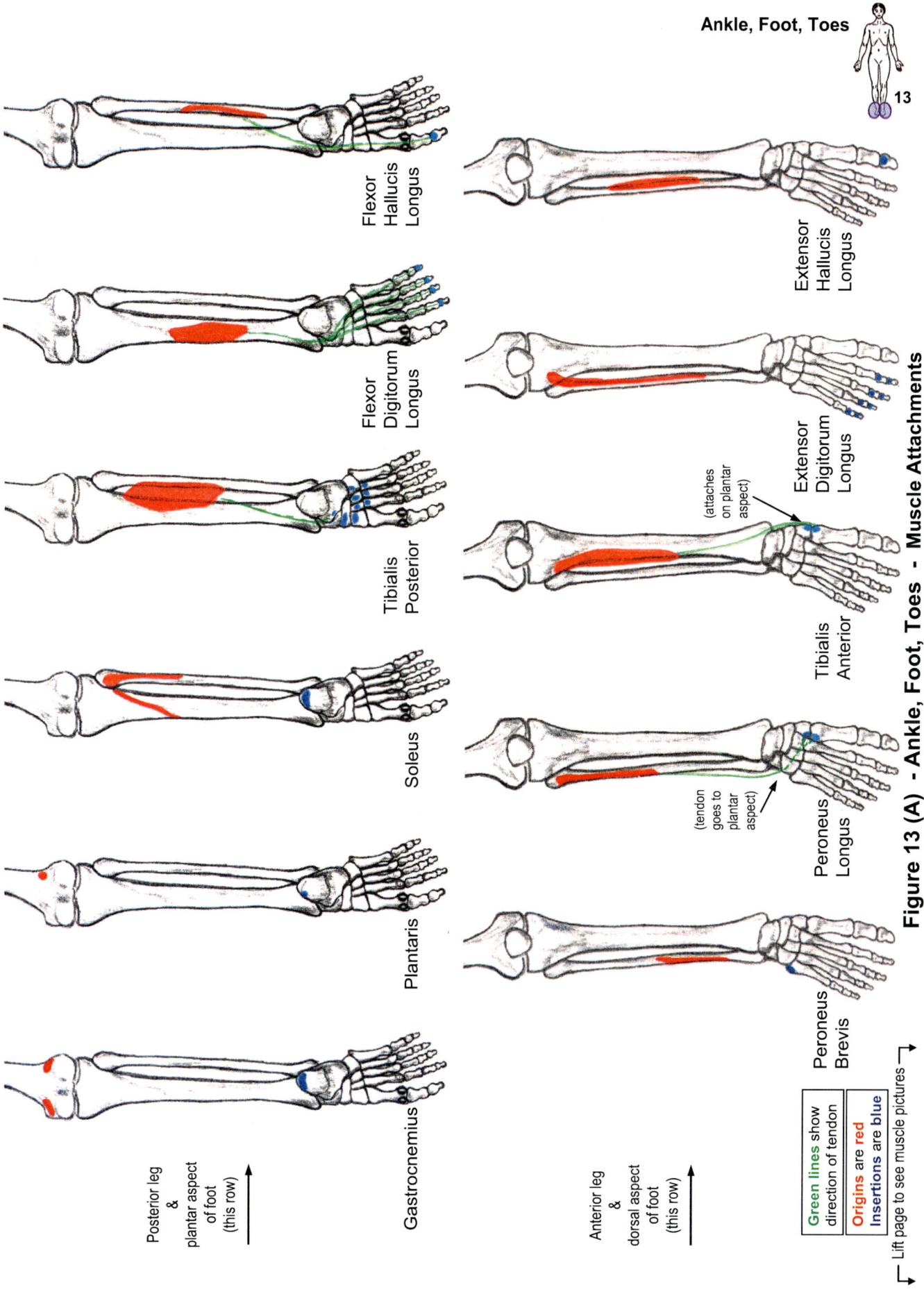

Figure 13 (A) - Ankle, Foot, Toes - Muscle Attachments

Ankle, Foot, Toes

#1-#5 indicate toe digit numbers (1=big toe, 5=little toe), TC jt.=Talocrural joint (ankle), TF jt.=Tibiofemoral joint (knee)

Muscles Acting On **Ankle, Foot, Toes**	Plantar flexion (=flexion) @ Ankle	Dorsiflexion (=extension) @ Ankle	Inversion (@ Subtalar joint)	Eversion (@ Subtalar joint)	Flexion of Toes	Extension of Toes	Flexion @ Knee (TF jt.)	Stabilization	Innervation	L4	L5	S1	S2
1. Gastrocnemius	X						X	Stabilizes knee	Tibial N. (S1, S2)			N	N
2. Plantaris	X weak		may assist				may assist		Tibial N. (L4, L5, S1)	N	N	N	
3. Soleus	X								Tibial N. (S1, S2)			N	N
4. Tibialis posterior	X		X					Stabilizer of ankle/foot	Tibial N. (L5, S1)		N	N	
5. Flexor digitorum longus	X		X		X #2-5				Tibial N. (L5, S1)		N	N	
6. Flexor hallucis longus	X		X		X #1 (hallux)				Tibial N. (L5, S1, S2)		N	N	N
7. Peroneus brevis	X assist			X				Helps stabilize foot	Superficial peroneal N. (L4, L5, S1)	N	N	N	
8. Peroneus longus	X assist			X				PL and TA form "Anatomical stirrup"...	Superficial peroneal N. (L4, L5, S1)	N	N	N	
9. Tibialis anterior		X	X					...helping to maintain balance & stabilize foot	Deep peroneal N. (L4, L5, S1)	N	N	N	
10. Extensor digitorum longus		X		X		X #2-5			Deep peroneal N. (L4, L5, S1)	N	N	N	
11. Extensor hallucis longus		X	weak assist			X #1 (hallux)			Deep peroneal N. (L4, L5, S1)	N	N	N	
(More muscles for the action) --->							see also Tables 11,12		Innervation				

Table 13 (B) - Ankle, Foot, Toes - Synergists & Antagonists

Ankle, Foot, Toes

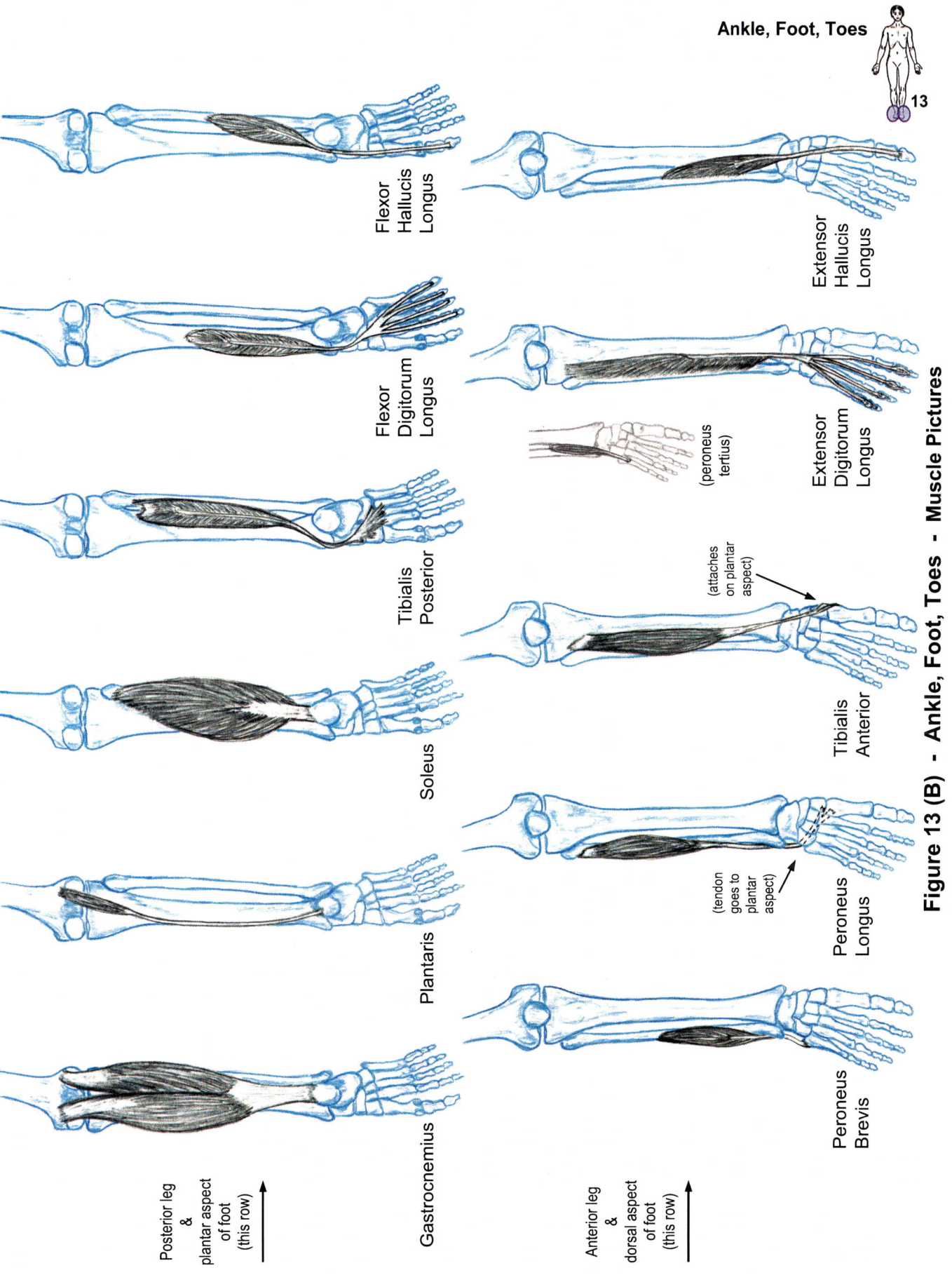

Figure 13 (B) - Ankle, Foot, Toes - Muscle Pictures

Chapter 6 - Muscles That Move the Lower Extremity

Ankle, Foot, Toes

Intrinsic Muscles of the Foot – Plantar Aspect

Muscle	Origin	Insertion	Action	Innervation
Plantar Layer #1 (superficial)				
Abductor Digiti Minimi	Tuberosity of the calcaneus	Proximal phalanx of the little toe (lateral base)	Abduction and flexion of the little toe	Lateral plantar N. (S2, S3)
Flexor Digitorum Brevis	Tuberosity of the calcaneus	Middle phalanges of toes #2-5 (sides)	Flexion of toes #2-5	Medial plantar N. (L5, S1)
Abductor Hallucis	Tuberosity of the calcaneus	Proximal phalanx of the big toe (medial base)	Abduction and flexion of the big toe	Medial plantar N. (L5, S1)
Plantar Layer #2 (intermediate)				
Lumbrical Muscles (4)	The four tendons of the flexor digitorum longus	The four tendons of the extensor digitorum longus (attach via the medial side of the dorsal digital expansions)	Flexion of toes #2-5 at the metatarsophalangeal joints, Extension of toes #2-5 at the interphalangeal joints	Lumbrical 1: Medial plantar N. (L5, S1) Lumbricals 2-4: Lateral plantar N. (S2, S3)
Quadratus Plantae	Plantar surface of the calcaneus	Tendon of the flexor digitorum longus (lateral margin, before it goes to the 4 toes)	Flexion of toes #2-5 (assists the FDL)	Lateral plantar N. (S2, S3)
Plantar Layer #3 (almost deepest)				
Flexor Digiti Minimi	Base of 5th metatarsal (& peroneus longus tendon)	Proximal phalanx of the little toe (plantar base)	Flexion of the little toe (at the MP joint)	Lateral plantar N. (S2, S3)
Adductor Hallucis	Oblique head: Bases of metatarsals #2-4, Transverse head: Metatarsophalangeal ligaments #3-5	Proximal phalanx of the big toe (lateral base)	Adduction of the big toe	Lateral plantar N. (S2, S3)
Flexor Hallucis Brevis	Cuboid and lateral cuneiform (plantar surfaces)	Proximal phalanx of the big toe (sides of base)	Flexion of the big toe (at the MP joint)	Medial plantar N. (L5, S1)
Plantar Layer #4 (deepest)				
Plantar Interossei (3)	3rd, 4th and 5th metatarsal bones (bases and medial side of shafts)	Bases of the proximal phalanges of toes #3-5 (and the dorsal digital expansions of toes #3-5)	Adduction of toes #3-5, Assist flexion of toes #3-5 at the metatarsophalangeal joints, Assist extension of toes #3-5 at the interphalangeal joints	Lateral plantar N. (S2, S3)

Intrinsic Muscles of the Foot – Dorsal Aspect

Muscle	Origin	Insertion	Action	Innervation
Dorsal Layer #1 (superficial)				
Extensor Digitorum Brevis	Dorsal surface of the calcaneus	Toes #2-4, via the tendons of the extensor digitorum longus (attach to the lateral side of the EDL tendons)	Extension of toes #2-4	Deep peroneal N. (L5, S1)
Extensor Hallucis Brevis	Dorsal surface of the calcaneus	Proximal phalanx of the big toe (dorsal surface of the base of the phalanx)	Extension of the big toe	Deep peroneal N. (L5, S1)
Dorsal Layer #2 (deep) Note: This layer is sometimes considered to be part of plantar layer #4				
Dorsal Interossei (4)	Shafts of metatarsal bones #1-5 (each muscle arises from the sides of two adjacent metatarsal bones)	Bases of the proximal phalanges of toes #2-4 (and the dorsal digital expansions of toes #2-4)	Abduction of toes #2-4, Assist flexion of toes #2-4 at the metatarsophalangeal joints, Assist extension of toes #2-4 the interphalangeal joints	Lateral plantar N. (S2, S3)

Intrinsic Muscles of the Foot

Ankle, Foot, Toes

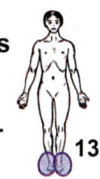

Plantar Aspect Has 4 Layers

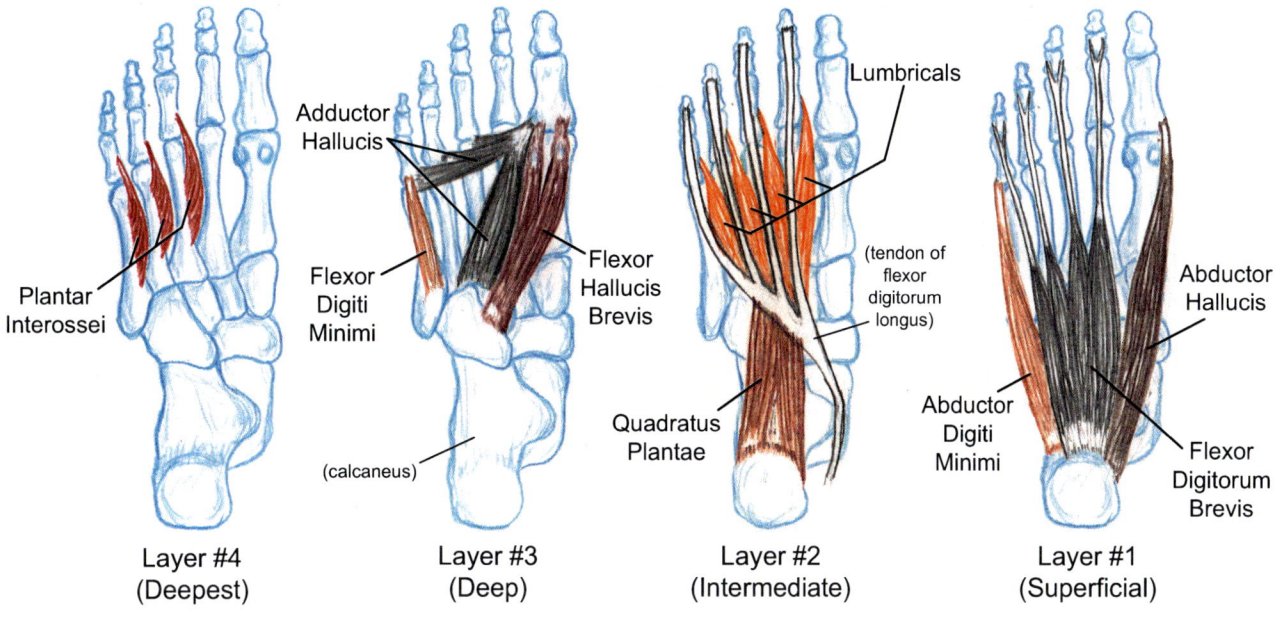

Building the Plantar Muscles One Layer at a Time →

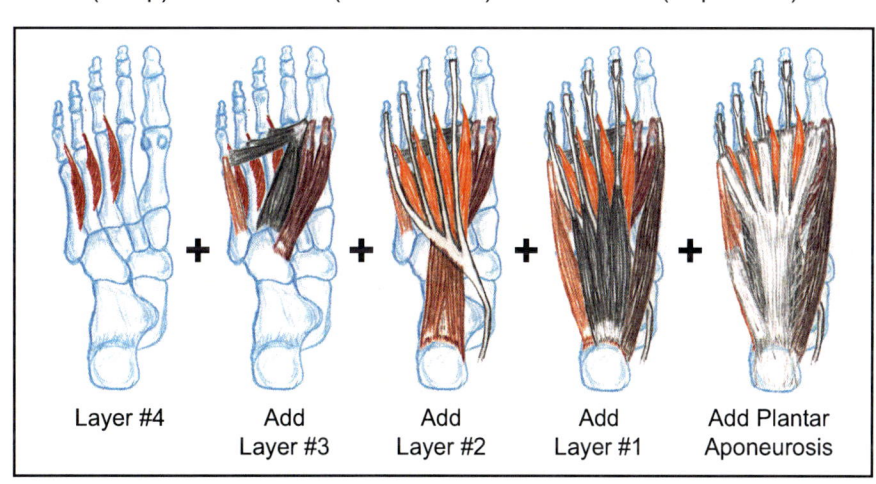

Dorsal Aspect Has 2 Layers

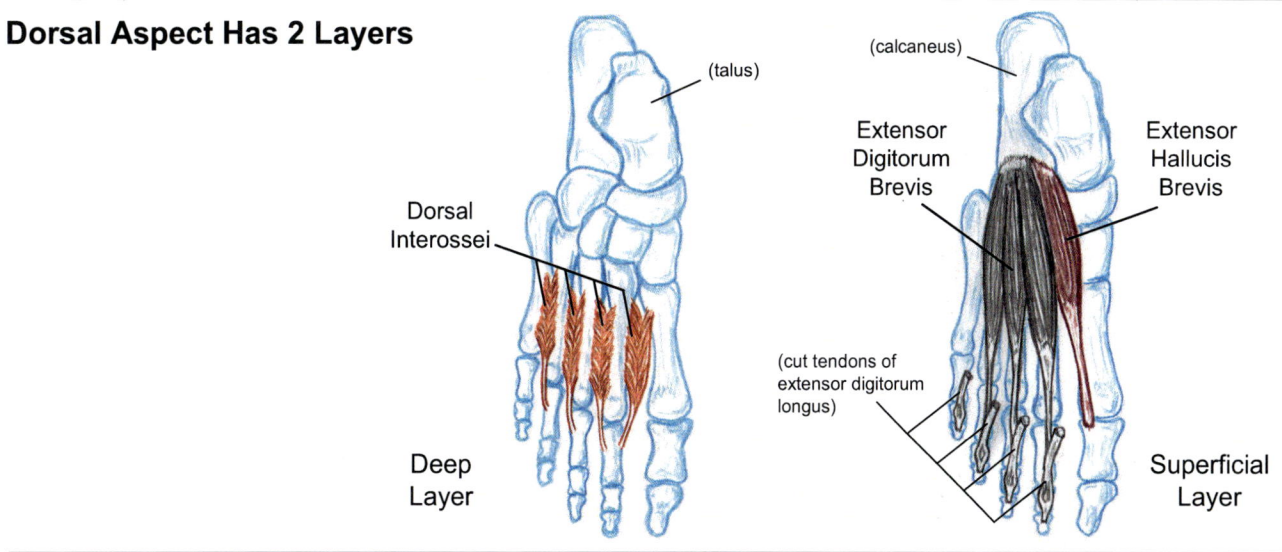

Ankle, Foot, Toes

Muscles of the Leg - by Compartment

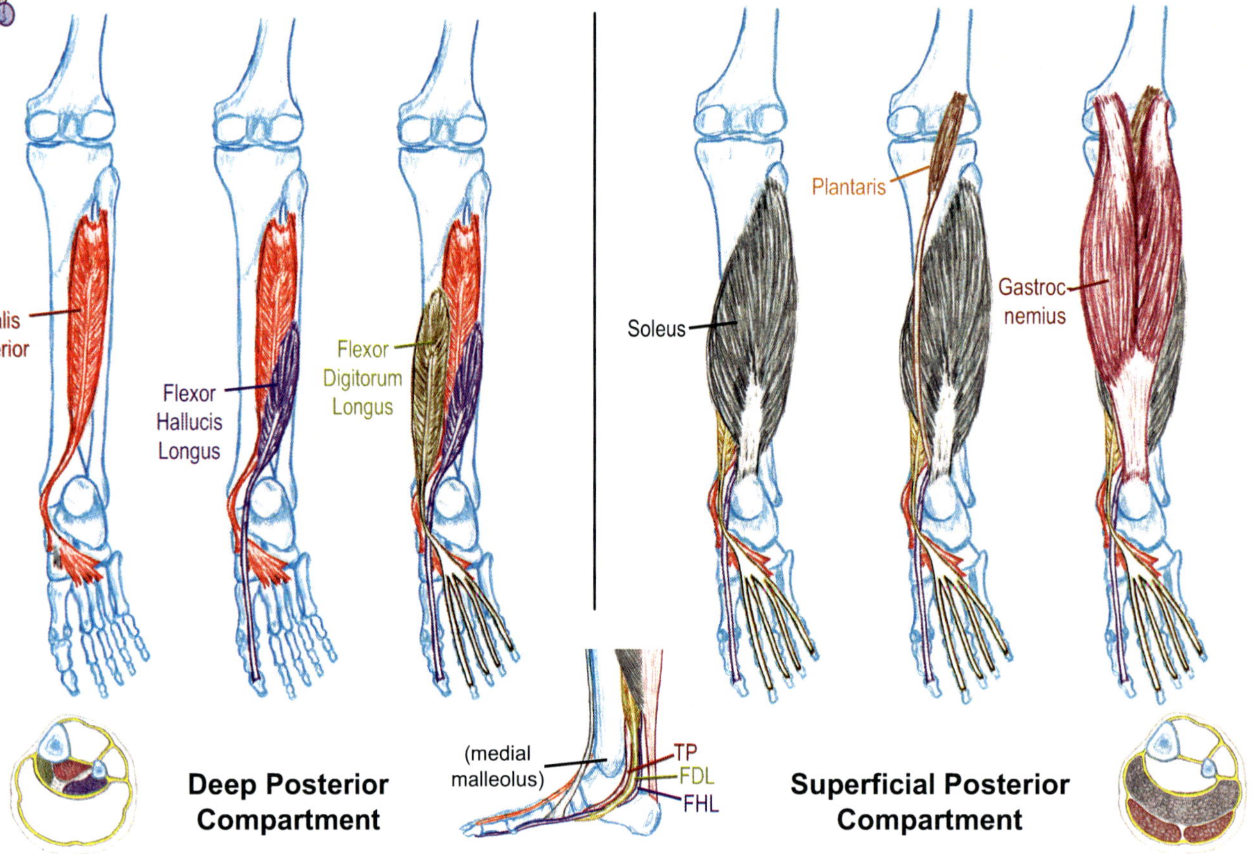

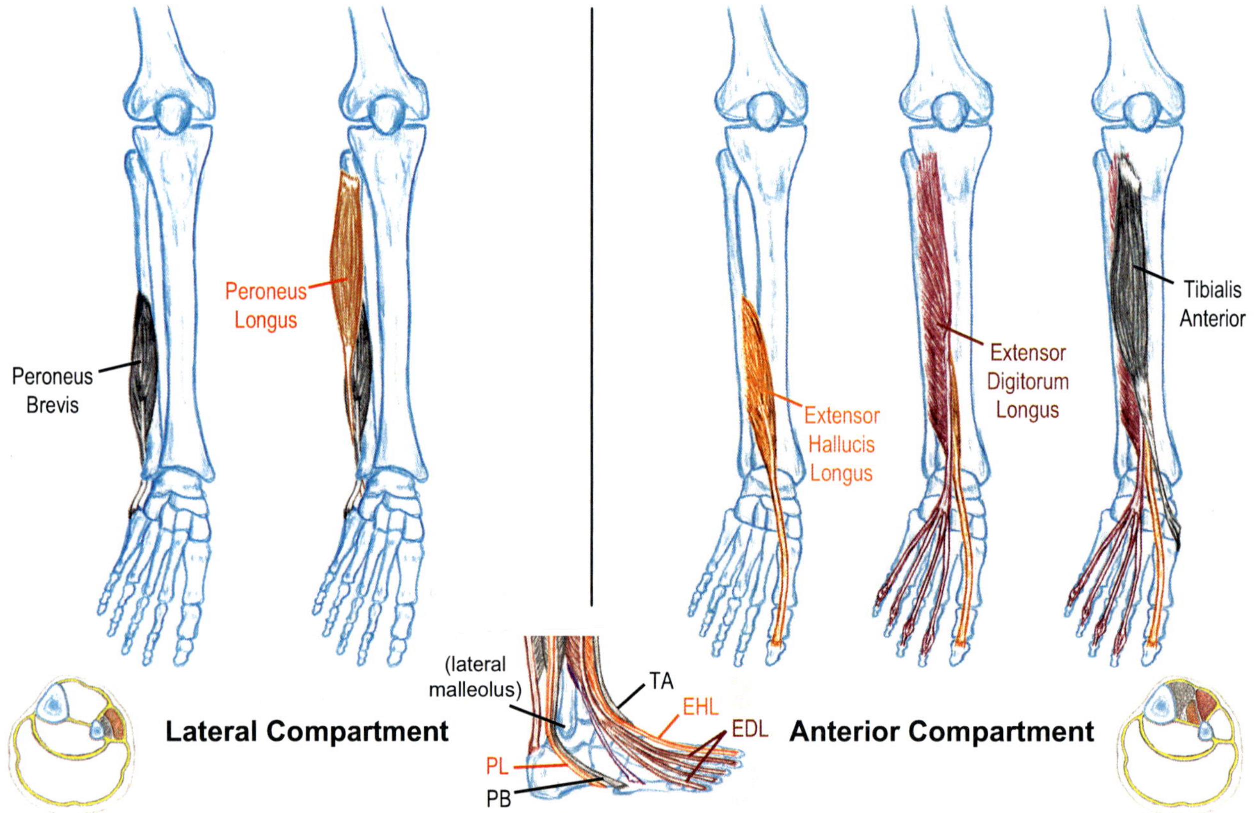

Tibialis Posterior

Flexor Hallucis Longus

Flexor Digitorum Longus

Plantaris

Soleus

Gastrocnemius

(medial malleolus) — TP, FDL, FHL

Deep Posterior Compartment

Superficial Posterior Compartment

Peroneus Brevis

Peroneus Longus

Extensor Hallucis Longus

Extensor Digitorum Longus

Tibialis Anterior

(lateral malleolus) — TA, EHL, EDL, PL, PB

Lateral Compartment

Anterior Compartment

186   Chapter 6 – Muscles That Move the Lower Extremity   Mastering Muscles & Movement   © 2007

Chapter 7

Summary Tables

Introduction **188**

Summary of Actions – Upper Extremity **189**
    Scapula / Clavicle 189
    Shoulder Joint 189
    Elbow, Forearm 190
    Wrist, Hand, Fingers 190

Summary of Actions – Axial Skeleton **191**
    Face, Jaw 191
    Head, Neck 191
    Spine, Trunk, Breathing 192

Summary of Actions – Lower Extremity **193**
    Hip Joint 193
    Knee 193
    Ankle, Foot, Toes 194

Joint Summary **195**

Ligament Summary **197**

Innervation Summary **198**

Introduction

Chapter 7 – Summary Tables provides a compact reference that can be quickly reviewed once you have learned all the muscles in Chapters 4, 5, and 6. The chapter includes "Summary of Actions" tables for each major structure of the body, and condensed reference tables for joints, ligaments, and innervation.

- Good for quick review before an exam
- Useful when you are assessing/analyzing a client's movement patterns

Summary of Actions Tables

Tables S-1, S-2 and S-3 use a format similar to the "B" Tables in Chapters 4-6, except that muscles are gathered from *multiple* B tables in cases where muscles from different groups move the same joint. This happens in Muscle Groups 1-13 because some muscles move multiple joints (e.g., biceps brachii moves both the elbow and the shoulder joints). Conversely, sometimes a single joint was presented over more than one group (e.g., Muscle Groups 11, 12, and 13 all have muscles that move the hip joint). The composite organization in this summary chapter provides a single unified table for each structure, to have a complete picture of the muscles affecting its movements.

Table S-1 – Summary of Actions – Upper Extremity: S-1 has four sub-tables: Scapula / Clavicle
Shoulder Joint
Elbow, Forearm
Wrist, Hand, Fingers

Table S-2 – Summary of Actions – Axial Skeleton: S-2 has three sub-tables: Head, Neck
Spine, Trunk, Breathing
Face, Jaw

Table S-3 – Summary of Actions – Lower Extremity: S-3 has three sub-tables: Hip Joint
Knee
Ankle, Foot, Toes

Other Tables

Tables S-4 through S-6 summarize information on joints, ligaments, and innervation.

Table S-4 – Joint Summary: A list of the joints of the body with pertinent information about each.

Table S-5 – Ligament Summary: A list of the major ligaments and their functions at the joints.

Table S-6 – Innervation Summary: A list of the major nerves of the body and the muscles they supply.

Table S-1 -- Summary of Actions – UPPER EXTREMITY

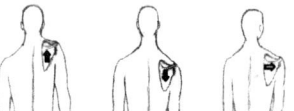

Scapula / Clavicle	Elevation	Depression	Protraction/ Abduction	Retraction/ Adduction	Lateral / Upward Rotation	Medial / Downward Rotation	Stabilization of Scapula	Ref. Table	Muscle also affects other joints:
Trapezius (upper fibers)	X				X		X	1	ROI: Head/Neck
Trapezius (middle fibers)				X			X	1	-
Trapezius (lower fibers)		X			X		X	1	-
Levator scapula	X					X		1	ROI: Neck
Rhomboid (major & minor)	X			X		X		1	-
Serratus anterior			X		X		X	1	-
Pectoralis minor		"anterior tilt"						1	ROI: Breathing
Subclavius		depresses clavicle						1	ROI: Breathing
Pectoralis Major		X (lower/ab fib)						2	Shoulder Joint
Latissimus Dorsi		X						2	Shoulder Joint

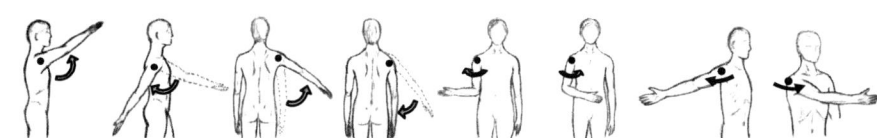

Shoulder Joint	Flexion @ GH jt.	Extension @ GH jt.	Abduction @ GH jt.	Adduction @ GH jt.	Lateral Rotation @ GH jt.	Medial Rotation @ GH jt.	Horizontal Abduction @ GH jt.	Horizontal Adduction @ GH jt.	Ref. Table	Muscle also affects other joints:
Deltoid (anterior fibers)	X		X			X		X	2	-
Deltoid (middle fibers)			X						2	-
Deltoid (posterior fibers)		X	X		X		X		2	-
Supraspinatus			X						2	-
Infraspinatus					X		may assist		2	-
Teres Minor					X				2	-
Subscapularis						X			2	-
Pectoralis Major (upper fibers)	X			X		X		X	2	-
Pectoralis Major (lower fibers)		X (from flexed pos.)		X		X			2	Scapula
Coracobrachialis	X			X				assist	2	-
Latissimus Dorsi		X		X		X			2	Spine/Trunk. Scapula
Teres Major		X		X		X			2	-
Biceps Brachii	X							assist (short head)	3	Elbow/ Forearm
Triceps Brachii (long head)		X		assist					3	Elbow

KEY: X – Muscle creates the action, Assist – Muscle assists the action, GH – Glenohumeral joint (shoulder joint)
ROI – Reversed Origin/Insertion (action moves O toward I)

Table S-1 -- Summary of Actions – UPPER EXTREMITY (continued)

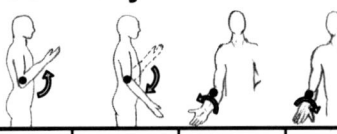

Elbow, Forearm	Flexion @ HU jt.	Extension @ HU jt.	Supination @ RU jt.	Pronation @ RU jt.	Stabilization	Ref. Table	Muscle also affects other joints:
Biceps brachii	X		X			3	Shoulder Joint
Brachialis	X					3	-
Brachioradialis	X (handshake)		assist (from sup.)	assist (from pron.)		3	-
Pronator teres	assist)			X		3	-
Pronator quadratus				X		3	-
Supinator			X			3	-
Triceps brachii		X				3	Shoulder Joint
Anconeus		assist			x	3	-
Extens. carpi radialis long.	Assist (handshake)					4	Wrist
All wrist/digit <u>flexors</u> that attach proximal to elbow	may assist					4	Wrist

KEY

X – Muscle creates the action
Assist – Muscle assists the acti
HU – Humeroulnar joint (elbow)
RU – Radioulnar joint
RC – Radiocarpal joint (wrist)

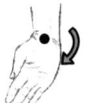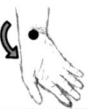

Wrist, Hand, Fingers	Flexion @ wrist	Extension @ wrist	Abduction/ Radial Deviation	Adduction/ Ulnar Deviation	Flexion Phalanges	Extension Phalanges	Abduction Phalanges	Adduction Phalanges	Ref. Table	Muscle also affects other joints:
Flexor carpi radialis	X		X						4	Elbow
Palmaris longus	X				Cups the hand				4	Elbow
Flexor carpi ulnaris	X			X					4	Elbow
Flexor digitorum superficialis	X				X fingers				4	Elbow
Flexor digitorum profundus	assist				X closed fist				4	-
Extensor carpi radialis longus		X	X						4	Elbow
Extensor carpi radialis brevis		X	X						4	-
Extensor carpi ulnaris		X		X					4	-
Extensor digitorum		X			X fingers	assist			4	-
Extensor indicis		assist			X index finger					-
Flexor pollicis longus	may assist				X thumb				5	-
Opponens pollicis					Thumb Opposition				5	-
Adductor pollicis								X thumb		-
Abductor pollicis longus			X		X thumb	X thumb			5	-
Extensor pollicis longus		assist	assist		X thumb				5	-
Extensor pollicis brevis			assist		X thumb				5	-

Thumb actions →

Table S-2 -- Summary of Actions – AXIAL SKELETON

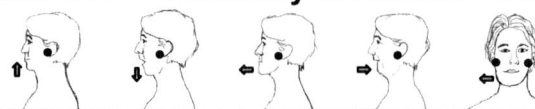

Face, Jaw	Elevation of mandible	Depression of mandible	Protrusion/ Protraction of mandible	Retrusion/ Retraction of mandible	Lateral Deviation of mandible	Other	Ref. Table	Muscle also affects other joints:
Masseter	X			assist (deep belly)			6	-
Temporalis	X			X (post. fibers)			6	-
Lateral Pterygoid		assist	X		UL to opp. side		6	-
Medial Pterygoid	assist		may assist		UL to opp. side		6	-
Occipitofrontalis						raise eyebrows	6	-
Platysma		assist				tighten skin of neck	6	-
Suprahyoids Group		X		assist		elev. hyoid (swallow)	6	-
Infrahyoids Group		assist				depress hyoid	6	-

Head, Neck	Flexion	Extension	Lateral Flexion	Rotation *Same* Side	Rotation *Opp.* Side	Comment	Ref. Table	Muscle also affects other joints:
Sternocleidomastoid	BL head/neck		UL head/neck		UL head/neck		7	Breathing
Scalenus Anterior	BL neck		UL neck			ROI use of muscle	7	Breathing
Scalenus Medius	BL neck		UL neck			ROI use of muscle	7	Breathing
Scalenus Posterior			UL neck			ROI use of muscle	7	Breathing
Longus Capitis	BL head/neck						7	-
Longus Colli	BL neck		may assist				7	-
Suboccipitals Group		BL head	UL head	UL (OCI only)			7	-
Splenius Capitis		BL head/neck	UL head/neck	UL head/neck			7	-
Splenius Cervicis		BL neck	UL neck	UL neck			7	-
Levator Scapula (scapula held fixed)		BL neck	UL neck	UL neck		ROI use of muscle	7	Scapula
Trapezius, upper (scapula held fixed)		BL head/neck	UL head/neck		UL head/neck	ROI use of muscle	7	Scapula
Spinalis (Cervicis)		X					8	Spine/Trunk
Longissimus (Capitis)		BL head/neck	UL head/neck	UL head/neck			8	Spine/Trunk
Longissimus (Cervicis)		BL neck	UL neck	UL neck			8	Spine/Trunk
Iliocostalis (Cervicis)		BL neck	UL neck	UL neck			8	Spine/Trunk
Semispinalis (Capitis)		BL head/neck					8	Spine/Trunk
Semispinalis (Cervicis)		BL neck			UL neck		8	Spine/Trunk
Multifidus (upper slips)		BL neck			UL neck		8	Spine/Trunk
Rotatores (upper slips)		BL neck			UL neck		8	Spine/Trunk

KEY: X = Muscle creates the action, UL = Unilateral contraction creates the action, BL = Bilateral contraction creates the action
ROI = Reversed Origin/Insertion (the action moves the origin toward the insertion)

Head/Neck = Muscle pulls the head, and the head & neck both move with a complex dynamic
Neck = Muscle pulls the neck, and the head goes along for the ride
Head = Muscle pulls the head from C1/C2, so the head moves on top of the spine, *without* the neck being moved

Table S-2 -- Summary of Actions – AXIAL SKELETON (continued)

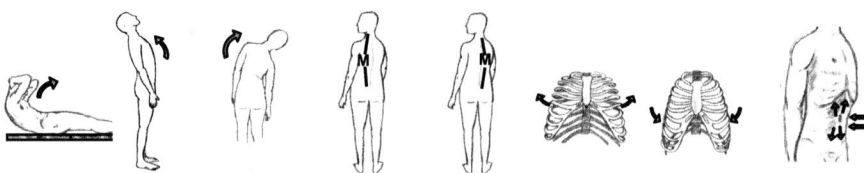

Spine, Trunk, Breathing	Flexion	Extension	Lateral Flexion	Rotation *Same* Side	Rotation *Opp.* Side	Inhalation/ Inspiration	Exhalation/ Expiration	Compress Abdominal Contents	Ref. Table	Muscle also affects other joints
Spinalis (thoracis)		BL							8	Head/Neck
Longissimus (thoracis)		BL	UL	UL (assist)					8	Head/Neck
Iliocostalis (thorac, lumb.)		BL	UL	UL (assist)			assist		8	Neck
Semispinalis (thoracis)		BL			UL				8	Neck
Multifidus		BL			UL				8	Neck
Rotatores		BL			UL				8	Neck
Quadratus Lumborum		BL (lumbar)	UL				X stabilize rib 12		8	Hip hike
Psoas Major	BL (↑ lordosis)		may assist						10	Hip Joint
Sternocleidomastoid						assist (ROI)			7	Head/Neck
Scalenes						X			7	Neck
Pectoralis Minor						assist (ROI)			1	Scapula
Subclavius						assist (ROI)			1	Clavicle
Diaphragm						X		X (downward)	9	-
Rectus Abdominis	X						X	X	9	-
External Oblique	BL		UL		UL		X	X	9	-
Internal Oblique	BL		UL	UL			X	X	9	-
Transverse Abdominis							X	X	9	-
External Intercostals						X			9	-
Internal Intercostals							X		9	-
Serratus Post. Superior						X			9	-
Serratus Post. Inferior							X		9	-
Levator Costae						X			9	-
Latissimus Dorsi		BL	UL						2	-

KEY: X = Muscle creates the action, UL = Unilateral contraction creates the action, BL = Bilateral contraction creates the action
ROI = Reversed Origin/Insertion (the action moves the origin toward the insertion)

Rotation *Same* side = Front of body moves to same side as the muscle contracting
Rotation *Opp.* side = Front of body moves to opposite side from the muscle contracting

Table S-3 -- Summary of Actions – LOWER EXTREMITY

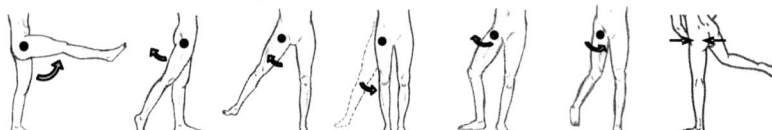

Hip Joint	Flexion @ Hip jt.	Extension @ Hip jt.	Abduction @ Hip jt.	Adduction @ Hip jt.	Lateral Rotation @ Hip jt.	Medial Rotation @ Hip jt.	Stabilization of Hip jt.	Ref. Table	Muscle also affects other joints:
Gluteus maximus		X	may assist (upper fib.)	X (lower fib.)	X			10	-
Gluteus medius	assist (anter. fib.)	assist (poster. fib.)	X		assist (poster. fib.)	assist (anter. fib.)	X	10	-
Gluteus minimus	may assist		X			X	X	10	-
Piriformis (& other 5 lateral rotators)					X			10	-
iliopsoas { Iliacus	X					may assist		10	-
iliopsoas { Psoas major	X					may assist		10	ROI: Trunk / Lumbar Spine
Sartorius	X		X		X			11	Knee
Tensor fascia latae	X		X			X	X	11	Knee
Pectineus	X			X		X		11	-
Adductor brevis	X			X		X		11	-
Adductor longus	X			X		X		11	-
Adductor magnus	X (anterior fib.)	X (poster. fib.)		X (all fibers)		X (anterior fib.)		11	-
Gracilis	may assist			X		may assist		11	Knee
Rectus femoris	X							12	Knee
Biceps femoris		X (long head)			X (long head)			12	Knee
Semitendinosus		X				X		12	Knee
Semimembranosus		X				X		12	Knee

Knee	Flexion @ TF jt.	Extension @ TF jt.	Lateral Rotation @ TF jt. (flexed)	Medial Rotation @ TF jt. (flexed)	Stabilization of TF jt.	Ref. Table	Muscle also affects other joints:
Rectus femoris		X				12	Hip Joint
Vastus medialis		X				12	-
Vastus lateralis		X				12	-
Vastus intermedius		X				12	-
Biceps femoris	X		X			12	Hip Joint
Semitendinosus	X			X		12	Hip Joint
Semimembranosus	X			X		12	Hip Joint
Popliteus	may assist			X		12	-
Sartorius	X			X		11	Hip Joint
Tensor fascia latae					X	11	Hip Joint
Gracilis	X			X	X	11	Hip Joint
Gastrocnemius	X				X	13	Ankle
Plantaris	may assist					13	Ankle

KEY

X – Muscle creates the action

Assist – Muscle assists the action

TF – Tibiofibular joint (the knee joint)

ROI – Reversed O/I action

Table S-3 -- Summary of Actions – LOWER EXTREMITY (continued)

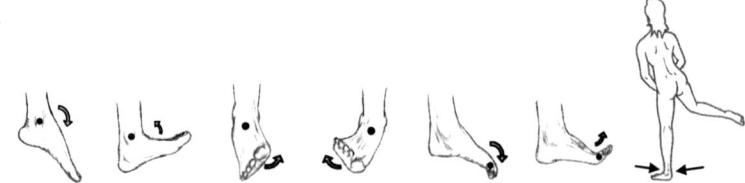

Ankle, Foot, Toes	Plantarflex (flexion @ TC jt.)	Dorsiflex (extension @ TC jt.)	Inversion (subtalar joint)	Eversion (subtalar joint).	Flexion of Phalanges	Extension of Phalanges	Stabilization of Ankle/Foot	Ref. Table	Muscle also affects other joints:
Gastrocnemius	X							13	Knee
Plantaris	may assist		may assist					13	Knee
Soleus	X							13	-
Tibialis posterior	X		X				X	13	-
Flexor digitorum longus	X		X		X #2-5			13	-
Flexor hallucis longus	X		X		X #1 (hallux)			13	-
Peroneus brevis	assist			X				13	-
Peroneus longus	assist			X			P.L. & T.A. create stirrup to stabilize foot/ankle	13	-
Tibialis anterior		X	X					13	-
Extensor digitorum longus		X		X		X #2-5		13	-
Peroneus tertius		X		assist				13	
Extensor hallucis longus		X	may assist			X #1 (hallux)		13	-

KEY: X – Muscle creates the action, Assist – Muscle assists the action, May assist – May help action under certain circumstances
TC – Talocrural joint (ankle), Hallux – Big toe (digit #1)

Table S-4 -- JOINT SUMMARY

SYNOVIAL JOINT	JOINT TYPE	COMMENTS ("◄►"=articulation)
Head		
Temporomandibular	condyloid/gliding/hinge	6 actions: elevation, depression protraction, retraction, R.&L. lateral deviation
Spine		
Atlantooccipital (Occipitoatlantal)	ellipsoid	2 points of contact (at facets) = 1 ellipsoid joint
Atlantoodontoid	pivot	C1/C2: anterior arch of atlas ◄► dens of axis
Atlantoaxial	gliding	C1/C2: inferior facets C1 ◄► superior facets C2
Intervertebral Facets	gliding	
Intervertebral Discs	(amphiarthrotic)	Cartilaginous joint (not synovial)
Sacroiliac	part gliding	Part fibrous (synarthrotic), part synovial
Trunk		
Sternocostal (rib 1)	(synarthrotic)	Fibrous joint (not synovial)
Sternocostal (ribs 2-7)	gliding	Ribs 2-7, sternum ◄► costal cartilage
Costochondral	(synarthrotic-fibrous)	Ribs 1-10, "junctions" of rib ◄► costal cartilage
Costovertebral	gliding	Head of rib ◄► costal facet on vertebral body
Costotransverse	gliding	Tubercle of rib ◄► costal facet on transverse process
Shoulder Complex		
Glenohumeral	ball and socket	This is the true shoulder joint, humerus ◄► glenoid fossa. Does 6 B&S actions+ Horiz. abduction, Horiz. adduction
Sternoclavicular (SC)	modif. ball & socket/hinge	SC, AC, & ST joints move together, with 6 actions:
Acromioclavicular (AC)	gliding	elevation, depression, protraction, retraction,
Scapulothoracic (ST)	false	upward rotation, downward rotation
Elbow		
Humeroulnar	hinge	The elbow joint
Radioulnar (Proximal)	pivot	Does supination & pronation (i.e., rotations)
Radioulnar (Distal)	pivot (but not much)	Pivots slightly to help accommodate supin. & pronation
Wrist, Hand, Fingers		
Radiocarpal	ellipsoid	Radius ◄► scaphoid, lunate, & triquetrum
Intercarpal, (aka mid-carpal)	gliding	Articulations of carpal bones to other carpal bones
Carpometacarpal (CM) #1 (thumb)	saddle	Trapezium ◄► 1st metacarpal
Carpometacarpal (CM) #2-5	gliding	
Metacarpophalangeal (MP)	condyloid	
Interphalangeal (PIP & DIP)	hinge	Fingers 1-4 have DIP & PIP, thumb has DIP only
Hip		
Coxal	ball and socket	Head of femur ◄► acetabulum of hip bone
Knee		
Tibiofemoral	modified hinge	The knee – "modified" because also rotates when flexed
Patellofemoral	gliding	Kneecap
Tibiofibular (Proximal)	gliding	
Tibiofibular (Distal)	(synarthrotic-fibrous)	Where tib./fib. connect to form upper part of ankle joint
Ankle, Foot, Toes		
Talocrural	hinge	Ankle joint (plantarflex=flexion, dorsiflex=extension)
Subtalar (=Talocalcaneal joint)	gliding/pivot	Inversion & eversion occur at the subtalar joint
Talocalcaneonavicular (TCN)	gliding, rotation	⎫
Transverse Tarsal	gliding	⎬ These create the springy arches of the foot
Tarsometatarsal (TM)	gliding	⎭
Metatarsophalangeal (MP)	condyloid	Ball of foot
Interphalangeal (PIP & DIP)	hinge	Toes 1-4 have DIP & PIP, big toe has DIP only

SYNOVIAL JOINTS (DIARTHROTIC) – TYPES & MOVEMENTS

Ball & Socket: flexion, extension, abduction, adduction, lateral rotation, medial rotation
Pivot: rotation   **Hinge:** flexion, extension   **Ellipsoid/Condyloid:** flexion, extension, abduction, adduction
Gliding: gliding   **Saddle:** same actions as ellipsoid, + opposition

Table S-5 -- LIGAMENT SUMMARY

Temporomandibular Joint
    Temporomandibular (Lateral)   Limits downward, lateral, and posterior motion, major jaw stabilizer
    Sphenomandibular   Limits anterior and downward motion
    Stylomandibular   Limits anterior motion and lateral deviation

Cervical Spine
    Ligamentum Nuchae (Nuchal)   Limits excessive flexion of the neck
    Apical   Maintains relative position of dens with occiput
    Alar   Limits lateral flexion of head on neck, holds position of dens
    Transverse (part of cruciate)   Limits posterior displacement of the dens, prevents dens from pushing into spinal cord

Spine (Lumbar/Thoracic)
    Supraspinous   Limits too much flexion
    Anterior Longitudinal   Prevents too much extension
    Posterior Longitudinal   Limits too much flexion
    Intertransverse   Limits lateral flexion
    Ligamentum flavum   Connects adjacent lamina on anterior side

Pelvis: Sacroiliac Joint & L5/S1
    Iliolumbar   Restrains movement at the lumbosacral junction (L5/S1)
    Sacroiliac   Limits rotation/gliding in sacroiliac (SI) joints, main stabilizer of SI joints
    Sacrotuberous   Limits forward rotation of the sacrum
    Sacrospinous   Stabilizes sacrum vs. hip bone

Shoulder: Sternoclavicular (SC) Joint
    Sternoclavicular   Reinforces the joint capsule, resists dislocation of clavicle from sternum
    Costoclavicular   Limits clavicular elevation
    Interclavicular   Limits clavicular depression

Shoulder: Acromioclavicular (AC) Joint
    Acromioclavicular (AC)   Reinforces the joint capsule, holds the AC joint together
    Coracoclavicular (CC)   Stability for the AC joint, and between scapula and clavicle
    Coracoacromial (CA)   Above the GH jt., forms a protective lid for rotator cuff tendons & SA bursa

Shoulder: Glenohumeral (GH) Joint
    Glenohumeral   Thickenings in the fibers of the capsule, reinforce anterior portion of capsule
    Coracohumeral   Strengthens upper part of the joint capsule

Elbow: Humeroulnar Joint
    Lateral (Radial) Collateral   Limits varus stress
    Medial (Ulnar) Collateral   Limits valgus stress
    Annular   Holds head of radius against radial notch of ulna (radioulnar joint)

Wrist: Radiocarpal Joint
    Lateral (Radial) Collateral   Prevents too much ulnar deviation and too much varus stress
    Medial (Ulnar) Collateral   Prevents too much radial deviation and too much valgus stress
    Palmar and Dorsal Radiocarpal   Limit wrist extension and flexion, respectively
    Flexor Retinaculum   Holds down the tendons of the flexor muscles, roof of carpal tunnel

Hand: Phalangeal Joints, etc.
    Carpometacarpal & Intercarpal Ligs   Reinforce joints in palmar hand
    Metacarpophalangeal Collateral   Resist stresses on the MP condyloid joints
    PIP and DIP Collateral Ligs   Limit varus and valgus stresses on the PIP and DIP hinge joints

Hip: Coxal Joint
    Iliofemoral   Limits extension and lateral rotation
    Ishiofemoral   Limits extension and medial rotation
    Pubofemoral   Limits extension and abduction
    Ligament of Head of Femur   Helps attach head of femur into the acetabulum, carries blood vessel

Knee: Tibiofemoral Joint
    Anterior Cruciate (ACL)   Limits hyperextension, prevents too much anterior movement of tibia on femur
        (Note: anterior movement of tibia on femur = posterior movement of femur on tibia)
    Posterior Cruciate (PCL)   Limits flexion, prevents too much posterior movement of tibia on femur
    Medial (Tibial) Collateral (MCL)   Medial stabilizer, prevents too much valgus stress
    Lateral (Fibular) Collateral (LCL)   Lateral stabilizer, prevents too much varus stress

Ankle, Foot: Talocrural Joint & Subtalar Joint
    Deltoid   Medial stabilizer, prevents too much valgus stress on the ankle
    Anterior Talofibular (ATFL)   Lateral stabilizer, prevents too much varus stress (most commonly injured lig.)
    Posterior Talofibular (PTFL)   Lateral stabilizer, prevents too much varus stress
    Calcaneonavicular (Spring)   Supports arch of foot, keeps head of talus from "falling down"

Table S-6 -- INNERVATION SUMMARY

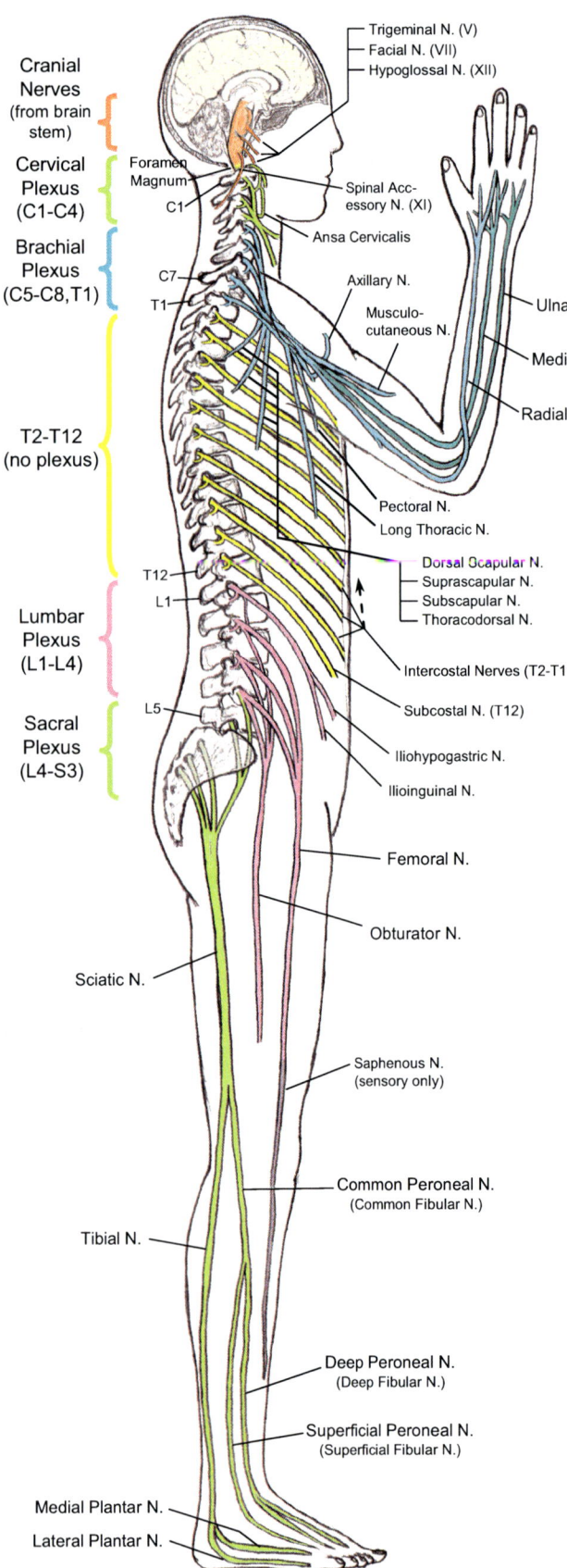

The set of tables on this and the following page summarize the nerves that supply the muscles of the body. For an overview description of the nervous system, please refer to pages 22-25 in Chapter 1 – Basic Information.

Cranial Nerves		
Nerve Name	**Muscle**	**Segment**
Trigeminal N.	Masseter	Cranial V
Trigeminal N.	Temporalis	Cranial V
Trigeminal N.	Lateral pterygoid	Cranial V
Trigeminal N.	Medial pterygoid	Cranial V
Trigeminal N.	Mylohyoid	Cranial V
Trigeminal N.	Digastric, anterior belly	Cranial V
Facial N.	Digastric, posterior belly	Cranial VII
Facial N.	Stylohyoid	Cranial VII
Facial N.	Occipitofrontalis	Cranial VII
Facial N.	Platysma	Cranial VII
Facial N.	Muscles of facial expression	Cranial VII

Cranial Nerves with	Upper Cervical Nerves	
Hypoglossal N. & C1	Geniohyoid	Cranial XII, and C1
Spinal Accessory N. & ventral rami C2,C3	Sternocleidomastoid	Cranial XI, and C2, C3
Spinal Accessory N. & ventral rami C3,C4	Trapezius	Cranial XI, and C3, C4

Cervical Plexus (Ventral Rami of C1-C4)		
Ansa cervicalis	Sternohyoid	C1,C2,C3
Ansa cervicalis	Sternothyroid	C1,C2,C3
Ansa cervicalis	Omohyoid	C1,C2,C3
Ansa cervicalis	Thyrohyoid	C1,C2,C3
Phrenic N.	Diaphragm	C3,C4,C5
Ventral rami C1-C3	Longus capitis	C1-C3
Ventral rami C2-C6	Longus colli *	C2-C6
Ventral rami C3-C8	Middle scalene *	C3-C8
Ventral rami C3-C5	Levator scapula *	C3-C5
	* muscle has nerve supply from both cervical plexus and brachial plexus	

Dorsal Rami of Cervical Nerves		
Suboccipital N.	Rectus capitis posterior major	C1
Suboccipital N.	Rectus capitis posterior minor	C1
Suboccipital N.	Oblique capitis superior	C1
Suboccipital N.	Oblique capitis inferior	C1
Dorsal rami of upper cervicals	Semispinalis capitis	C1-C5
Dorsal rami of middle cervicals	Splenius capitis	C3-C6
Dorsal rami of lower cervicals	Splenius cervicis	C5-C8

Paraspinal Muscles (Dorsal Rami of Spinal Nerves)		
Dorsal rami (segmental)	Spinalis	C5-T12
Dorsal rami (segmental)	Longissimus	C3-L4
Dorsal rami (segmental)	Iliocostalis	C6-L2
Dorsal rami (segmental)	Semispinalis	C1-T10
Dorsal rami (segmental)	Multifidus	C3-L5
Dorsal rami (segmental)	Rotatores	C2-L5
Dorsal rami (segmental)	Levator costae	C8-T11

Brachial Plexus (Ventral Rami of C5-T1)

Nerve Name	Muscle	Segments
Ventral rami C5,C6	Scalene, anterior	C5,C6
Ventral rami C3-C8	Scalene, middle *	C3-C8
Ventral rami C6-C8	Scalene, posterior	C6-C8
Ventral rami C2-C6	Longus colli *	C2-C6
Dorsal scapular N. (C5) & ventral rami C3, C4	Levator scapula *	C5 & C3,C4
Dorsal scapular N.	Rhomboid major	C5
Dorsal scapular N.	Rhomboid minor	C5
	* muscle has nerve supply from both cervical plexus and brachial plexus	
Subclavian N.	Subclavius	C5,C6
Suprascapular N.	Supraspinatus	C5
Suprascapular N.	Infraspinatus	C5,C6
Upper Subscapular N.	Subscapularis, upper part	C5,C6
Lower Subscapular N.	Subscapularis, lower part	C5,C6
Lower Subscapular N.	Teres major	C5,C6
Long thoracic N.	Serratus anterior	C5,C6,C7
Thoracodorsal N.	Latissimus dorsi	C6,C7,C8
Lateral pectoral N.	Pectoralis major, upper part	C5,C6,C7
Medial pectoral N.	Pectoralis major, lower part	C8,T1
Medial pectoral N.	Pectoralis minor	C8,T1
(Below are the five major "terminal branches" of the brachial plexus)		
Axillary N.	Deltoid	C5,C6
Axillary N.	Teres minor	C5
Musculocutaneous N.	Coracobrachialis	C6,C7
Musculocutaneous N.	Biceps brachii	C5,C6
Musculocutaneous N.	Brachialis	C5,C6
Radial N.	Brachioradialis	C5,C6
Radial N.	Triceps brachii	C7,C8
Radial N.	Anconeus	C7,C8,T1
Radial N.	Supinator	C6
Radial N.	Extensor carpi radialis longus	C6,C7
Radial N.	Extensor carpi radialis brevis	C6,C7
Radial N.	Extensor carpi ulnaris	C6,C7,C8
Radial N.	Extensor digitorum	C6,C7,C8
Radial N.	Extensor indicis	C7,C8
Radial N.	Abductor pollicis longus	C7,C8
Radial N.	Extensor pollicis longus	C6,C7,C8
Radial N.	Extensor pollicis brevis	C7,C8
Median N.	Pronator teres	C6,C7
Median N.	Pronator quadratus	C8,T1
Median N.	Flexor carpi radialis	C6,C7
Median N.	Palmaris longus	C6,C7
Median N.	Flexor digitorum superficialis	C7,C8,T1
Median N.	Flexor pollicis longus	C8,T1
Median N.	Opponens pollicis	C8,T1
Median N.	Abductor pollicis brevis	C8,T1
Median N.	Flexor pollicis brevis, sup. hd	C8,T1
Median N.	Flexor digitorum profundus, to digits 2 & 3	C8,T1
Median N.	Lumbrical muscles, digits 2-3	C8,T1
Ulnar N.	Flexor pollicis brevis, deep hd	C8,T1
Ulnar N.	Flexor digitorum profundus, to digits 4 & 5	C8,T1
Ulnar N.	Lumbrical muscles, digits 4-5	C8,T1
Ulnar N.	Flexor carpi ulnaris	C8,T1
Ulnar N.	Dorsal interossei	C8,T1
Ulnar N.	Palmar interossei	C8,T1
Ulnar N.	Abductor digiti minimi	C8,T1
Ulnar N.	Flexor digiti minimi	C8,T1
Ulnar N.	Opponens digiti minimi	C8,T1
Ulnar N.	Adductor pollicis	C8,T1

Ventral Rami of Thoracic Nerves

Nerve Name	Muscle	Segments
Intercostal nerves	External intercostals	T1-T11
Intercostal nerves	Internal intercostals	T1-T11
Ventral rami T1-T4	Serratus posterior superior	T1-T4
Ventral rami T9-T12	Serratus posterior inferior	T9-T12

Abdominal Muscles (Ventral Rami of T6-T12 and L1)

Nerve Name	Muscle	Segments
Intercostal nerves	Rectus abdominis	T6-T12
Intercostal nerves	External oblique	T7-T12
Intercostal N., Iliohypogastric N. & Ilioinguinal N.	Internal oblique	T8-T12, T12-L1, & L1,
Intercostal N., Iliohypogastric N. & Ilioinguinal N.	Transverse abdominis	T7-T12, T12-L1, & L1

Lumbar Plexus (Ventral Rami of L1-L4)

Nerve Name	Muscle	Segments
Lumbar plexus	Quadratus lumborum	T12,L1-L3
Lumbar plexus	Psoas major	L2,L3,L4
Lumbar plexus	Psoas minor	L1
Femoral N.	Iliacus	L2,L3
Femoral N.	Sartorius	L2,L3
Femoral N.	Rectus femoris	L2,L3,L4
Femoral N.	Vastus lateralis	L2,L3,L4
Femoral N.	Vastus intermedius	L2,L3,L4
Femoral N.	Vastus medialis	L2,L3,L4
Femoral N. (& sometimes Obturator N.)	Pectineus	L2,L3 (L3,L4)
Obturator N.	Adductor brevis	L2,L3,L4
Obturator N.	Adductor longus	L2,L3,L4
Obturator N.	Gracilis	L2,L3
Obturator N.	Adductor magnus, anterior part	L2,L3,L4
Obturator N.	Obturator externus	L3,L4

Sacral Plexus (Ventral Rami of L4-S3)

Nerve Name	Muscle	Segments
Inferior gluteal N.	Gluteus maximus	L5,S1,S2
Superior gluteal N.	Gluteus medius	L4,L5,S1
Superior gluteal N.	Gluteus minimus	L4,L5,S1
Superior gluteal N.	Tensor fascia latae	L4,L5,S1
Sacral Plexus	Piriformis	S1,S2
Sacral Plexus	Gemellus superior	L5,S1,S2
Sacral Plexus	Obturator internus	L5,S1,S2
Sacral Plexus	Gemellus inferior	L4,L5,S1
Sacral Plexus	Quadratus femoris	L4,L5,S1
Sciatic N.	Adductor magnus, posterior part	L4,L5,S1
Sciatic N., tibial part	Semitendinosus	L5,S1,S2
Sciatic N., tibial part	Semimembranosus	L5,S1,S2
Sciatic N., tibial part	Biceps femoris, long head	S1,S2,S3
Sciatic N., peroneal part	Biceps femoris, short head	L5,S1,S2
Deep peroneal N.	Extensor digitorum longus	L4,L5,S1
Deep peroneal N.	Extensor hallucis longus	L4,L5,S1
Deep peroneal N.	Tibialis anterior	L4,L5,S1
Deep peroneal N.	Peroneus tertius	L5,S1
Deep peroneal N.	Extensor digitorum brevis	L5,S1
Deep peroneal N.	Extensor hallucis brevis	L5,S1
Superficial peroneal N.	Peroneus brevis	L4,L5,S1
Superficial peroneal N.	Peroneus longus	L4,L5,S1
Tibial N.	Popliteus	L4,L5,S1
Tibial N.	Gastrocnemius	S1,S2
Tibial N.	Plantaris	L4,L5,S1
Tibial N.	Soleus	S1,S2
Tibial N.	Tibialis posterior	L5,S1
Tibial N.	Flexor digitorum longus	L5,S1
Tibial N.	Flexor hallucis longus	L5,S1,S2
Medial plantar N.	Flexor digitorum brevis	L5,S1
Medial plantar N.	Abductor hallucis	L5,S1
Medial plantar N.	Flexor hallucis brevis	L5,S1
Medial plantar N.	Lumbrical muscle #1	L5,S1
Lateral plantar N.	Lumbrical muscles #2-4	S2,S3
Lateral plantar N.	Abductor digiti minimi	S2,S3
Lateral plantar N.	Quadratus plantae	S2,S3
Lateral plantar N.	Flexor digiti minimi	S2,S3
Lateral plantar N.	Adductor hallucis	S2,S3
Lateral plantar N.	Plantar interossei	S2,S3
Lateral plantar N.	Dorsal interossei	S2,S3

Chapter 8

Study Tools

Introduction	**202**
MusclePlus⁺ Flashcards	202
Bony Landmark Practice Sheets	202
Muscle Tickets	202
Erasable Skeleton Pictures	203
Action Table Practice Pages	203
Muscle Group Summary Cards	203
Bony Landmark Flashcards	203
Muscle Action Card Games	203
How To Use the MusclePlus⁺ Flashcards	**204**
Sample Flashcard	205
How To Use the Bony Landmark Practice Sheets	**206**
Sample – Type 1	207
Sample – Type 2	208
Sample Muscle Tickets	**209**
Sample Erasable Skeleton Pictures	**210**
Sample Action Table Practice Pages	**212**

Introduction

The **MMM Study Tools** packet associated with **Mastering Muscles & Movement** provides many resources for studying and practicing the information in the book. This chapter describes the study tools available and provides a sample of each tool. **MMM Study Tools** are purchased separately from the book.

How to Order:
Please use the order form provided in the back of this book, or visit www.bodylightbooks.com.

The following MMM Study Tools are available:

MusclePlus⁺ Flashcards

Description: These unique full-color flashcards gather all drawings and information for each muscle in the book. The front of each card gives *both* a muscle drawing and a red/blue origin/insertion drawing, and the reverse side has all pertinent muscle information. The cards include body-location icons, muscle ID tags (e.g., G1-4, as described below), and page references to the book Mastering Muscles & Movement.

A special booklet is included in the box that contains bone and joint drawings, illustrations of actions at each joint, study tips, and a list of the cards organized by Muscle Groups. Please refer to the section "How To Use the MusclePlus⁺ Flashcards" on page 204 for more information.

Sample: A sample MusclePlus⁺ Flashcard is included on page 205.

Bony Landmark Practice Sheets

Description: These pages have the bone drawings from Chapter 2 with the labels or words removed to facilitate repetitive practice to memorize bony landmarks. Please refer to the section "How to Use the Bony Landmark Practice Sheets" on page 206 for more information.

Sample: Samples are included for the bones of the foot on pages 207-208.

Muscle Tickets

Description: These ticket-sized cards have the muscle names on them. As you study each of the 13 groups of muscles, cut up the muscle tickets and use them to draw out of a hat and randomly test yourself. Each ticket has a small label at the lower right corner, for example, "G1-4". This label tells which muscle group to go to in the book to read the origin, insertion, etc., for the muscle. The "G" is for "Group", so "G1-4" indicates the 4th muscle in Muscle Group 1.

Sample: A sample of Muscle Tickets is included on page 209.

Erasable Skeleton Pictures

Description: Two skeleton drawings - upper body and lower body - are inserted in plastic sleeves (or laminated). On the back of each is a Key that lists all the muscle names to go with the small labels at the bottom of the flashcards. These skeleton pictures can be marked and erased many times using fine point dry erase markers (preferably red and blue to match the origin/insertion convention used in this book).

Sample: Sample Erasable Skeleton Pictures are included on pages 210-211.

Action Table Practice Pages

Description: These are the Summary Tables from Chapter 7 in a special "practice" format. The X's are removed so the learner can test their knowledge of synergists for each action of each joint of the body, and then check their answers in Chapter 7 of Mastering Muscles & Movement.

Sample: A sample Action Table Practice Page is included on page 212.

Muscle Group Summary Cards

Description: There are thirteen large 5" x 8" cards, one card for each Muscle Group in the book. Each card has an overview drawing of all muscles in the group on one side, and the "A" Table from the book on the reverse side. This provides a handy take-it-with-you summary for each of the 13 muscle groups.

(In development – available soon)

Bony Landmark Flashcards

(In development – available soon)

Muscle Action Card Games

(In development – available soon)

How To Use the MusclePlus⁺ Flashcards

MusclePlus⁺ Flashcards can, of course, be used as you would any flashcards to drill yourself on recognizing muscles and memorizing their names, origins, insertion, actions, and innervation.

In addition, the special red/blue origin/insertion drawings on the cards provide extra ways to focus your study efforts and really master the muscles.

Here are a few suggestions on ways to use MusclePlus⁺ cards. These methods assume you are using the laminated skeleton pictures and red and blue dry erase markers provided in the **MMM Study Tools** packet. You could, however, do these exercises using any picture of a skeleton on paper and red and blue colored pencils or markers.

Method 1: To practice in a **verbal-to-visual** direction

1. Shuffle the cards and turn them so the side with *words* is facing up and you can't see the picture.
2. Select a card, keeping the name of the muscle at the top of the card covered.
3. Read the words for **Origin**, and use a *red* dry erase marker to color the skeleton where the words describe.
4. Read the words for **Insertion** and use a *blue* dry erase marker to color the skeleton.
5. By now you may be able to guess what muscle is being described. If not, read the **Action** part of the card to get more clues.
6. Turn the card over and check if you colored the skeleton accurately.
7. If you want to further study the muscle in your Mastering Muscles & Movement book, turn to the page number given at the lower left of the card.

Method 2: To practice in a **visual-to-verbal** direction

1. Shuffle the cards and turn them so the side with the *picture* is facing up.
2. Select a card.
3. Look at the red areas colored on the card and name the bony landmarks indicated. Check if the words you said match the **Origin** on the back of the card.
4. Look at the blue areas colored on the card and name the bony landmarks indicated. Check if the words you said match the **Insertion** on the back of the card.
5. Read the **Action**, and visualize the blue spot on the moveable bone being pulled toward the red spot on the stable bone.

Method 3: To study muscle shapes and fiber directions

1. Draw the muscle on the bone picture connecting origin to insertion and showing the shape and fiber direction.
2. Visualize the muscle belly getting shorter (contracting) and pulling the blue insertion toward the red origin, and recite the actions of the muscle.

Method 4 - ∞: Make up other game/drills

Drawing of muscle's attachments to the bones

Origin: Shown in Red
Insertion: Shown in Blue
Action: Visualize the blue being pulled toward the red

MusclePlus⁺ Flashcards

www.musclepluscards.com

- Full color double-illustration flashcards describing all the major muscles of the body
- Includes illustrated booklet with bones, actions, muscle group lists, and study tips.

Muscle Illustration - Lines show direction of fibers

Muscle Identifier
G# - n
G# = Group # (1-13)
n = Muscle within that group

© 2007 Bodylight Books   Anterior   G2 - 6

Aspect of the body you are viewing

(Cards are shown actual size - 3¼" x 4¼")

Front of card contains:

- Illustration of muscle
- Illustration of origin & insertion
- Organizational tags at the bottom

Back of card contains:

- Name of muscle
- Muscle attachments (origin & insertion)
- Actions
- Innervation
- Organizational tags at the bottom

Pectoralis Major

Origin:
Clavicular head: Medial half of clavicle
Sternocostal part: Sternum and cartilages of ribs 1-6

Insertion: Intertubercular groove of the humerus (lateral lip)

Action:
All fibers: Adduction and medial rotation of humerus
Upper fibers: Flexion and horizontal adduction of humerus
Lower fibers: Extension of humerus (from a flexed position), and depression of shoulder girdle

Nerve: Lateral Pectoral N. (C5-C7) & Medial Pectoral N. (C8,T1)

MMM, p. 80   G2 - 6

Page reference in the book "Mastering Muscles & Movement"

Icon indicating the joint(s) moved by the muscle

Muscle Identifier

Mastering Muscles & Movement © 2007   Chapter 8 – Study Tools   **205**

How to Use the Bony Landmark Practice Sheets

Chapter 2 of **Mastering Muscles & Movement** contains fully labeled bony landmark drawings. Each page of bone drawings is organized with the bones in one area of the page and a list of bone names, bony landmarks, and joints in a separate area of the page. The arrangement allows you to cover the list of names and use the labels on the drawings to test yourself as you memorize the names. This facilitates learning the landmarks from a visual direction, that is, you *see* a place on a bone and you recall its bony landmark name.

To fully learn the bones and bony landmarks, you should be able to recall the information from both visual and verbal directions. Recalling from the verbal direction means you *read* the name of a landmark and you recall and visualize where it is on the bone.

The **MMM Study Tools** packet includes dual versions of the bone drawings from Chapter 2 of the book. One version has the labels removed from the drawings and the list of landmarks is left intact. The other version has the list of landmarks removed and the drawing labels are left intact. With these opposite arrangements, you can memorize the information from both visual and verbal directions. To illustrate how this works for the bones of the foot, sample drawings from the Study Tools packet are shown on pages 207 and 208.

The Bony Landmark Practice Sheets can be slipped into a plastic sleeve and marked using fine point dry erase markers to allow multiple practice sessions.

The following are included in the **MMM Study Tools** packet:

Skeleton
    Anterior - Bones
    Lateral and Posterior - Bones
    Anterior - Joints

Upper Extremity
    Scapula
    Shoulder Girdle
    Upper Arm (Humerus)
    Forearm (Radius, Ulna)
    Hand

Axial Skeleton
    Skull and Hyoid
    Spine (Vertebral Column)
    Vertebrae – Special Features by Section
    Thorax (Ribs, Sternum)

Lower Extremity
    Pelvis
    Thigh (Femur)
    Leg (Tibia, Fibula)
    Foot

Samples

The following pages show two types of Bony Landmark Practice Sheets from the **MMM Study Tools** packet.

Right Foot

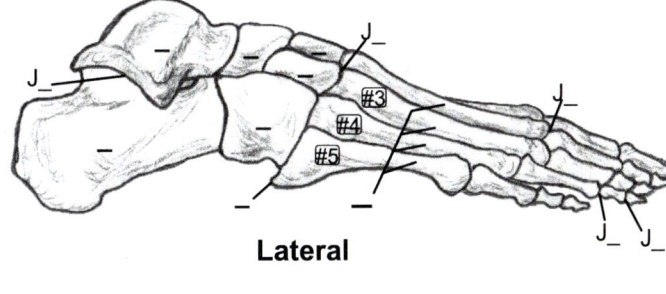

Plantar   **Dorsal**

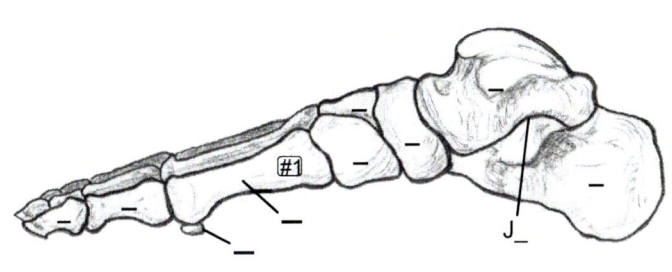

Lateral

Medial

A. Tarsal Bones (7)

1. Talus
2. Calcaneus
3. Navicular
4. Medial Cuneiform (1st cuneiform)
5. Middle Cuneiform (2nd cuneiform)
6. Lateral Cuneiform (3rd cuneiform)
7. Cuboid

B. Metatarsal Bones (5)

8. Head
9. Base
10. Tuberosity of 5th metatarsal

#1 medial side to #5 lateral side

C. Phalanges (14)

11. Head
12. Base

    C1 – Proximal phalanx
    C2 – Middle phalanx
    C3 – Distal phalanx

Digit #1 = Big toe (Hallux)
Digits #2 - #5 = toes

D. Sesamoid bones

Joints

J1. Subtalar (talocalcaneal)
J2. Intertarsals
J3. Tarsometatarsal (TM)
J4. Metatarsophalangeal (MP)
J5. Proximal Interphalangeal (PIP)
J6. Distal Interphalangeal (DIP)

Bony Landmarks - Practice Page – Foot

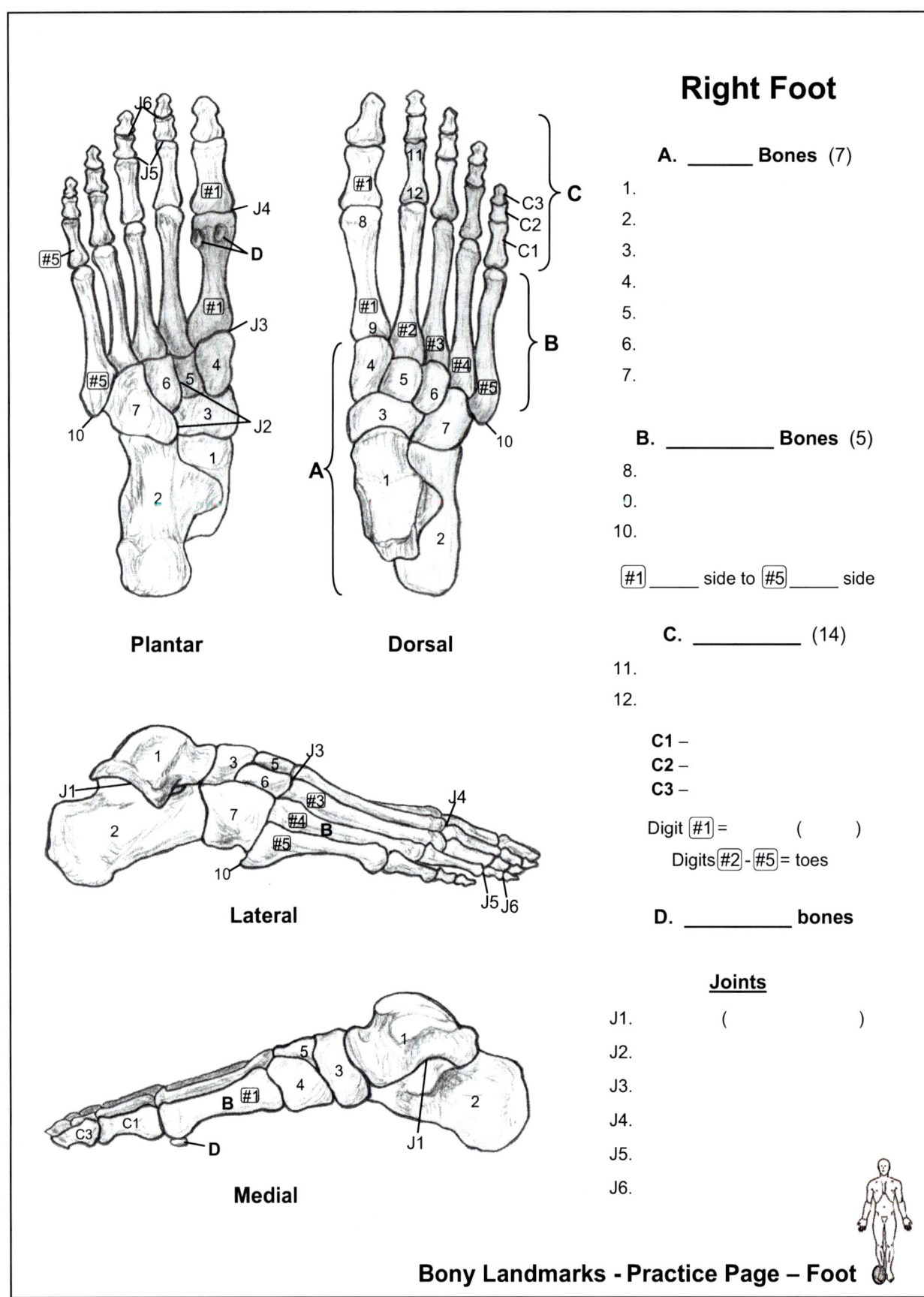

Muscle Tickets – Upper Extremity

Trapezius G1 - 1	**Levator Scapula** G1 - 2	**Rhomboid Major and Minor** G1 - 3	**Serratus Anterior** G1 - 4
Pectoralis Minor G1 - 5	**Subclavius** G1 - 6	-	**Deltoid** G2 - 1
Supraspinatus G2 - 2	**Infraspinatus** G2 - 3	**Teres Minor** G2 - 4	**Subscapularis** G2 - 5
Pectoralis Major G2 - 6	**Coracobrachialis** G2 - 7	**Latissimus Dorsi** G2 - 8	**Teres Major** G2 - 9
Biceps Brachii G3 - 1	**Brachialis** G3 - 2	**Brachioradialis** G3 - 3	**Pronator Teres** G3 - 4
Pronator Quadratus G3 - 5	**Triceps Brachii** G3 - 6	**Anconeus** G3 - 7	**Supinator** G3 - 8
Flexor Carpi Radialis G4 - 1	Palmaris Longus G4 - 2	Flexor Carpi Ulnaris G4 - 3	Flexor Digitorum Superficialis G4 - 4
Flexor Digitorum Profundus G4 - 5	Extensor Carpi Radialis Longus G4 - 6	Extensor Carpi Radialis Brevis G4 - 7	Extensor Carpi Ulnaris G4 - 8
Extensor Digitorum G4 - 9	Extensor Indicis G4 - 10	-	-
Flexor Pollicis Longus G5 - 1	**Flexor Pollicis Brevis** G5 - 2	**Opponens Pollicis** G5 - 3	**Adductor Pollicis** G5 - 4
Abductor Pollicis Longus G5 - 5	**Abductor Pollicis Brevis** G5 - 6	**Extensor Pollicis Longus** G5 - 7	**Extensor Pollicis Brevis** G5 - 8

(This is a sample from the **MMM Study Tools** packet – sold separately)

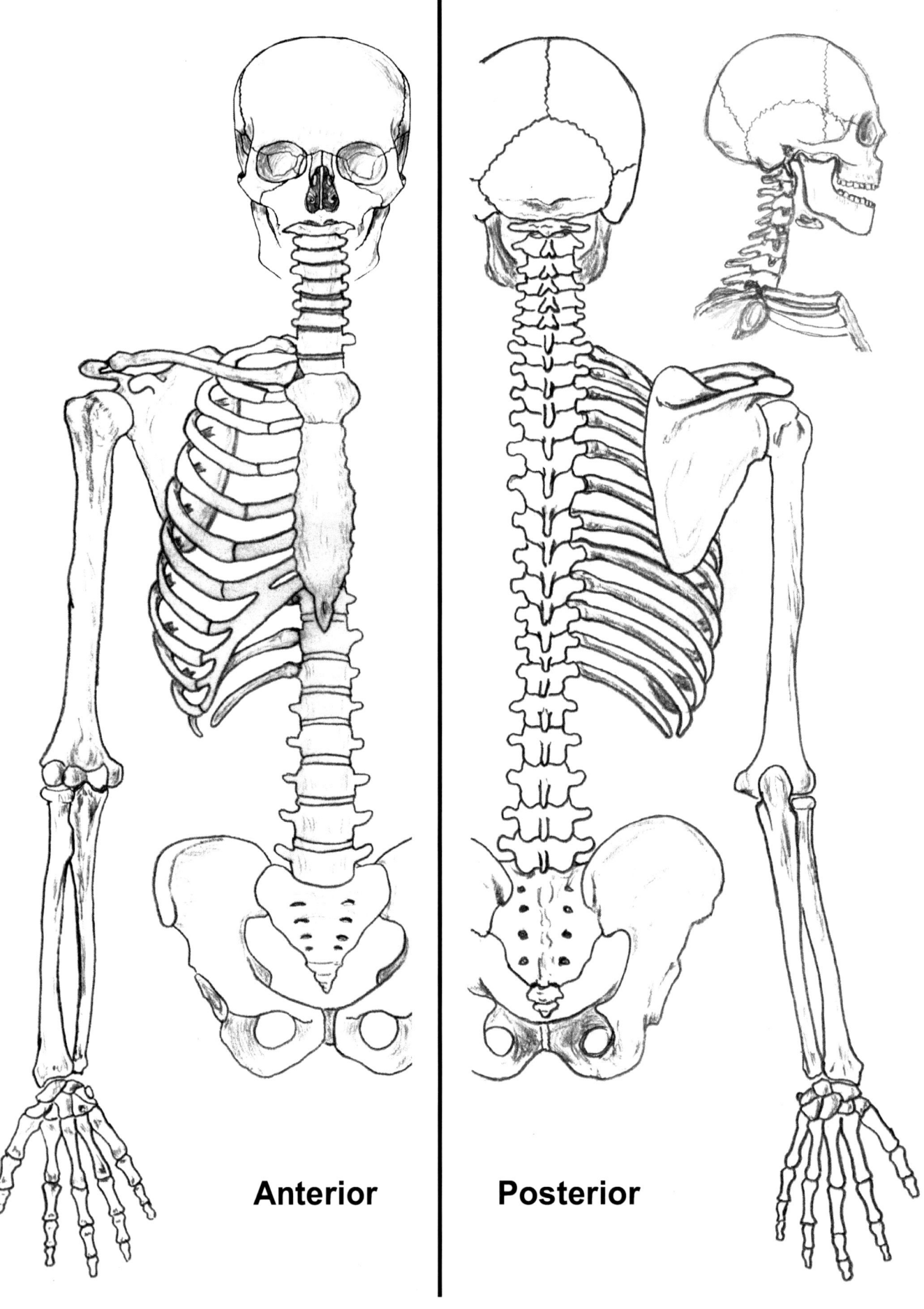

Anterior **Posterior**

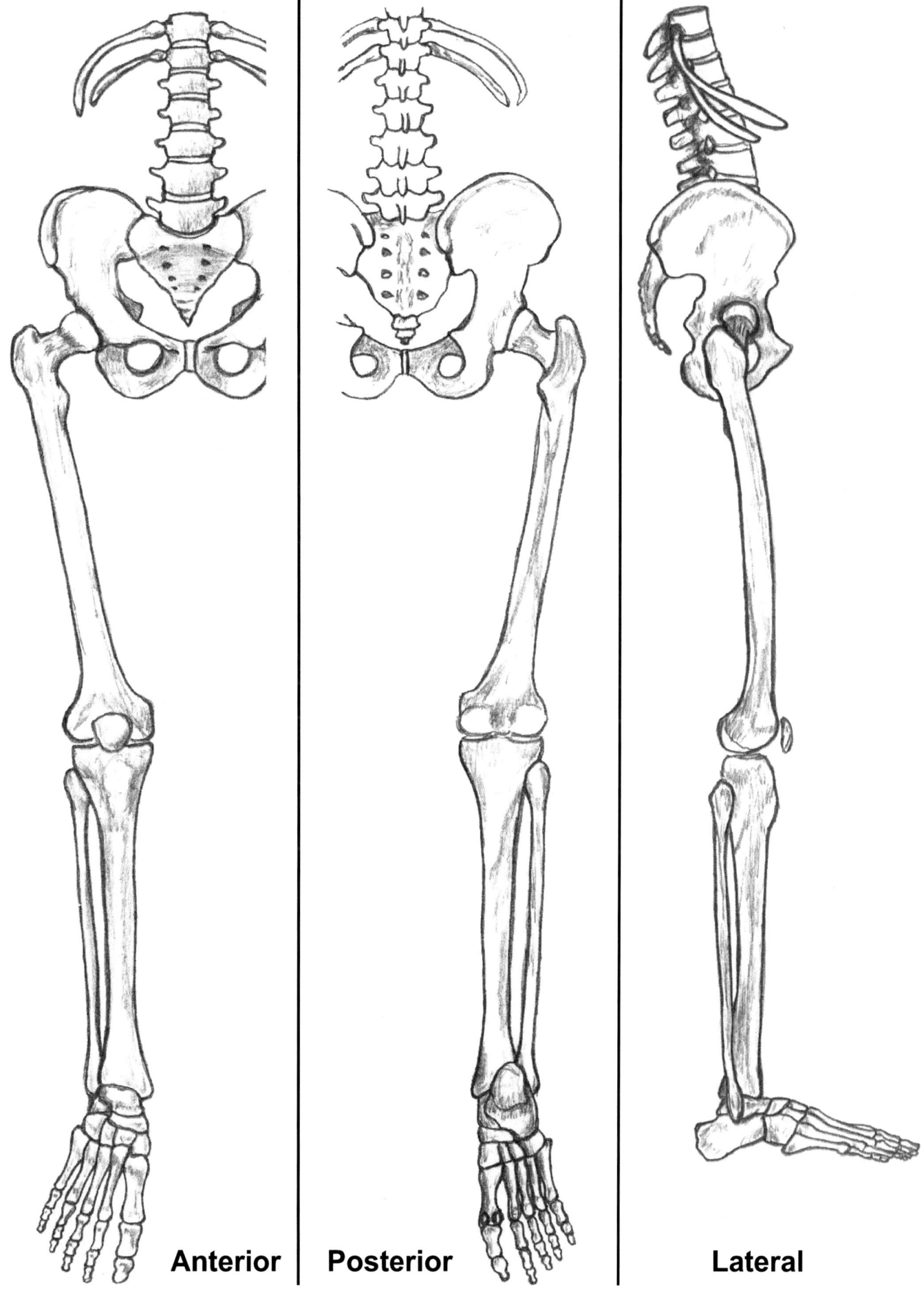

Anterior | **Posterior** | **Lateral**

Table S-1 -- Actions Practice Page – UPPER EXTREMITY

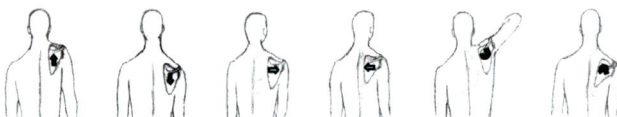

Scapula / Clavicle	Elevation	Depression	Protraction/ Abduction	Retraction/ Adduction	Lateral / Upward Rotation	Medial / Downward Rotation	Stabilization of Scapula	Ref. Table	Muscle also affects other joints:
Trapezius (upper fibers)								1	ROI: Head/Neck
Trapezius (middle fibers)								1	-
Trapezius (lower fibers)								1	-
Levator scapula								1	ROI: Neck
Rhomboid (major & minor)								1	-
Serratus anterior								1	-
Pectoralis minor								1	ROI: Breathing
Subclavius								1	ROI: Breathing
Pectoralis Major								2	Shoulder Joint
Latissimus Dorsi								2	Shoulder Joint

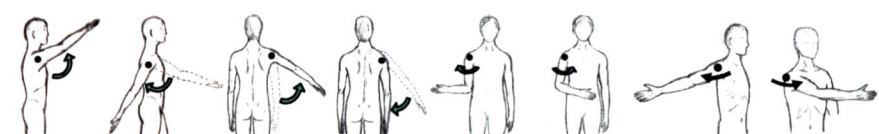

Shoulder Joint	Flexion @ GH jt.	Extension @ GH jt.	Abduction @ GH jt.	Adduction @ GH jt.	Lateral Rotation @ GH jt.	Medial Rotation @ GH jt.	Horizontal Abduction @ GH jt.	Horizontal Adduction @ GH jt.	Ref. Table	Muscle also affects other joints:
Deltoid (anterior fibers)									2	-
Deltoid (middle fibers)									2	-
Deltoid (posterior fibers)									2	-
Supraspinatus									2	-
Infraspinatus									2	-
Teres Minor									2	-
Subscapularis									2	-
Pectoralis Major (upper fibers)									2	-
Pectoralis Major (lower fibers)									2	Scapula
Coracobrachialis									2	-
Latissimus Dorsi									2	Spine/Trunk. Scapula
Teres Major									2	-
Biceps Brachii									3	Elbow/ Forearm
Triceps Brachii (long head)									3	Elbow

KEY: X – Muscle creates the action, Assist – Muscle assists the action, GH – Glenohumeral joint (shoulder joint)
ROI – Reversed Origin/Insertion (action moves O toward I)

(This is a sample from the **MMM Study Tools** packet – sold separately)

Bibliography

The following list gives the main resources used in the development of this book.

1. Barcsay, Jeno, *Anatomy for the Artist*, New York: Barnes & Noble Books, 2002
2. Biel, Andrew, *Trail Guide to the Body*, 2nd ed., Boulder: Books of Discovery, 2001
3. Calais-Germain, Blandine, *Anatomy of Movement*, Seattle: Eastland Press, 1993
4. Clemente, Carmine D., *Anatomy – A Regional Atlas of the Human Body*, 4th ed., Baltimore: Williams & Wilkins, 1997
5. Goldfinger, Eliot, *Human Anatomy for Artists*, New York: Oxford University Press, 1991
6. Goss, CM, ed. *Gray's Anatomy of the Human Body*, 28th ed. Philadelphia: Lea & Febiger, 1966
7. Juhan, Deane, *Job's Body – A Handbook for Bodywork*, 3rd ed., Barrytown, NY: Station Hill Press, 2003
8. Kapandji, I.A., *The Physiology of the Joints, Volume 1 Upper Limb*, 5th ed., London: Churchill Livingstone, 1982
9. Kapandji, I.A., *The Physiology of the Joints, Volume 2 Lower Limb*, 5th ed., London: Churchill Livingstone, 1985
10. Kapandji, I.A., *The Physiology of the Joints, Volume 3 The Trunk and the Vertebral Column*, 2nd ed., London: Churchill Livingstone, 1974
11. Kapit, Wynn, and Elson, Lawrence M., *The Anatomy Coloring Book*, 3rd ed., San Francisco: Benjamin Cummings, 2002
12. Kendall, F.P., McCreary, E.K., and Provance, P.G., *Muscles, Testing and Function*, Baltimore: Williams & Wilkins, 1993
13. Lippert, Lynn S., *Clinical Kinesiology for Physical Therapy Assistants*, 3rd ed., Philadelphia: F. A. Davis Company, 2000
14. McMinn, R.M.H., and Hutchings, R.T., Color Atlas of Human Anatomy, Chicago: Year Book Medical Publishers, Inc., 1977
15. Muscolino, Joseph E., *The Muscular System Manual*, JEM Publications, 2002
16. Netter, Frank H., M.D., *Atlas of Human Anatomy*, New Jersey: CIBA-GEIGY Corporation, 1987
17. Netter, Frank H., M.D., *The CIBA Collection of Medical Illustrations, Volume 8: Musculoskeletal System*, New Jersey: CIBA-GEIGY Corporation, 1993
18. Platzer, Werner, *Color Atlas/Text of Human Anatomy, Vol. 1: Locomotor System*, 4th ed., New York: Thieme Medical Publishers, 1991
19. Pugh, Maureen Barlow, Ed., *Stedman's Medical Dictionary*, 27th ed., Baltimore: Lippincott Williams & Wilkins, 2000
20. Rohen, Yokochi and Lutjen-Drecoll, *Color Atlas of Anatomy*, 4th ed., Baltimore: Williams & Wilkins, 1998
21. Simons, David G., M.D., Travell, Janet G., M.D., and Simons, Lois S., P.T., *Myofascial Pain and Dysfunction: The Trigger Point Manual*, Volume 1. Upper Half of Body, Second Edition, Baltimore: Lippincot Williams & Wilkins, 1999
22. Thibodeau, Gary A., and Patton, Kevin T., *The Human Body In Health & Disease*, 2nd ed., St. Louis: Mosby, 1997
23. Thompson, Clem W., and Floyd, R. T., *Manual of Structural Kinesiology*, 14th ed., New York: McGraw-Hill, 2001
24. Tortora, Gerard J., and Grabowski, Sandra Reynolds, *Principles of Anatomy and Physiology*, 7th ed., New York: Harper Collins, 1993
25. Travell, Janet G., M.D., and Simons, David G., M.D., *Myofascial Pain and Dysfunction: The Trigger Point Manual*, Volume 2. The Lower Extremities, Baltimore: Lippincot Williams & Wilkins, 1983

Index

A
Abdominal muscles, 137
Actions of all body parts, 6-8
Actions of all muscles (tables), 189-194
    axial skeleton, 191-192
    lower extremity, 193-194
    upper extremity, 189-190
Actions of,
    abdomen, 137
    ankle, 177
    elbow, 85
    face, 113
    fingers, 93
    foot, 177
    forearm, 85
    hand, 93
    head on neck, 122
    hip joint, 153, 161, 169
    jaw, 113
    knee, 169
    neck, 121
    respiration, 137
    ribs, 137
    scapula/clavicle, 69
    shoulder girdle, 69
    shoulder joint, 77
    spine, 129
    thumb, 101
    toes, 177
    trunk, 137
    vertebral column, 129
    wrist, 93
Actions practice page, 212
Active insufficiency, 94
Adductor group of muscles, 164
Anatomical position, 1
Anatomical snuffbox, 102
Anatomical stirrup, 178
Aponeurosis, 21
Appendicular skeleton, 12, 68, 152
Arches of the foot, 178
Articular system, 14-18
Atlas and axis, 45, 121
Axial skeleton, 12, 110
Axis, 9

B
Bilateral action, 112
Biomechanics, definition, 29
Bones of the body, 11, 33-35, 40-49
Bones, classification by shape, 13
Bony landmarks of the,
    clavicle, 40
    coccyx, 45
    femur, 48
    fibula, 48
    foot, 49
    hand, 42
    hip (coxal) bone, 47
    humerus, 41
    mandible, 43
    occiput, 43
    pelvis, 47
    radius, 41
    ribs, 46
    sacrum, 45
    scapula, 40
    shoulder girdle, 40
    skull, 43
    sphenoid, 43
    spine, 44
    sternum, 46
    temporal bone, 43
    tibia, 48
    ulna, 41
    vertebrae, 44, 45
Bony landmarks, common terms, 13
Bony landmarks, definition, 13
Bony landmark practice page, 206-208
Brachial plexus, 24, 198
Bursae, 15

C
Calcaneal (Achilles) tendon, 179
Carpal tunnel, 94, 108
Cervical plexus, 24, 198
Common extensor tendon, 95
Common flexor tendon, 95
Compartments of the leg, 179, 186
Concentric contraction, 28
Costal cartilage, 46
Cranial nerves, 24, 198
Cranial sutures, 43

D
Deglutition, 113
DIP, 93, 177
Directional terms, 3
Disc, intervertebral, 44, 111

E
Eccentric contraction, 28
Erector Spinae Group, 132

F
Facial expression muscles, 114
Flashcards, MusclePlus+, 204-205

G
Galea aponeurotica, 115

H
Hallux, 49, 178
Hip lateral rotator s -"Deep 6", 155
Hypothenar eminence, 102, 108

I
Iliotibial tract (band), 155, 163, 167
Inguinal ligament, 139, 155, 159
Innervation of all muscles, 198-199
Innervation, 25, 198
Intervertebral joint, 44,111
Intrinsic muscles of the foot, 184-185
Intrinsic muscles of the hand, 108
Isometric contractions, 28
Isotonic contractions, 28

J
Joint,
    Acromioclavicular joint, 69
    Atlantoaxial joint, 121
    Atlantooccipital joint, 121
    Atlantoodontoid joint, 121
    Carpometacarpal joint #1, 101
    Carpometacarpal joints #2-#5, 93
    Costochondral joint, 137
    Costotransverse joint, 137
    Costovertebral joint, 137
    Coxal (hip) joint, 153
    Glenohumeral joint, 77, 85
    Humeroulnar joint, 85
    Intercarpal joint, 93
    Interphalangeal joint - distal, 93, 177
    Interphalangeal joint - proximal, 93, 177
    Intervertebral joint, 121, 129
    Metacarpophalangeal joint, 93, 101
    Metatarsophalangeal joint, 177
    Occipitoatlantal joint, 121
    Patellofemoral joint, 169
    Radiocarpal joint, 93, 101
    Radiohumeral joint, 85
    Radioulnar joint, 85
    Sacroiliac joint, 129, 153
    Scapulothoracic joint
    Scapulothoracic joint, 69
    Sternoclavicular joint, 69
    Sternocostal joint, 137
    Subtalar joint, 177
    Talocalcaneaonavicular joint, 178
    Talocrural joint, 177
    Tarsometatarsal joint, 177
    Temporomandibular joint, 113
    Tibiofemoral joint, 169
    Tibiofibular joint, 168
    Transverse tarsal joint, 178
Joints of the body - all, 17, 36-39
Joints of the upper extremity, 16
Joints of the,
    ankle, 177
    elbow, 85
    fingers, 93
    foot, 177
    hand, 42, 93, 101
    hip, 153, 161
    knee, 169
    leg, 169
    neck, 45, 121
    pelvis, 47
    ribs, 46, 137

Index

scapula/clavicle, 40, 69
shoulder girdle, 40, 69
shoulder, 77
skull, 43, 113
thumb, 93
toes, 177
vertebral column, 44, 111, 121, 129
wrist, 93
Joints, overview 14-18
   accessory structures, 15
   amphiarthrotic (cartilagenous), 14
   definitions, 14
   degrees of freedom of, 18
   diarthrotic (synovial), 14
   six types of synovial joints, 15
   synarthrotic (fibrous), 14
   synovial joint structure, 14
   uniaxial/biaxial/triaxial, 18

K
Kyphotic curve, 44, 110, 130

L
Levers, 26
Ligaments, 15, 197
Ligamentum Nuchae, 121
Linea alba, 139
Lordotic curve, 44, 110, 130
Lumbar plexus, 24, 198
Lumbosacral plexus, 24

M
Mastication, 113
Muscle anatomy,
   endomysium, 19
   epimysium, 19
   fascicle, 19
   muscle attachment, 19, 27
   muscle belly, 19
   muscle fiber, 19, 20
   perimysium, 19
   periosteum, 19
   tendon, 19, 21
Muscles and,
   action, 27
   agonist, 29
   opposing movements, 21
   antagonist, 29
   attachments, 19, 21, 27
   fiber arrangements, 20
   line of pull, 20
   origin/insertion, 27
   reversed O/I action, 27
   stabilizer, 29
   synergist, 29
Muscle Tickets, 209
MusclePlus+ Flashcards, 204-205

Muscles,
   Abductor digiti minimi – foot, 185
   Abductor digiti minimi – hand, 108
   Abductor hallucis, 185
   Abductor pollicis brevis, 104
   Abductor pollicis longus, 104
   Adductor brevis, 164
   Adductor hallucis, 185
   Adductor longus, 164
   Adductor magnus, 164
   Adductor pollicis, 104
   Anconeus, 88
   Biceps brachii, 88
   Biceps femoris, 172
   Brachialis, 88
   Brachioradialis, 88
   Buccinator, 114
   Coracobrachialis, 80
   Deltoid, 80
   Deep Six lateral rotators of hip, 155
   Depressor anguli oris, 114
   Depressor labii inferioris, 114
   Diaphragm, 140
   Digastric, 116
   Dorsal interossei of the foot, 185
   Dorsal interossei of the hand, 108
   Erector spinae group, 132
   Extensor carpi radialis brevis, 96
   Extensor carpi radialis longus, 96
   Extensor carpi ulnaris, 96
   Extensor digiti minimi, 99
   Extensor digitorum (fingers), 96
   Extensor digitorum brevis (toes), 185
   Extensor digitorum longus (toes), 180
   Extensor hallucis brevis, 185
   Extensor hallucis longus, 180
   Extensor indicis, 96
   Extensor pollicis brevis, 104
   Extensor pollicis longus, 104
   External abdominal oblique, 140
   External intercostals, 140
   Fibularis, see peroneus
   Flexor carpi radialis, 96
   Flexor carpi ulnaris, 96
   Flexor digiti minimi of the foot, 185
   Flexor digiti minimi of the hand, 108
   Flexor digitorum brevis, 185
   Flexor digitorum longus, 180
   Flexor digitorum profundus, 96
   Flexor digitorum superficialis, 96
   Flexor hallucis brevis, 185
   Flexor hallucis longus, 180
   Flexor pollicis brevis, 104
   Flexor pollicis longus, 104
   Gastrocnemius, 180
   Gemellus inferior & superior, 156
   Geniohyoid, 116
   Gluteus maximus, 156
   Gluteus medius, 156
   Gluteus minimus, 156
   Gracilis, 164
   Hamstrings, 172

Muscles, continued
   Hypothenar eminence muscles, 108
   Iliacus, 156
   Iliocostalis, 132
   Iliopsoas, 156
   Infrahyoids group, 116
   Infraspinatus, 80
   Intercostals, external, 140
   Intercostals, internal, 140
   Internal abdominal oblique, 140
   Internal intercostals, 140
   Interspinales, 130
   Intertransversarii, 130
   Latissimus dorsi, 80
   Levator anguli oris, 114
   Levator costae, 140
   Levator labii superioris, 114
   Levator labii superioris alaeque nasi, 114
   Levator scapula, 72
   Levator scapula (reversed O/I), 124
   Longissimus, 132
   Longus capitis, 124
   Longus colli, 124
   Lumbricales of the foot, 185
   Lumbricales of the hand, 108
   Masseter, 116
   Mentalis, 114
   Multifidus, 132
   Mylohyoid, 116
   Nasalis, 114
   Oblique, abdominal, 140
   Oblique capitis inferior, 124
   Oblique capitis superior, 124
   Obturator externus, 156
   Obturator internus, 156
   Occipitofrontalis, 116
   Omohyoid, 116
   Opponens digiti minimi, 108
   Opponens pollicis, 104
   Orbicularis oculi, 114
   Orbicularis oris, 114
   Palmar interossei, 108
   Palmaris longus, 96
   Pectineus, 164
   Pectoralis major, 80
   Pectoralis minor, 72
   Peroneus brevis, 180
   Peroneus longus, 180
   Peroneus tertius, 183
   Piriformis, 156
   Plantar interossei, 185
   Plantaris, 180
   Platysma, 116
   Popliteus, 172
   Procerus, 114
   Pronator quadratus, 88
   Pronator teres, 88
   Psoas major, 156
   Psoas minor, 155
   Pterygoid, lateral, 116
   Pterygoid, medial, 116

Index

Muscles, continued
    Quadratus femoris, 156
    Quadratus lumborum, 132
    Quadratus plantae, 185
    Quadriceps, 172
    Rectus abdominis, 140
    Rectus capitis anterior, 122
    Rectus capitis lateralis, 122
    Rectus capitis posterior major, 124
    Rectus capitis posterior minor, 124
    Rectus femoris, 172
    Rhomboid major, 72
    Rhomboid minor, 72
    Risorius, 114
    Rotator cuff muscles, 80
    Rotatores, 132
    Sartorius, 164
    Scalenes group, 124
    Semimembranosus, 172
    Semispinalis, 132
    Semispinalis capitis, 124
    Semitendinosus, 172
    Serratus anterior, 72
    Serratus posterior superior, 140
    Serratus posterior inferior, 140
    Soleus, 180
    Spinalis, 132
    Splenius capitis, 124
    Splenius cervicis, 124
    Sternocleidomastoid, 124
    Sternohyoid, 116
    Sternothyroid, 116
    Stylohyoid, 116
    Subclavius, 72
    Suboccipital group (4), 124
    Subscapularis, 80
    Supinator, 88
    Suprahyoids group (4), 116
    Supraspinatus, 80
    Temporalis, 116
    Tensor fascia latae, 164
    Teres major, 80

Muscles, continued
    Teres minor, 80
    Thenar eminence muscles, 108
    Thyrohyoid, 116
    Tibialis anterior, 180
    Tibialis posterior, 180
    Transverse abdominis, 140
    Transversospinal group, 132
    Trapezius, 72
    Trapezius, upper fibers only, 124
    Triceps brachii, 88
    Vastus intermedius, 172
    Vastus lateralis, 172
    Vastus medialis, 172
    Zygomaticus major, 114
    Zygomaticus minor, 114
Muscles/Actions Summary Tables, 189-194
Muscular system, 19-21

N
Nervous system, 22-25
    cranial nerves, 22, 24
    innervation, 25
    motor units, 25
    muscle innervation summary, 198-199
    names of major nerves, 24
    overall organization, 22
    plexuses, 24
    primary rami, 23
    spinal nerve root, 23
    spinal nerve segments, 22
Nuchal ligament, 121

O
Opposite side rotation, 112

P
Palmar Aponeurosis, 108
Paraspinal muscles, 129
Passive insufficiency, 94
Patellar ligament (tendon), 171
Pelvic tilt, 47
PIP, 93, 177
Plane, 5
Plane/axis pairs, 10
Plantar aponeurosis, 189
Pollux, 42, 101
Positional terms, 3
Prevertebral muscles, 122
Primary curve of spine, 44, 110, 130
Pubic symphysis, 47

R
Range of motion, 18, 30
Rectus sheath, 139
Regions of the body, 3
Respiration, 138
Ribs, true/false/floating, 46
Rotator cuff muscles, 80

S
Sacral plexus, 24, 198
Sacrotuberous ligament, 159
Same side rotation, 112
Secondary curve of spine, 44, 110, 130
Shoulder complex, 77
Skeletal system, 11
Skull, sutures of, 43
Spine, curves of, 110, 130
Sutures, cranial, 43

T
Tendon, 19, 21
Tendons of medial malleolus, 178, 186
Tendons of lateral malleolus, 178, 186
Thenar eminence, 102, 108
Transversospinal Group, 132

U
Unilateral action, 112

Products from Bodylight Books

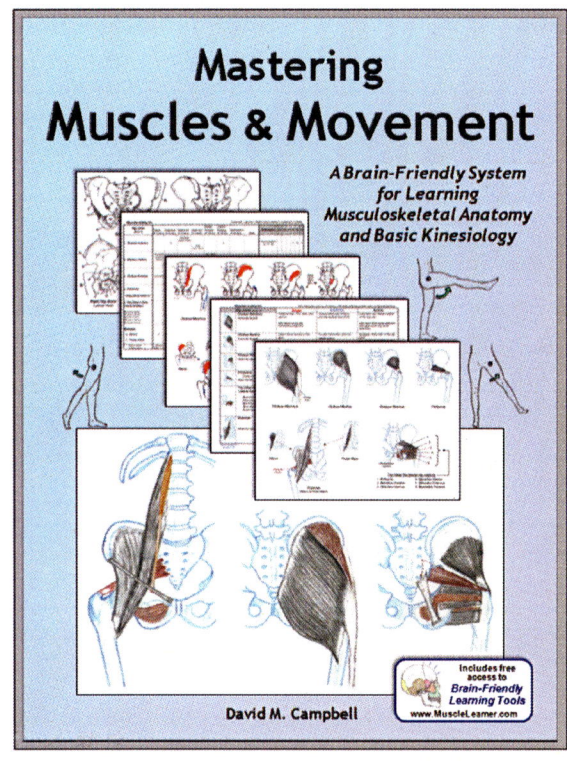

(Book)

Mastering Muscles & Movement

A Brain-Friendly System
for Learning
Musculoskeletal Anatomy
and Basic Kinesiology

ISBN: 978-0-9788664-0-2

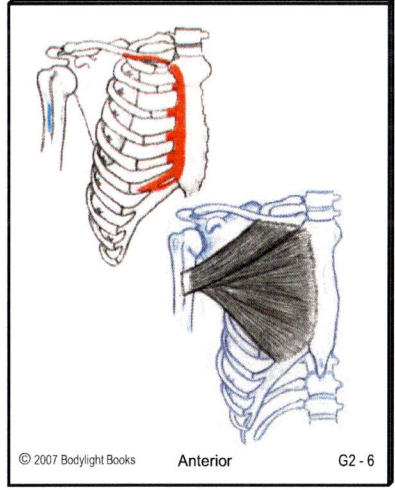 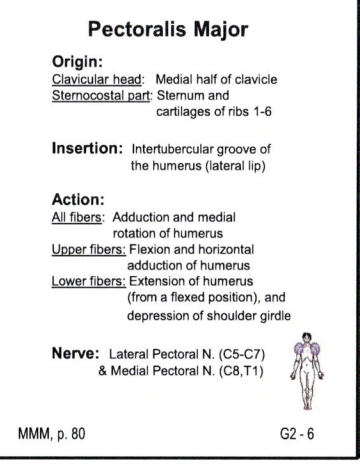

(actual card size is 3 ¼ " x 4 ¼ ")

(Flashcards)

MusclePlus⁺ Flashcards

Full color
double-illustration
study cards using figures
from the book

(see pages 204-205 for more info)

ISBN: 978-0-9788664-1-X

For more
information
or to order:

Bodylight Books
1707 F Street
Bellingham, WA 98225
(360) 734-1560
www.bodylightbooks.com

Muscles – Alphabetic List

Abductor digiti minimi of the foot .. 185
Abductor digiti minimi of the hand. 108
Abductor hallucis 185
Abductor pollicis brevis 104
Abductor pollicis longus 104
Adductor brevis 164
Adductor hallucis 185
Adductor longus 164
Adductor magnus 164
Adductor pollicis 104
Anconeus 88

Biceps brachii 88
Biceps femoris 172
Brachialis 88
Brachioradialis 88
Buccinator 114

Coracobrachialis 80

Deltoid 80
Deep Six lateral rotators of the hip .. 155
Depressor anguli oris 114
Depressor labii inferioris 114
Diaphragm 140
Digastric 116
Dorsal interossei of the foot 185
Dorsal interossei of the hand 108

Erector spinae group 132
Extensor carpi radialis brevis 96
Extensor carpi radialis longus 96
Extensor carpi ulnaris 96
Extensor digiti minimi 99
Extensor digitorum (fingers) 96
Extensor digitorum brevis (toes) 185
Extensor digitorum longus (toes) ... 180
Extensor hallucis brevis 185
Extensor hallucis longus 180
Extensor indicis 96
Extensor pollicis brevis 104
Extensor pollicis longus 104
External abdominal oblique 140
External intercostals 140

Fibularis, see peroneus
Flexor carpi radialis 96
Flexor carpi ulnaris 96
Flexor digiti minimi of the foot 185
Flexor digiti minimi of the hand 108
Flexor digitorum brevis 185
Flexor digitorum longus 180
Flexor digitorum profundus 96
Flexor digitorum superficialis 96
Flexor hallucis brevis 185
Flexor hallucis longus 180
Flexor pollicis brevis 104
Flexor pollicis longus 104

Gastrocnemius 180
Gemellus inferior & superior 156
Geniohyoid 116
Gluteus maximus 156
Gluteus medius 156
Gluteus minimus 156

Gracilis 164
Hamstrings 172
Hypothenar eminence muscles 108

Iliacus 156
Iliocostalis 132
Iliopsoas 156
Infrahyoids group 116
Infraspinatus 80
Intercostals, external 140
Intercostals, internal 140
Internal abdominal oblique 140
Internal intercostals 140
Interspinales 130
Intertransversarii 130

Latissimus dorsi 80
Levator anguli oris 114
Levator costae 140
Levator labii superioris 114
Levator labii superioris alaeque nasi 114
Levator scapula 72
Levator scapula (reversed O/I) 124
Longissimus 132
Longus capitis 124
Longus colli 124
Lumbricales of the foot 185
Lumbricales of the hand 108

Masseter 116
Mentalis 114
Multifidus 132
Mylohyoid 116

Nasalis 114

Oblique, abdominal 140
Oblique capitis inferior 124
Oblique capitis superior 124
Obturator externus 156
Obturator internus 156
Occipitofrontalis 116
Omohyoid 116
Opponens digiti minimi 108
Opponens pollicis 104
Orbicularis oculi 114
Orbicularis oris 114

Palmar interossei 108
Palmaris longus 96
Pectineus 164
Pectoralis major 80
Pectoralis minor 72
Peroneus brevis 180
Peroneus longus 180
Peroneus tertius 183
Piriformis 156
Plantar interossei 185
Plantaris 180
Platysma 116
Popliteus 172
Procerus 114
Pronator quadratus 88
Pronator teres 88
Psoas major 156

Psoas minor 155
Pterygoid, lateral 116
Pterygoid, medial 116

Quadratus femoris 156
Quadratus lumborum 132
Quadratus plantae 185
Quadriceps 172

Rectus abdominis 140
Rectus capitis anterior 122
Rectus capitis lateralis 122
Rectus capitis posterior major 124
Rectus capitis posterior minor 124
Rectus femoris 172
Rhomboid major 72
Rhomboid minor 72
Risorius 114
Rotator cuff muscles 80
Rotatores 132

Sartorius 164
Scalenes group 124
Semimembranosus 172
Semispinalis 132
Semispinalis capitis 124
Semitendinosus 172
Serratus anterior 72
Serratus posterior superior 140
Serratus posterior inferior 140
Soleus 180
Spinalis 132
Splenius capitis 124
Splenius cervicis 124
Sternocleidomastoid 124
Sternohyoid 116
Sternothyroid 116
Stylohyoid 116
Subclavius 72
Suboccipital group 124
Subscapularis 80
Supinator 88
Suprahyoids group 116
Supraspinatus 80

Temporalis 116
Tensor fascia latae 164
Teres major 80
Teres minor 80
Thenar eminence muscles 108
Thyrohyoid 116
Tibialis anterior 180
Tibialis posterior 180
Transverse abdominis 140
Transversospinal group 132
Trapezius 72
Trapezius, upper fibers only 124
Triceps brachii 88

Vastus intermedius 172
Vastus lateralis 172
Vastus medialis 172

Zygomaticus major 114
Zygomaticus minor 114

Muscles – List by Group

Muscles are placed in groups based on the bones and joints they *move* as they contract.

----- (Chapter 4) -----

Group 1 – Scapula / Clavicle
Trapezius p. 69-76
Levator scapula
Rhomboid major & minor
Serratus anterior
Pectoralis minor
Subclavius

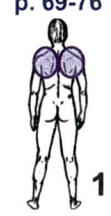

Group 2 – Shoulder Joint
Deltoid p. 77-84
Supraspinatus
Infraspinatus
Teres minor
Subscapularis
Pectoralis major
Coracobrachialis
Latissimus dorsi
Teres major

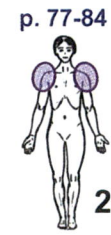

Group 3 – Elbow, Forearm
Biceps brachii p. 85-92
Brachialis
Brachioradialis
Pronator teres
Pronator quadratus
Triceps brachii
Anconeus
Supinator

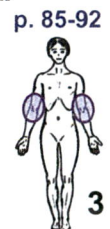

Group 4 – Wrist, Hand, Fingers
Flexor carpi radialis
Palmaris longus p. 93-100
Flexor carpi ulnaris
Flexor digitorum superficialis
Flexor digitorum profundus
Extensor carpi radialis longus
Extensor carpi radialis brevis
Extensor carpi ulnaris
Extensor digitorum
Extensor indicis

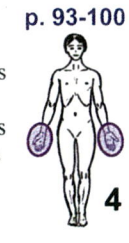

Group 5 – Thumb
Flexor pollicis longus
Flexor pollicis brevis p. 101-108
Opponens pollicis
Adductor pollicis
Abductor pollicis longus
Abductor pollicis brevis
Extensor pollicis longus
Extensor pollicis brevis
Intrinsic muscles of the hand

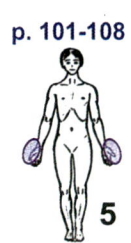

----- (Chapter 5) -----

Group 6 – Face, Jaw
Masseter
Temporalis p. 113-120
Lateral pterygoid
Medial pterygoid
Occipitofrontalis
Platysma
Suprahyoids Group
    Geniohyoid, Mylohyoid,
    Stylohyoid, Digastric
Infrahyoids Group
    Sternohyoid, Sternothyroid,
    Omohyoid, Thyrohyoid
Muscles of facial expression

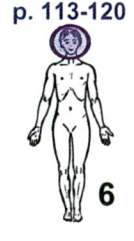

Group 7 – Neck, Head
Sternocleidomastoid
Scalenes group p. 121-128
Longus capitis & longus colli
Suboccipital group
    Rectus Capitis Posterior Major
    Rectus Capitis Posterior Minor
    Oblique Capitis Superior
    Oblique Capitis Inferior
Splenius capitis
Splenius cervicis
Semispinalis capitis
Levator scapula*
Trapezius, upper fibers*
    *(revisited for reversed O/I actions)

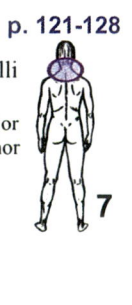

Group 8 – Spine
Spinalis p. 129-136
Longissimus
Iliocostalis
Semispinalis
Multifidus
Rotatores
Quadratus lumborum
Interspinales & Intertransversarii

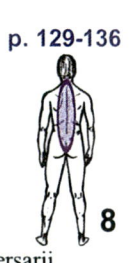

Group 9 – Thorax, Abdomen, Breathing
Rectus abdominis
External oblique p. 137-144
Internal oblique
Transverse abdominis
Diaphragm
External intercostals
Internal intercostals
Serratus posterior superior
Serratus posterior inferior
Levator costae

----- (Chapter 6) -----

Group 10 – Hip Joint (Part 1)
Gluteus maximus
Gluteus medius p. 153-160
Gluteus minimus
Piriformis (1st lateral rotator)
The other 5 lateral rotators
    Gemellus Superior
    Obturator Internus
    Gemellus Inferior
    Obturator Externus
    Quadratus Femoris
Iliopsoas
    (Iliacus & Psoas major)

Group 11 – Hip Joint (Part 2)
Sartorius
Tensor fascia latae p. 161-168
Pectineus
Adductor brevis
Adductor longus
Adductor magnus
Gracilis

Group 12 – Knee (& Hip Joint, Part 3)
Rectus femoris
Vastus lateralis p. 169-176
Vastus intermedius
Vastus medialis
Biceps femoris
Semitendinosus
Semimembranosus
Popliteus

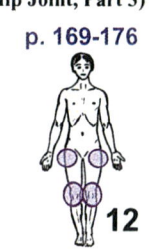

Group 13 – Ankle, Foot, Toes
Gastrocnemius
Plantaris
Soleus p. 177-186
Tibialis posterior
Flexor digitorum longus
Flexor hallucis longus
Peroneus longus
Peroneus brevis
Tibialis anterior
Extensor digitorum longus
Extensor hallucis longus
Intrinsic muscles of the foot

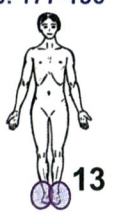

Mastering Muscles & Movement © 2007

Don't just memorize muscles...

The innovative design of this book provides a *brain-friendly* environment for learning the muscles, bones, joints, and movements of the human body.

- Promotes learning from many directions – visual, verbal, relational
- Clearly models how to study, compare, and contrast the information
- Accelerates the learning process and enhances long-term functional memory
- Facilitates easy, repetitive self-testing while studying

... master them!

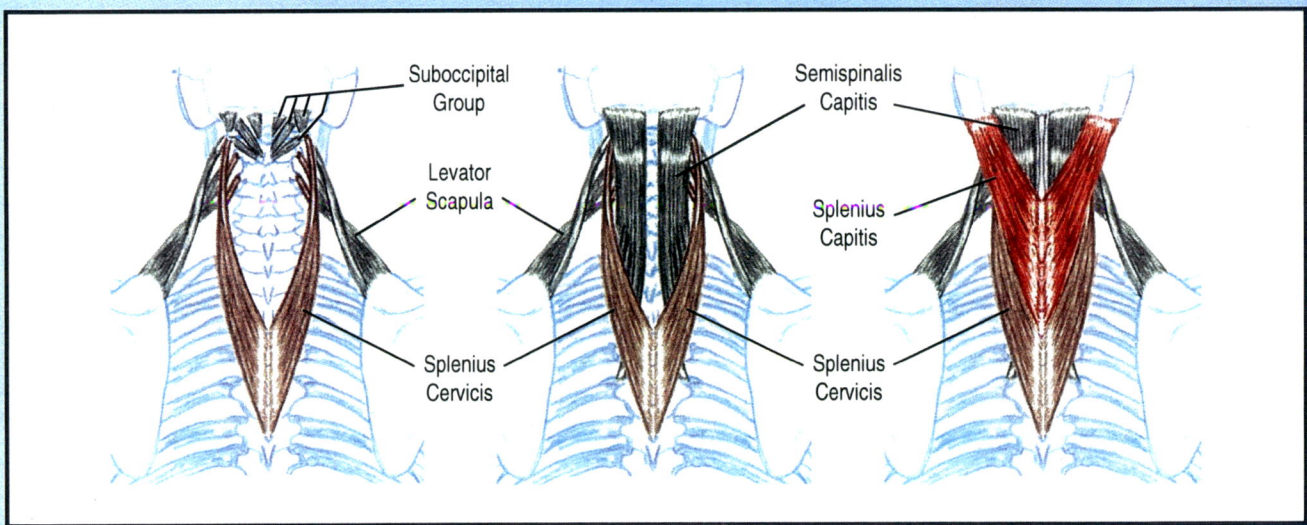

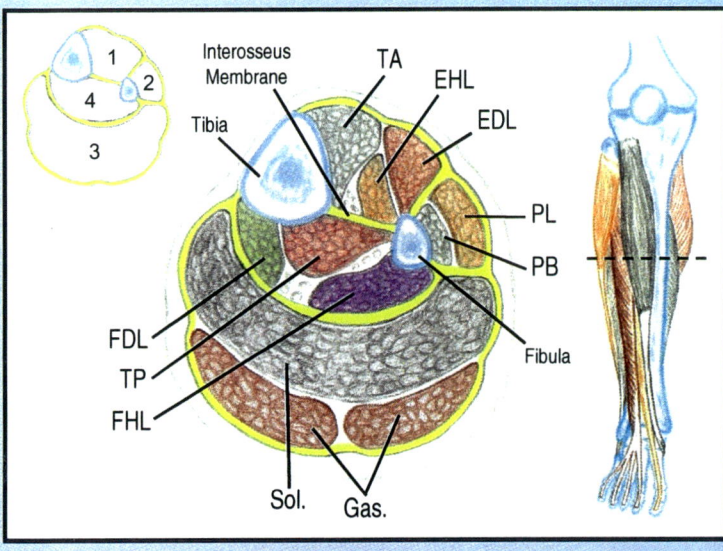

About the author/illustrator

David M. Campbell (LMP, NCTMB, CTP) is a licensed massage therapist and instructor of kinesiology. He has been in practice for over 20 years and has an enduring fascination with the intricacies of the human body, mind, and spirit. He holds a B.S. in Mathematics from the University of California and is a certified Trager® practitioner. Campbell, a former key engineer at the Fairchild Research Labs in Palo Alto, CA, developed his brain-friendly approach to teaching muscles and movement based on years of experience in the study of human perception, artificial intelligence, and graphic representation of scientific information.

Mastering Muscles & Movement

A Brain-Friendly System
for Learning
Musculoskeletal Anatomy
and Basic Kinesiology

Bodylight Books ~ www.bodylightbooks.com